12

Springer Series on Chemical Sensors and Biosensors

Methods and Applications

Series Editor: G. Urban

For further volumes:
http://www.springer.com/series/5346

Springer Series on Chemical Sensors and Biosensors

Series Editor: G. Urban

Recently Published and Forthcoming Volumes

Designing Receptors for the Next Generation of Biosensors
Volume Editors: S.A. Piletsky, M.J. Whitcombe
Vol. 12, 2013

Solid State Gas Sensors – Industrial Application
Volume Editors: M. Fleischer, M. Lehmann
Vol. 11, 2012

Optical Nano- and Microsystems for Bioanalytics
Volume Editors: W. Fritzsche, J. Popp
Vol. 10, 2012

Mathematical Modeling of Biosensors
An Introduction for Chemists and Mathematicians
Volume Authors: R. Baronas, F. Ivanauskas, J. Kulys
Vol. 9, 2010

Optical Guided-wave Chemical and Biosensors II
Volume Editors: M. Zourob, A. Lakhtakia
Vol. 8, 2010

Optical Guided-wave Chemical and Biosensors I
Volume Editors: M. Zourob, A. Lakhtakia
Vol. 7, 2010

Hydrogel Sensors and Actuators
Volume Editors: Gerlach G., Arndt K. -F.
Vol. 6, 2009

Piezoelectric Sensors
Volume Editors: Steinem C., Janshoff A.
Vol. 5, 2006

Surface Plasmon Resonance Based Sensors
Volume Editor: Homola J.
Vol. 4, 2006

Frontiers in Chemical Sensors
Novel Principles and Techniques
Volume Editors: Orellana G., Moreno-Bondi M. C.
Vol. 3, 2005

Ultrathin Electrochemical Chemo- and Biosensors
Technology and Performance
Volume Editor: Mirsky V. M.
Vol. 2, 2004

Optical Sensors
Industrial, Environmental and Diagnostic Applications
Volume Editors: Narayanaswamy R., Wolfbeis O. S.
Vol. 1, 2003

Designing Receptors for the Next Generation of Biosensors

Volume Editors:
Sergey A. Piletsky
Michael J. Whitcombe

With contributions by

M.R. Andreu · P.A. Bertin · R.E. Carlson · A.M. Costero · S.N. Dean · S. Gil · F. Hof · K. Karim · H.Y. Kuchelmeister · P.A. Lieberzeit · G. Mustafa · S.H. North · M. Parra · S.A. Piletsky · T. Pinter · A. Poma · R.W. Roska · C. Schmuck · L.C. Shriver-Lake · S. Subrahmanyam · C.R. Taitt · R. Tel-Vered · C. Thomas · M. Whitcombe · I. Willner

Springer

Editors
Prof. Dr. Sergey A. Piletsky
Dr. Michael J. Whitcombe
Cranfield University
Cranfield Health
Cranfield, Bedfordshire, UK

ISSN 1612-7617
ISBN 978-3-642-32328-7 ISBN 978-3-642-32329-4 (eBook)
DOI 10.1007/978-3-642-32329-4
Springer Heidelberg New York Dordrecht London

Library of Congress Control Number: 2012947654

Printed on acid-free paper

Springer is part of Springer Science+Business Media (www.springer.com)

Series Editor

Prof. Dr. Gerald Urban

IMTEK - Laboratory for Sensors
Institute for Microsystems Engineering
Albert-Ludwigs-University
Georges-Köhler-Allee 103
79110 Freiburg
Germany
urban@imtek.de

Aims and Scope

Chemical sensors and biosensors are becoming more and more indispensable tools in life science, medicine, chemistry and biotechnology. The series covers exciting sensor-related aspects of chemistry, biochemistry, thin film and interface techniques, physics, including opto-electronics, measurement sciences and signal processing. The single volumes of the series focus on selected topics and will be edited by selected volume editors. The *Springer Series on Chemical Sensors and Biosensors* aims to publish state-of-the-art articles that can serve as invaluable tools for both practitioners and researchers active in this highly interdisciplinary field. The carefully edited collection of papers in each volume will give continuous inspiration for new research and will point to existing new trends and brand new applications.

Preface

Synthetic Receptors: Design Versus Evolution

Despite the considerable progress achieved to date in the application of enzymes, antibodies, and biological receptors to diagnostics and in separations, the last two decades have witnessed the emergence of a clutch of alternative technologies based on their synthetic analogues. Research and development in this area is expanding, as can be seen from a steady increase in the number of relevant publications (Fig. 1). The move to the use of artificial receptors in place of natural molecules has been driven by practical considerations, in particular the need for stable, sterilizable, and non-immunogenic materials. Novel diagnostics based on synthetic receptors are not only suitable for clinical applications but have the potential to fill niches in, e.g., online monitoring of industrial (pharmaceutical manufacturing) processes or continuous environmental monitoring, where the technology can benefit from materials that combine excellent specificity and affinity with durability at low cost.

Most publications in this area, however, are related to fundamental aspects of materials design with only a small proportion being concerned with practical applications of the technology.

The challenges facing the adoption of synthetic receptors in "real life" applications are substantial and stem from two reasons: The first is related to the fact that tremendous progress has already been achieved in improving the performance of biological molecules through, e.g., site-directed mutagenesis and by combinatorial chemistry. The second reason is the difficulty of translating the fundamental knowledge of molecular recognition mechanisms gleaned from biological systems into designs for the creation of synthetic receptors. Basically it is easier to develop completely new synthetic recognition systems than to mimic the corresponding biological analogue. Synthetic receptor molecules prepared by design do not, however, always possess superior characteristics that can justify their use in, e.g., biosensors. Thus the important questions are whether there is a real need to develop a new generation of synthetic

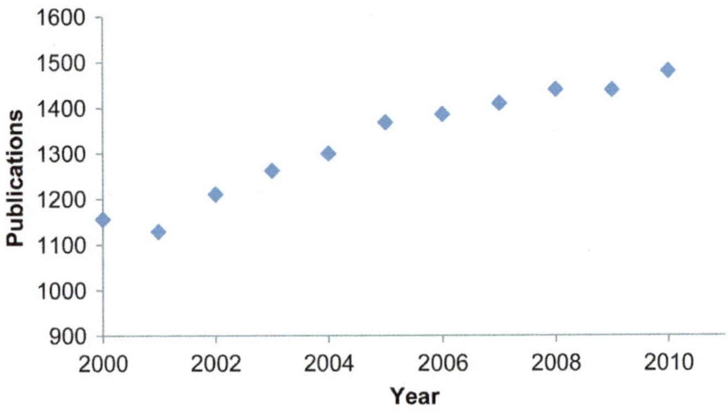

Fig. 1 Number of publications dedicated to synthesis and investigation of synthetic receptors (calculations made from the analysis of information available at ISI Web of Knowledge)

receptors which can successfully compete with biological molecules in diagnostics, and how this can be achieved? The answer to the first question is firmly positive. Enzymes, antibodies, and receptors are products of natural evolution and consequently are relatively unstable under non-physiological conditions. They also tend to be expensive and difficult to integrate with the device using standard manufacturing tools and protocols such as photolithography. There is a fundamental limit, imposed by nature, to what can be achieved by modification of biological molecules to suit their practical applications. Synthetic chemistry, however, is not limited to 20 or so amino acids and 4 or 5 nucleotides (the building blocks of peptides, proteins, and aptamers) and can be used to construct its products from a huge range of functional motifs and chemical entities. Bespoke molecules can be designed with particular functional characteristics for complex and diverse applications such as diagnostics.

The present book intends to answer the second question – how synthetic receptors can be designed with characteristics that make them attractive alternatives to biological molecules. To start with we would like to suggest a definition of synthetic receptors: these can be considered to be: *synthetic, semisynthetic, or rationally designed biological molecules created with the purpose of selectively binding a single compound or group of structurally similar compounds*. Although the structural and chemical diversity of chemical entities which might have affinity for a particular compound or groups of compounds is very large, the most prominent groups of synthetic receptors include the following (Fig. 2):

- Synthetic peptides with receptor properties
- Macrocycles
- Molecularly imprinted polymers (MIPs)
- Aptamers
- Combinatorially selected compounds
- Supramolecular receptors

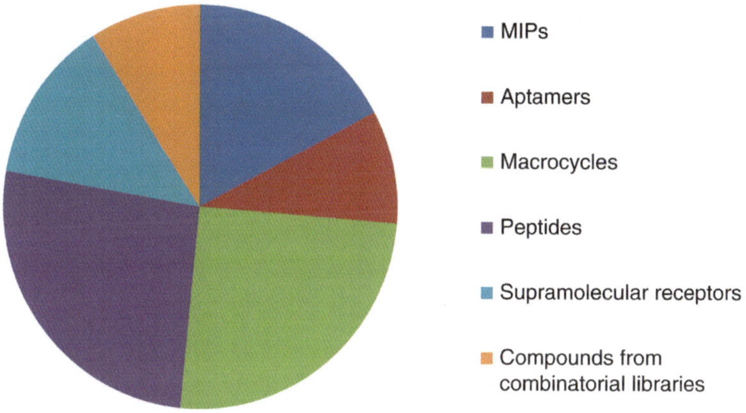

Fig. 2 The relative share of research publications dedicated to various groups of synthetic receptors (calculations made from the analysis of information available at Web of Knowledge)

For this collection we have invited contributions from top experts in the area of synthetic receptors whose work is relevant to diagnostic applications. The particular focus of all chapters is on practical aspects, either in the development process or on the applications of the synthesized materials. We hope this will differentiate our book from other, recently published volumes and make it an interesting and important reference work for business leaders and technology experts in the sensors and diagnostics sectors. In conclusion, we would like to express our deepest gratitude to our colleagues who found the time in their busy schedules to share their results, thoughts, and visions with you, the reader.

Cranfield, Bedfordshire, UK — Sergey A. Piletsky
Cranfield, Bedfordshire, UK — Michael J. Whitcombe

Contents

Multichannel Sensors Based on Biphenyl and Cyclohexane Conformational Changes

Ana M. Costero, Margarita Parra, Salvador Gil, and M. Rosario Andreu

Abstract Conformational changes as a transduction mechanism are considered in both biphenyl- and cyclohexane-based chemosensors. The sensors under study have been designed by following the binding site-signaling unit approach. Modifications in the dihedral angle of the biphenyl and bipyridine system induced by complexation with target molecules can give rise to major changes in the sensor's UV–Vis, fluorescence, and electrochemical properties. This behavior has been used to detect both cations and anions in different solvents. 1,2,4,5-Tetrasubtituted cyclohexane derivatives with a *trans-transoid-trans* configuration and appropriate binding sites have also been used in carboxylate and dicarboxylate sensing. Diastereoisomeric and enantiomeric selectivity has been explored in addition to the discrimination between α,ω-dicarboxylates of different lengths.

Keywords Anion, Biphenyl, Cation, Chemosensor, Cyclohexane, Fluorescence

Contents

A.M. Costero (✉), M. Parra, S. Gil, and M.R. Andreu
Centro de Reconocimiento Molecular y Desarrollo Tecnológico (IDM), Unidad Mixta Universidad de Valencia-Universidad Politécnica de Valencia, Doctor Moliner 50, 46100 Burjassot, Valencia, Spain
e-mail: ana.costero@uv.es

S.A. Piletsky and M.J. Whitcombe (eds.), *Designing Receptors for the Next Generation of Biosensors*, Springer Series on Chemical Sensors and Biosensors (2013) 12: 1–32
DOI 10.1007/5346_2012_14,
Published online: 27 June 2012

Abbreviations

CD	Circular dichroism
CV	Cyclic voltammetry
DC2	Oxalate
DC3	Malonate
DC4	Succinate
DC5	Glutarate
DC6	Adipate
DMSO	Dimethyl sulfoxide
HBA	Hydrogen-bond acceptor
HBD	Hydrogen-bond donor
MeCN	Acetonitrile
SCE	Standard carbon electrode
SQWV	Square wave voltammetry
TBA	Tetrabutylammonium
TBA(*t*-BuO)	Tetrabutylammonium *t*-butoxide
TBAOH	Tetrabutylammonium hydroxide
THF	Tetrahydrofuran
TMA	Tetramethylammonium
TMB	Tetramethylbenzidine

1 Introduction

Molecular systems that combine binding ability and photophysical or electrochemical properties are of much interest for designing chemosensors [1–13]. There are different transduction mechanisms for transmitting the information of the complexation event to the macroscopic world. Among these mechanisms, conformational changes can be used to modify the ligand's photophysical or electrochemical properties. In line with this, useful probes should be rigid enough to guarantee one main conformation under the conditions of measurement. Additionally, those probes that can be studied through two channels are more interesting as this helps enhance selectivity and/or sensitivity.

Herein we summarize the use of several sensors based on biphenyl and cyclohexane scaffolds (Fig. 1). Biphenyl can be used both as signaling unit and simultaneously as transduction moiety. Thus the dihedral angle formed between both aromatic rings in the 2,2′-disustituted biphenyl system plays an important role in the photophysical or electrochemical properties not only of the free ligand but also of the corresponding complexes. So it was that both N. S. Finney [14–16] and A. C. Benniston [17] demonstrated how conformational restriction is a feasible mechanism for transducing ion binding into an enhanced fluorescence emission in organic fluorophores. On the other hand, in appropriately substituted biphenyls,

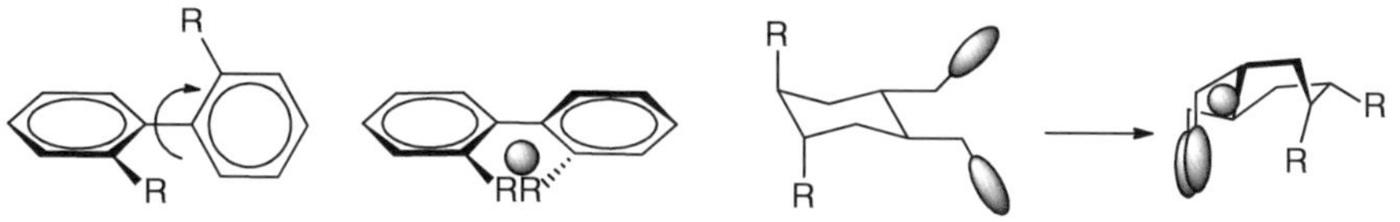

Fig. 1 Recognition event transduction through conformational changes in biphenyl and cyclohexane derivatives

conformational changes can also give rise to marked changes in oxidation/reduction potentials.

Moreover, the rigidity of cyclohexane derivatives can be employed to design photophysical sensors, where the ligand's stereochemistry plays a very important role because it can modify the signaling unit properties. Here we describe the use of *trans*-1,2-disustituted and *trans-transoid-trans*-1,2,4,5-tetrasustituted cyclohexanes which are capable of modifying their conformation after complexation to give rise to marked changes in their photophysical properties.

2 Biphenyl Derivatives

2.1 Recognition and Sensing of Cations

N,N,N′,N′-tetramethylbenzidine (TMB) is a very interesting antenna for the design of both electrochemical and fluorescent chemosensors based on conformational changes. In fact, electrochemical oxidation of TMB systems leads to a radical cation in which both aromatic rings are on the same plane allowing for unpaired electron delocalization (Scheme 1). When this moiety has bonded a binding site at the 2,2′-positions, such as a crown ether or podand chains, the oxidation process can be modified by the presence of different analytes.

Additionally, modification of TMB in its dihedral angle induced by substituents or by the presence of different target molecules (Fig. 2) also leads to modifications in absorption and fluorescence emission spectra (Fig. 3).

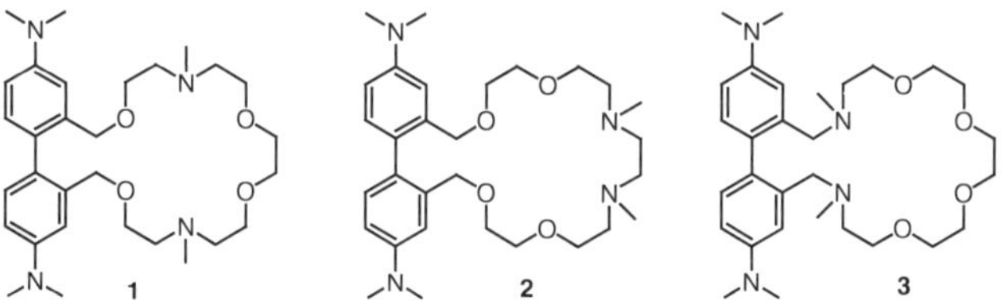

Thus, the fluorescent properties of compounds **1–3** were clearly modified in the presence of transition metal cations [18]. Free ligands show a UV–Vis spectrum in acetonitrile with bands in the UV region centered at 270 nm, which can also give rise to an emission band at 370 nm ($\lambda_{exc} = 270$ nm). For ligands **1** and **2**, the

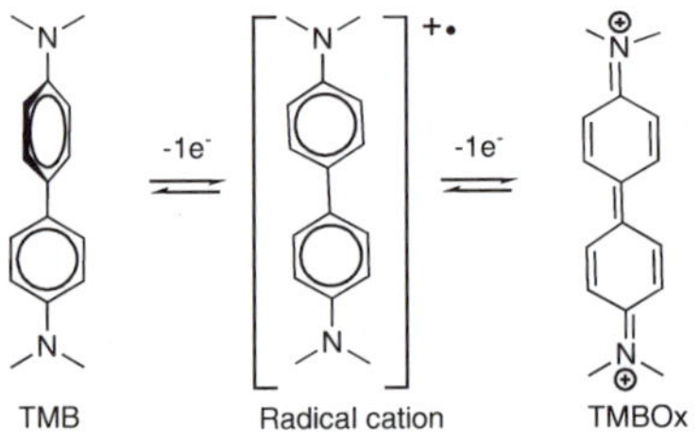

Scheme 1 Oxidation of *N,N,N′,N′*-tetramethylbenzidine (TMB)

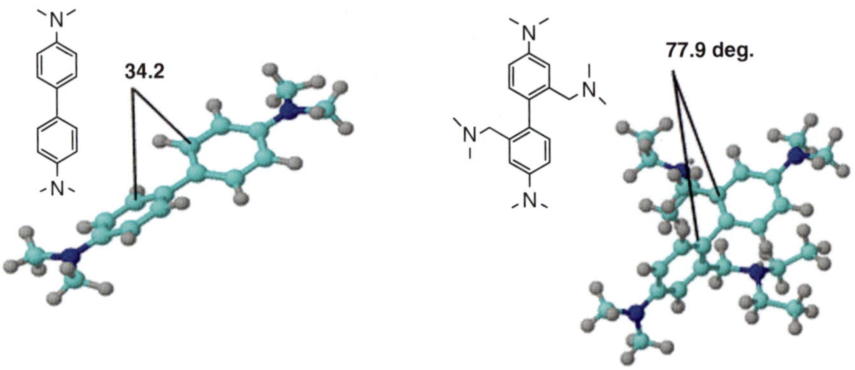

Fig. 2 Dihedral angle in TMB and 2,2′-disustituted TMB

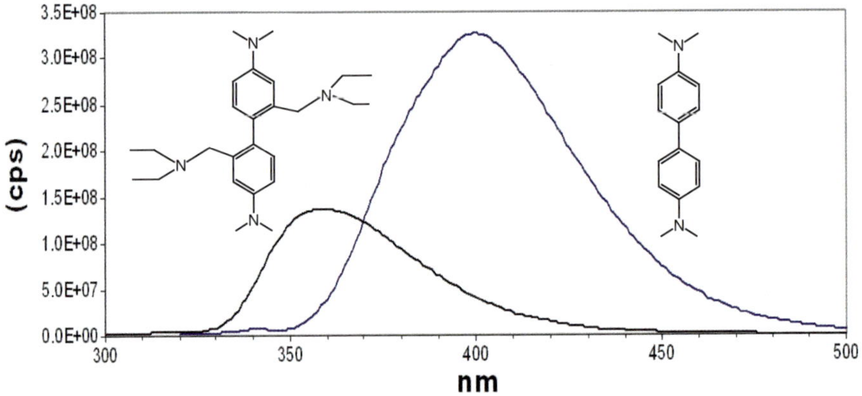

Fig. 3 Influence of the dihedral angle in the fluorescence emission of TMB derivatives

addition of Ni^{2+}, Cu^{2+}, Zn^{2+}, Cd^{2+}, and Hg^{2+} metal ions induces a partial quenching of the emission band centered at 370 nm. Conversely in the presence of Pb^{2+}, both ligands undergo a selective red shift of the emission band of ca. 50 nm (Fig. 4). Compound **3** also displays selective behavior. Thus, the presence of Cu^{2+} produces

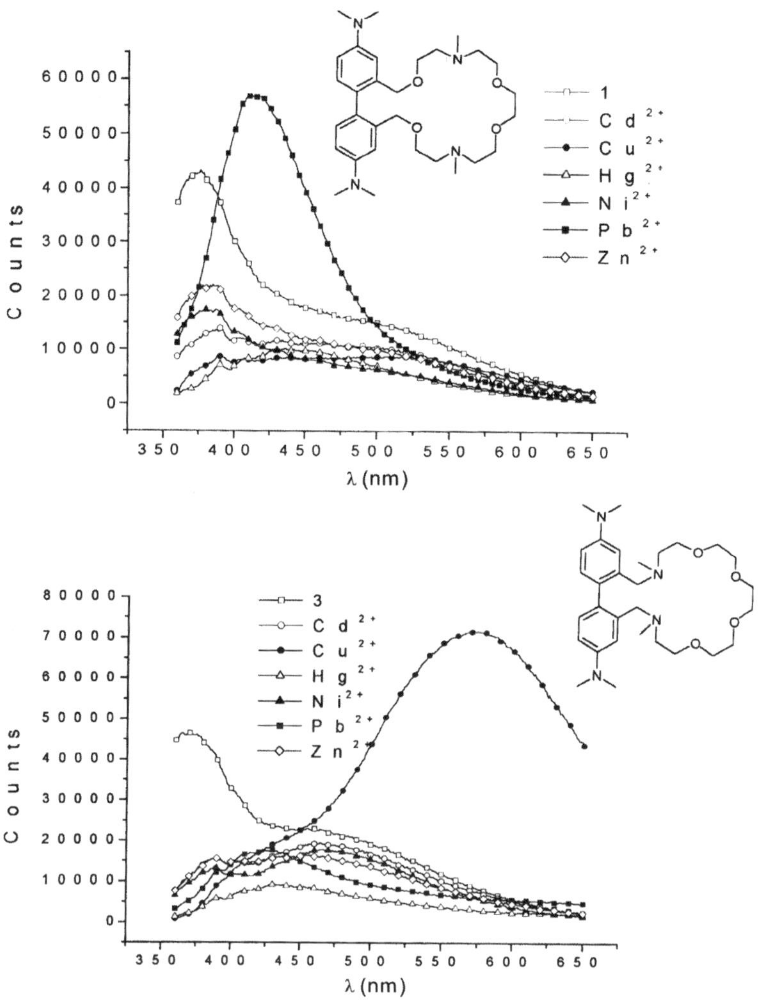

Fig. 4 Emission spectrum of (*left*) **1**, and (*right*) **3** in acetonitrile in the presence of Ni^{2+}, Cu^{2+}, Zn^{2+}, Cd^{2+}, Hg^{2+}, and Pd^{2+}

a new band at ca. 570 nm, shifted 200 nm from the emission band of the free receptor. This new band is obtained only in the presence of Cu^{2+}; other metal ions such as Ni^{2+}, Zn^{2+}, Cd^{2+}, Hg^{2+}, and Pb^{2+} merely produce a quenching of the 367 nm emission band. Studies carried out with ligands **1** and **3** have shown a linear correlation between changes in the optical spectrum and metal ion concentration. On the other hand, the detection limits were 0.7 ppm for Pb^{2+} with ligand **1**, and 0.2 ppm for Cu^{2+} with ligand **3**. Transition metal ions are known to quench fluorescence very effectively. However, it has also been shown that Cu^{2+}, Ni^{2+}, and Pb^{2+} can cause fluorescence enhancement in a cryptand-based fluorophore due to the strong cryptate effect. In addition, similar to the results observed by

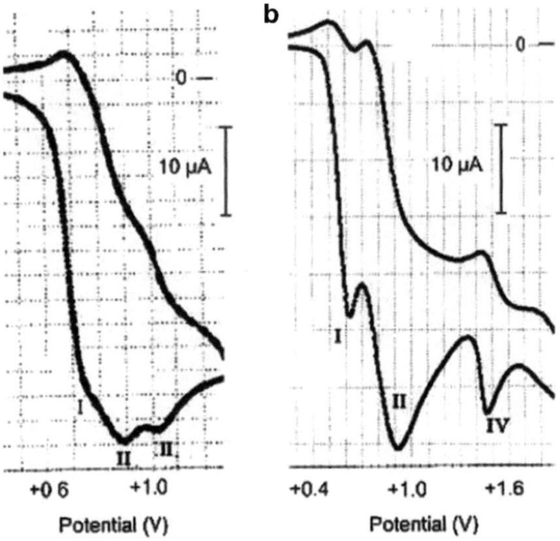

Fig. 5 CVs of (**a**) a 2.0 mM **4** solution in 0.10 M Bu_4NPF_6/MeCN. (**b**) A 2.0 mM **5** solution in 0.10 M Bu_4NPF_6/MeCN. Potential scan rate of 100 mV/s

Bharadwaj et al. [19], no significant reduction wave in the cyclic voltammogram of Cu^{2+}·**3** in the −1.0 to 1.0 range (vs. SCE) was observed and the absence of the redox process $Cu^{2+} \rightarrow Cu^{+}$ in the complex could explain the enhanced fluorescence observed in Cu^{2+}·**3**.

On the other hand, the red shift of the fluorescence emission is too large ($\Delta\lambda = 200$ nm) to be due only to a metal ion-induced change in polarity around the fluorophore. This shift in emission could well result from the modification in the dihedral angle induced by the complexation process. This possibility is supported by the complex's great stability (log $K = 6.30 \pm 0.2$ in acetonitrile (MeCN) for a 1:1 complex). There are similar reasons that could explain the shift observed for Pb^{2+}·**1**.

Related compounds **4** and **5** are able to act as electrochemical sensors for Zn^{2+} and Cd^{2+} [20].

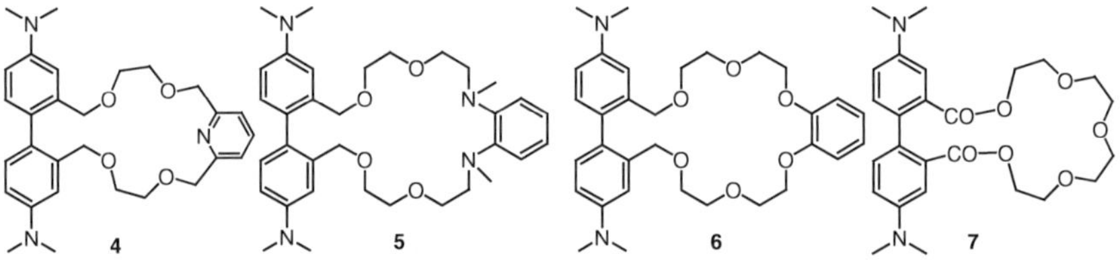

The CV response of compound **4** (10^{-3} mM solution in 0.10 M Bu_4NPF_6/MeCN) at the platinum electrode consists of three anodic peaks which overlap at +0.74 (Ia), +0.92 (IIa), and +1.09 V (IIIa), coupled with their cathodic counterparts at +0.72 (Ic), +0.84 (IIc) (Fig. 5a).

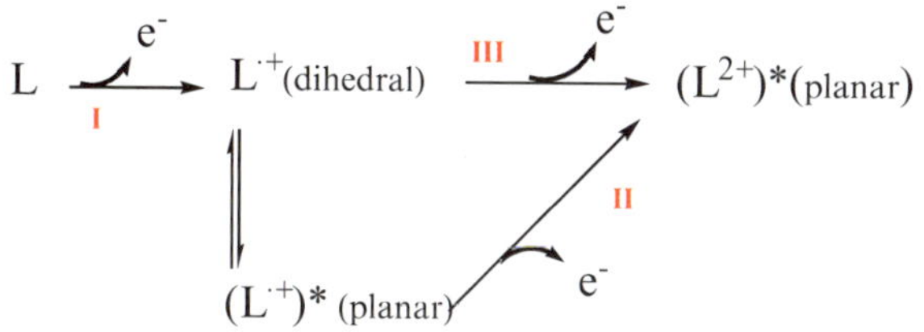

Scheme 2 Mechanism of oxidation of ligand **4**

All three Ia/Ic, IIa/IIc, and IIIa/IIIc pairs can be described as essentially reversible one-electron-transfer processes, as judged from the variation of the anodic-to-cathodic peak potential separation with the Epa–Epc potential scan rate.

Thus, parent neutral ligand, L, is reversibly oxidized to the corresponding radical cation $L^{\cdot +}$ and dication, L^{2+} in two successive one-electron-transfer steps. These correspond to the Ia/Ic and IIa/IIc pairs. The presence of an additional IIIa/IIIc pair is rationalized if we consider that the overall oxidation process is accompanied by a significant stereochemical modification: a transition takes place from the dihedral neutral molecule to the planar dication. Accordingly, the first electron-transfer step yields a nonplanar cation radical ($L^{\cdot +}$), which undergoes a relatively slow pre-organization process to some extent (Scheme 2). Under similar conditions, ligand **5** shows the Ia/Ic (0.62 V) and IIa/IIc (0.91 V) pairs, but not the IIIa/IIIc pair. The absence of these peaks suggests that the oxidation process of ligands exclusively occurs through the nonplanar dication. In addition, a new pair (IVa/IVc) around 1.5 V appears due to the oxidation of *o*-phenylenediamine (Fig. 5b).

In the presence of Cd^{2+} or Zn^{2+}, ligand **4**'s preferred form of oxidation is through the nonplanar radical cation. Thus, peak III totally disappears and overlaps peaks II and I, as observed (Fig. 6). This behavior is due to the small size of this ligand which precludes rotation toward the planar radical cation. In contrast, ligand **5**'s larger size allows the transition metal cations to be located close to the softer nitrogen atoms of the *o*-phenylenediamine moiety and then oxidation through planar geometry is made possible.

These results indicate that both compounds **4** and **5** can be used as electrochemical sensors, their electrochemical response being directly related to possibility of each ligand being able to adopt a coplanar conformation under oxidation conditions. Unfortunately, these ligands are unable to distinguish between $Zn^{2+}$and Cd^{2+}.

On the other hand, compound **6**, which has a related structure, is able to act as a selective fluorescent chemosensor for Hg^{2+}. This ligand shows an emission band at 372 nm, yet a new emission band at 464 nm appears in the presence of Hg^{2+}. This band was attributed to intermolecular excimer formation. Selectivity experiments carried out with mixtures of Zn^{2+}, Cd^{2+}, and Hg^{2+} demonstrate that the presence of Zn^{2+} and Cd^{2+} does not bring about any changes in the response (Fig. 7).

Additionally, the fluorescence studies carried out with compound **7** demonstrate that this compound is able to distinguish between different mercury salts. Thus,

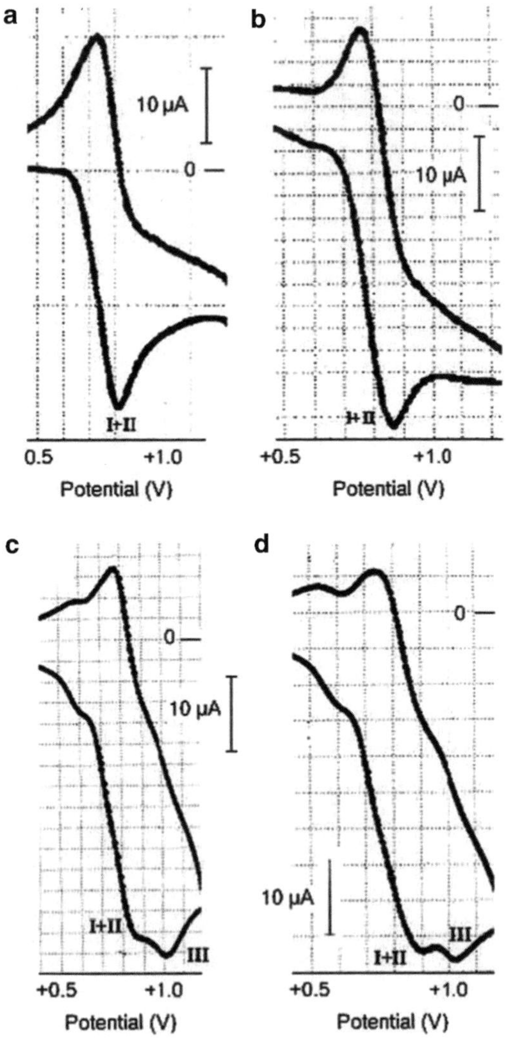

Fig. 6 (From *left* to *right*) CVs of a 2.0 mM **4** + 4 times excess of Zn^{2+} solution in 0.10 M Bu_4NPF_6/MeCN. A 2.0 mM **4** + 4 times excess of Cd solution in 0.10 M Bu_4NPF_6/MeCN. A 2.0 mM **5** + 4 times excess of Zn^{2+} solution in 0.10 M Bu_4NPF_6/MeCN. A 2.0 mM **5** + 4 times excess of Cd^{2+} solution in 0.10 M Bu_4NPF_6/MeCN. Potential scan rate of 100 mV/s

ionic mercury salts, such as $Hg(ClO_4)_2$, give rise to fluorescence quenching, whereas covalent derivatives, such as $Hg(CN)_2$, enhance fluorescence (Fig. 8). These results are related to the different types of interactions between host and guest. Hence, ionic salts place their cations inside the crown cavity, whereas covalent mercury compounds probably interact with the external diamine groups [21].

Compounds **8** and **9**, with related structures yet of different sizes and binding atoms, behave as double channel sensors for some transition metal cations (Zn^{2+} and Cd^{2+}). Thus, these compounds give rise to fluorescence quenching in the presence of transition metal cations due to their coordination inside the coronand cavity.

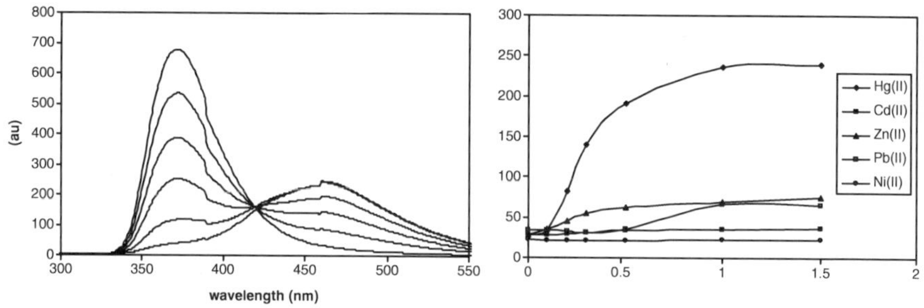

Fig. 7 (*left*) Complexation experiments with ligand **3** in CH_3CN with Hg^{2+} as triflate salts, (*right*). Effect of 1 equiv. of metal ions on the emission at 470 nm for solutions of **6** (3.0×10^{-5} M) in acetonitrile

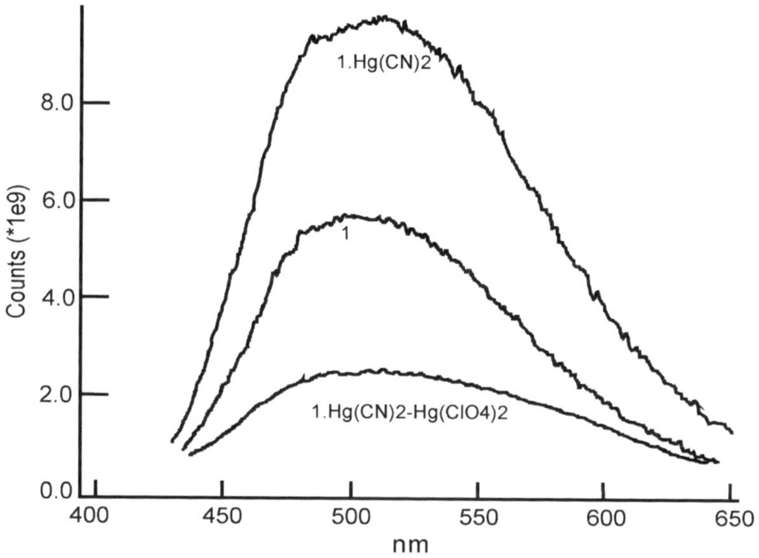

Fig. 8 Selective sensing of ionic and covalent mercury derivatives with ligand **7**

As expected, this coordination modifies the dihedral angle between both biphenyl aromatic rings, with the corresponding modifications in fluorescent properties (Fig. 9).

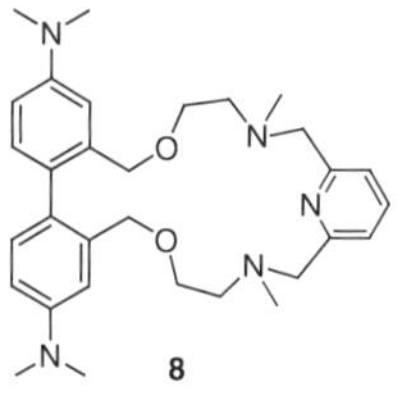

8

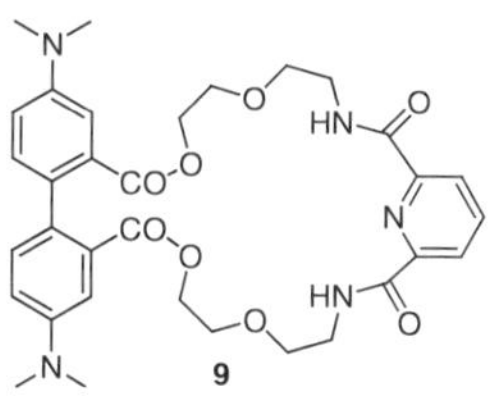

9

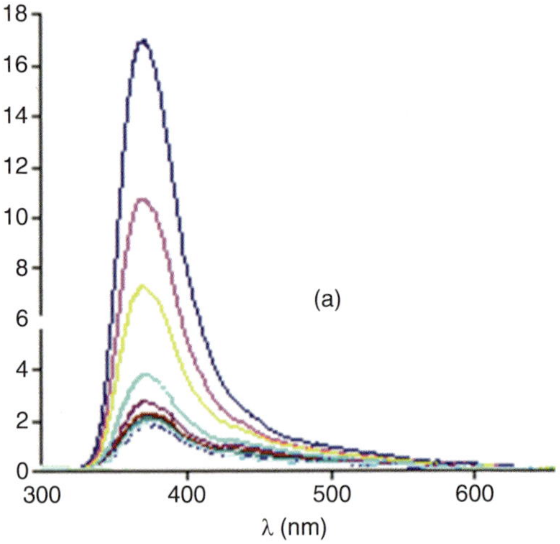

Fig. 9 Fluorescence spectra of **8** after addition of Zn^{2+} at 20°C, $\lambda_{exc} = 300$ nm. The initial ligand concentration was 10^{-5} M in CH_3CN and salt was added until high excess was achieved

Figure 9 shows the quenching experimented by ligand **8** in the presence of increasing amounts of Zn^{2+}. The results obtained with Cd^{2+} are very similar.

From the electrochemical point of view, these ligands exhibit clear modifications in the cyclic voltammetric (CV) data in the presence of transition metal cations. These modifications are reflected in Fig. 10 and are similar to those observed with compounds **4** and **5** [22].

Thus, oxidation takes place through the nonplanar radical cation; consequently, peak III totally disappears and peaks II and I overlap. These results demonstrate that these compounds are able to act as electrochemical sensors for Zn^{2+} and Cd^{2+}. However in this case, selectivity is not observed as the behavior with both cations is very similar.

In order to find selectivity, polyazapodands deriving from TMB were prepared. These ligands offer the advantage of greater flexibility which, despite diminishing complex strength, enables a clamp-type complex formation (Scheme 3).

Thus, compounds **10–12** (Scheme 3) show marked perturbations in their fluorescence spectra, which are associated with complexation by aliphatic amine groups. Thus, ligand **10** displays a quenching of emission intensity of around 90% after the addition of 1 equiv. of metal, and little dependence on metal identity. Given the 1:1 stoichiometry of the binding, the most likely geometry of the complex formation is (a) in Fig. 11, in which both chains are involved in binding the metal.

Since this complexation hinders the rotation of the two phenyl rings, it accounts for the fluorescence quenching observed in the presence of metal ions and the similar behavior noted for all the metals studied (Fig. 12).

On the other hand, compound **11**'s behavior strongly depends on pH. For instance in the case of Zn^{2+}, which is a post-transition metal with no ligand field stabilization energy effects, complexation of the metal is not sufficiently favorable

Fig. 10 CVs of (**a**) a 2.0 mM **9** solution in 0.10 M Bu_4NPF_6/MeCN; (**b**) **9** + four times excess of Zn^{2+}. Potential scan rate of 100 mV/s

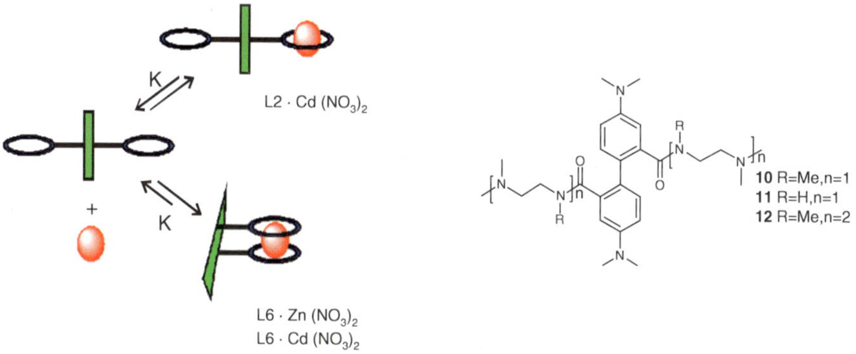

Scheme 3 Possible complex formed by ligands **10–12**

in energetic terms to be accompanied by amidic hydrogen removal. For this very reason, when no base is present in the medium, it interacts with Zn^{2+} in the same way as ligand **10** does, resulting in 80% quenching. When an excess of base (tetrabutylammonium hydroxide) is added, amidic hydrogen is removed and, in this situation, Zn^{2+} is chelated by both nitrogens of the same branch (Fig. 11b). This disposition does not constrain the torsion angle between the aromatic biphenyl rings and quenching is not induced.

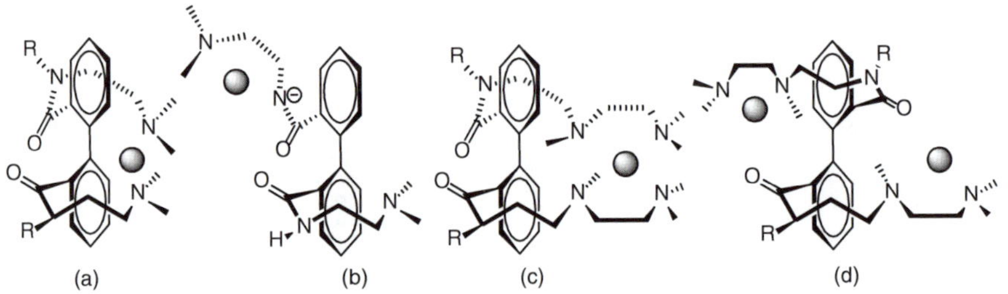

Fig. 11 Possible complexing arrangements suggested to account for the fluorescence results obtained for the different ligands and cations

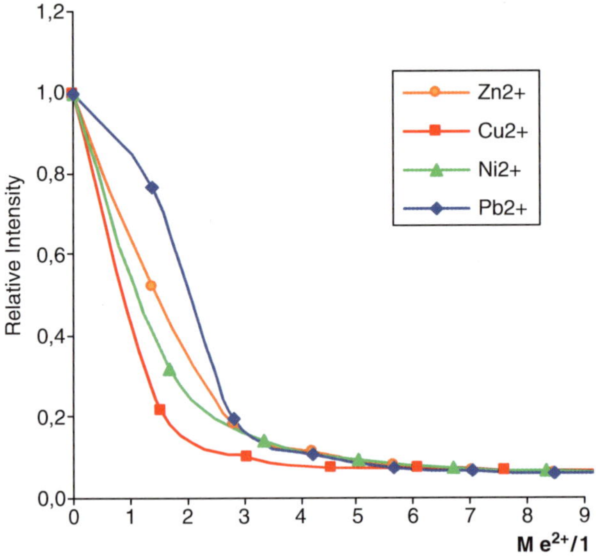

Fig. 12 Fluorescence emission ($\lambda_{exc} = 340$ nm) of **10** in the presence of different transition metal cations in acetonitrile ($T = 20°C$). Concentration of **10** 6×10^{-6} M, TMAOH 1.3×10^{-5} M and $TBAPF_6$ of 1×10^{-4} M

Ligand **12** provides more interesting results with two possible, yet distinct, binding stoichiometries and depends on the identity of the metal ion. Cu^{2+}, Ni^{2+}, and Pb^{2+} give a 1:1 complex, even in the presence of a high cation concentration. Fluorescence quenching is observed in each case (Fig. 13) along with a considerable shift of the wavelength of the maximum emission for Cu^{2+} and Pb^{2+}. In this case, the geometry involves both amino chains in complexation, as indicated above, with this type of complex influencing the rotation of the biphenyl moiety, which gives rise to the corresponding fluorescence quenching. In contrast, addition of Cd^{2+} and Zn^{2+} initially leads to fluorescence quenching with a concomitant red shift, but it subsequently results in enhanced fluorescence due to the formation of new complexes with the 1:2 stoichiometries (Fig. 11d) [23].

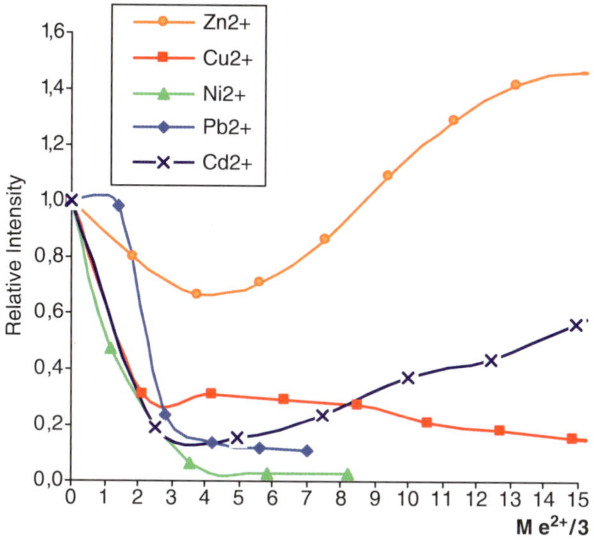

Fig. 13 Relative intensity emission of fluorescence ($\lambda_{exc} = 340$ nm) in tritiation of **12** in the presence of different transition metal cations in acetonitrile ($T = 20$°C). Concentration of **12** 4.4×10^{-6} M, TMAOH 1.3×10^{-5} M and $TBAPF_6$ 1×10^{-4} M

Reducing the amide groups to the corresponding amino derivatives gives rise to compounds **13–15**.

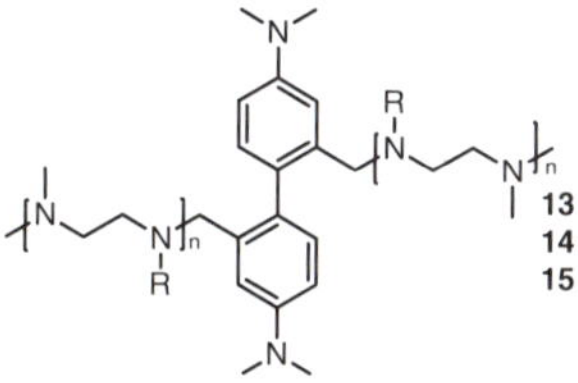

The sensing properties of these compounds were studied with Zn^{2+}, Cd^{2+}, Pb^{2+}, Ni^{2+}, and Cu^{2+} as triflate salts. The results obtained with Cd^{2+}, Pb^{2+}, and Ni^{2+} are similar for all three ligands, and fluorescence quenching occurs (Fig. 14). This behavior may be due not only to the presence of the transition metal cation in the solution but also to the modification in the dihedral angle between aromatic rings or the conformational restriction induced by the complexation event.

In contrast, Cu^{2+} and Zn^{2+} display a completely different behavior when complexation experiments are carried out with ligands **13–15**. The behavior observed with these two cations depends on salt concentration. Thus in the case of Zn^{2+}, when the cation is in solution, a quenching of the main emission band is observed, which also occurs with other studied cations. Nonetheless, when the presence of Zn^{2+} is above 1 equiv., the broadband centered at ca. 550 nm increases in intensity. Similar behavior is noted for Cu^{2+} but, in this case, the band starts growing even with smaller amounts of the cation (0.7 equiv.). A very large amount of Zn^{2+} (about 16 equiv.) is required to achieve a band of a similar intensity to that obtained with

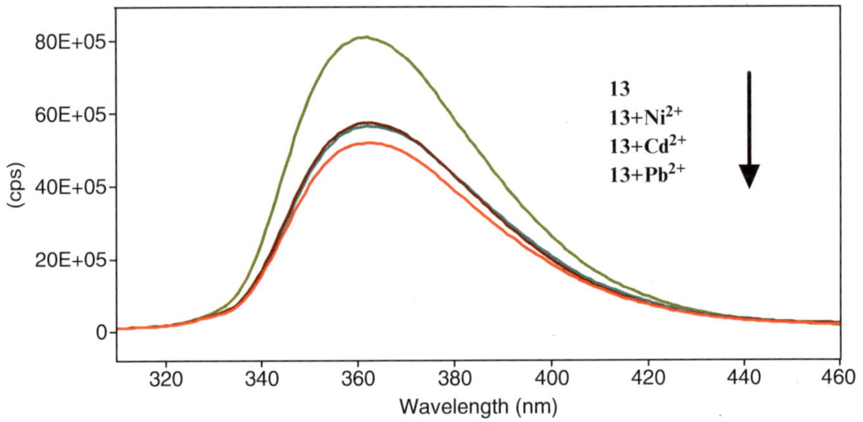

Fig. 14 Fluorescence spectra of **13** after addition of ca. 4.5 equiv. of Ni^{2+}, Cd^{2+}, and Pb^{2+} at 20°C, $\lambda_{exc} = 300$ nm. The initial concentration of ligand was ca. 5.6×10^{-6} M in CH_3CN

only 2 equiv. of Cu^{2+}. The results obtained with ligand **13** and copper are reflected in Fig. 15. These new bands are apparently due to intermolecular excimers. So after adding 2.93 equiv. of Cu^{2+}, the excitation spectrum of **14** at $\lambda = 372$ nm resembles the corresponding UV spectrum, while a shift in the maximum of the corresponding excitation spectrum at $\lambda = 494$ nm is obtained. This finding, therefore, reveals that the species emitting at these two wavelengths are not similar in nature. Even though dependence between complex concentration and the intensity of the new band is observed, intensity values do not directly relate to the expected square of concentration for intermolecular excimer formation [24].

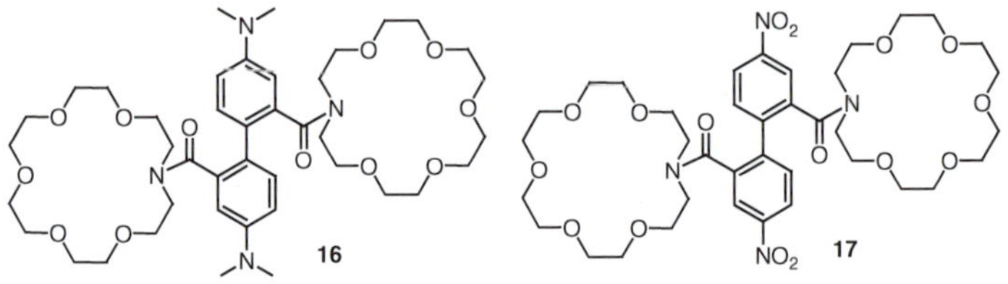

16 17

Compound **16** is a bis-crown ether deriving from TMB, which combines the possibility of forming clamp complexes with the macrocyclic effect, and prove most important in complexation processes [25]. This compound is able to recognize both Zn^{2+} and Cd^{2+} with a 1:1 stoichiometry. The UV–visible absorbance spectrum of ligand **16** in acetonitrile displays strong absorbance in the UV region, centered at 295 nm ($\varepsilon = 29{,}600$) and with a shoulder at 333 nm ($\varepsilon = 5{,}680$). Upon excitation at 340 nm, compound **16** exhibits an intense fluorescence emission band centered at 474 nm with a quantum yield of 0.08 (Fig. 16). These values compare with a measured quantum yield of 0.04 for tetramethylbenzidine under the same conditions, indicating significantly enhanced fluorescence when these substituents are introduced.

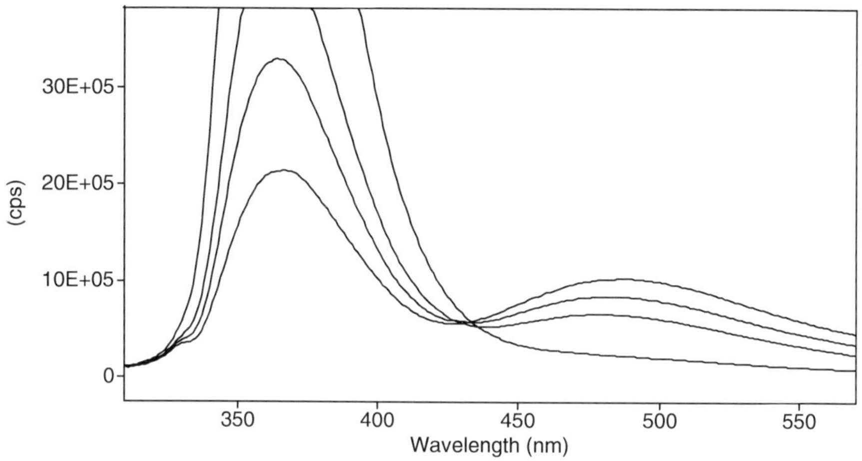

Fig. 15 Fluorescence spectra of **13** after addition of Cu^{2+} at 20°C, $\lambda_{exc} = 300$ nm. Initial concentration of ligand is 5.7×10^{-6} M in CH_3CN. Spectra correspond to additions of 0, 0.70, 1.40, and 2.10 equiv. of Cu^{2+}, respectively

Titration of ligand **16** against Zn^{2+} or Cd^{2+} (Fig. 16) in acetonitrile leads to a pronounced and red-shifted shoulder in the UV spectra, and a well-defined isosbestic point at 340 nm. Addition of Zn^{2+} or Cd^{2+} (as triflate salts) to the acetonitrile solutions of ligand **16** gives rise to a red shift in the emission spectrum (e.g., the emission maximum shifted by 17 nm after the addition of 2 equiv. of Zn^{2+}, and by 12 nm in the case of Cd^{2+}; $\lambda_{exc} = 340$ nm). Use of the corresponding nitrate salts in acetonitrile offers similar results. Additionally, both cations induce partial fluorescence quenching, with Zn^{2+} displaying a more marked effect (76% quenching after 2 equiv. of metal) than Cd^{2+} (38% quenching under the same conditions) (Fig. 16). The significant changes noted in the fluorescence spectrum of ligand **16** upon complexation of Zn^{2+} or Cd^{2+} are indicative of a profound interaction and effect on conformation.

These results are consistent with a clamp complex whose formation is accompanied by a substantial modification of the value in the dihedral angle between both aromatic rings. This change is responsible for the perturbations observed in absorption and emission bands. The greater extent of quenching observed with zinc compared to cadmium is consistent with the larger equilibrium constant associated with the clamp-type complex formation for zinc. When competitive experiments are carried out, ligand **16** selectively responds to Zn^{2+} in the presence of Cd^{2+}. This ligand's greater affinity for zinc over cadmium seems to be related to energetic factors: although clamp compound formation certainly results in loss of entropy, it is, on the other hand, accompanied by a gain in enthalpy due to a larger number of oxygen–metal interactions. Thus for both metal complexes, the enthalpic factor must outweigh the entropic one when the clamp complex is formed. Zn^{2+} is a significantly harder metal ion than Cd^{2+} due to its smaller size and, for this reasons, a more marked increase in enthalpy is expected when it forms a clamp

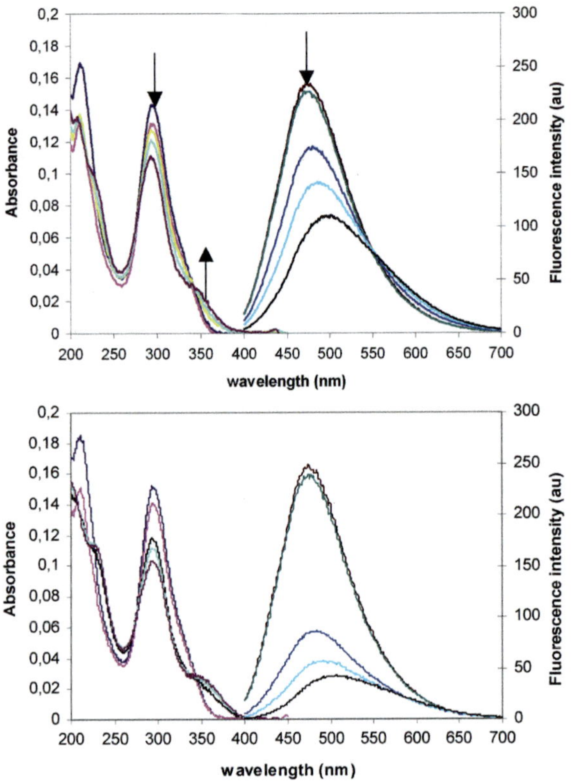

Fig. 16 UV–visible absorbance (from 200 to 400 nm) and fluorescence emission ($\lambda_{exc} = 340$ nm) spectra (from 400 to 700 nm) of **16** (4.76×10^{-6} M in acetonitrile) and upon addition of increasing concentrations of (*left*) Cd$(Tf)_2$ (1–10 equiv.), (*right*) Zn$(Tf)_2$ (1–10 equiv.) both in the presence of Bu_4NClO_4 (1.41×10^{-4} M)

complex if compared to Cd^{2+}, and also as a result of a better "hard-to-hard" interaction with oxygen.

In order to investigate the influence of the substituents in the biphenyl moiety on sensing properties, compound **17** was studied under similar conditions. This ligand is also able to form 1:1 complexes with Zn^{2+} and Cd^{2+} as triflate salts. In this case however, the geometries of the complexes differ from those observed in ligand **16**. Thus, addition of either Zn^{2+} or $Cd^{2+}$to **17** brings about absolutely no change in either the intensity or the wavelength of the maximum, indicating that the dihedral angle in the TMB unit is not changed. This fact disproves the formation of a clamp complex involving both crown cavities. On the other hand, the ^{1}H NMR studies carried out with ligand **17** agree with a symmetrical structure for these complexes; for all these reasons, a fast interchange, which makes both cavities equivalent on the NMR time scale, seems to be the most likely complexation (Scheme 4).

Compound **17** also displays interesting electrochemical behavior. The cyclic voltammetric (CV) response of the receptor dissolved in acetonitrile (0.10 M Bu_4NPF_6) is illustrated in Fig. 17a. In the initial cathodic scan, two well-defined reduction peaks appear at -1.02 (C1) and -1.94 V (C2), and the former is preceded by a weak shoulder near -0.74 V (C3). In the subsequent anodic scan, an ill-defined shoulder appears at -1.55 V (A2), followed by two overlapping peaks at -0.88

Scheme 4 Type of complexes proposed

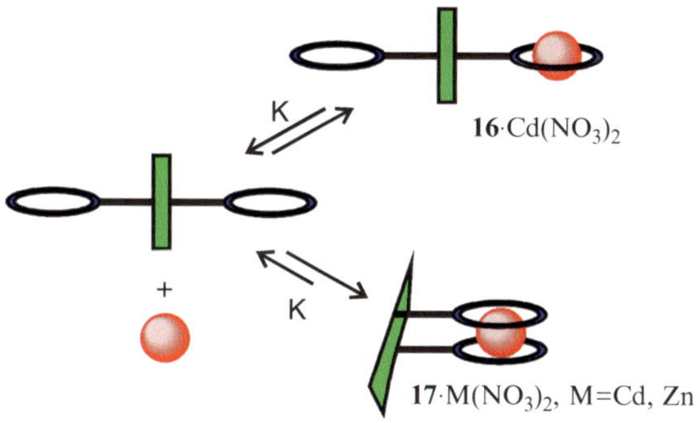

(A1) and −0.61 V (A3). The latter peak disappears in CVs if the potential is reversed at potentials close to −1.25 V, suggesting that peak A3 corresponds to the oxidation of any species generated during electrode process C2 (Fig. 17b).

The observed response is described on the basis of the well-known electrochemistry of the aromatic and nitroaromatic compounds in aprotic solvents, consisting of two successive one-electron-transfer processes yielding an anion radical and a dianion. The voltammetric profile, however, depends largely on the intermediate radical anion's stability. If this is unstable if compared with the disproportionateness into dianion and the parent aromatic compound, the voltammogram resembles a single two-electron wave. This is the case of the studied receptors; here the obtained half-peak width value of process C1, measured in square wave voltammetries (SQWVs) as 115 mV, is clearly lower than that expected for a reversible one-electron process (126 mV). The overall electrochemical process can be represented as:

$$\mathbf{L} + 2^{e^-} \rightarrow \mathbf{L}^{2-}.$$

As the SQWVs show in Fig. 18, noncoordinated **17** displays isolated peaks C_1 and C_2. On addition of Cd^{2+}, additional peaks appear at −0.49 (C_5) and −1.64 V (C_6), whereas peak C_1 is resolved in two peaks overlapping at −0.92 and −1.03 V. Electrode process C_5 is described in terms of the biphenyl-centered reduction of the complex. This is represented as:

$$\mathbf{L} + xM^{2+} + 2^{e^-} \rightarrow (M^{2+})_x(\mathbf{L}^{2-}).$$

Electrode process C_6 is ascribed to metal-centered reduction:

$$(M^{2+})_x(\mathbf{L}^{2-}) + 2x^{e^-} \rightarrow xM^0 + \mathbf{L}^{2-}.$$

This process takes place at considerably more negative potentials than reduction of uncomplexed Cd^{2+}, which is as expected for the electrochemical reduction of

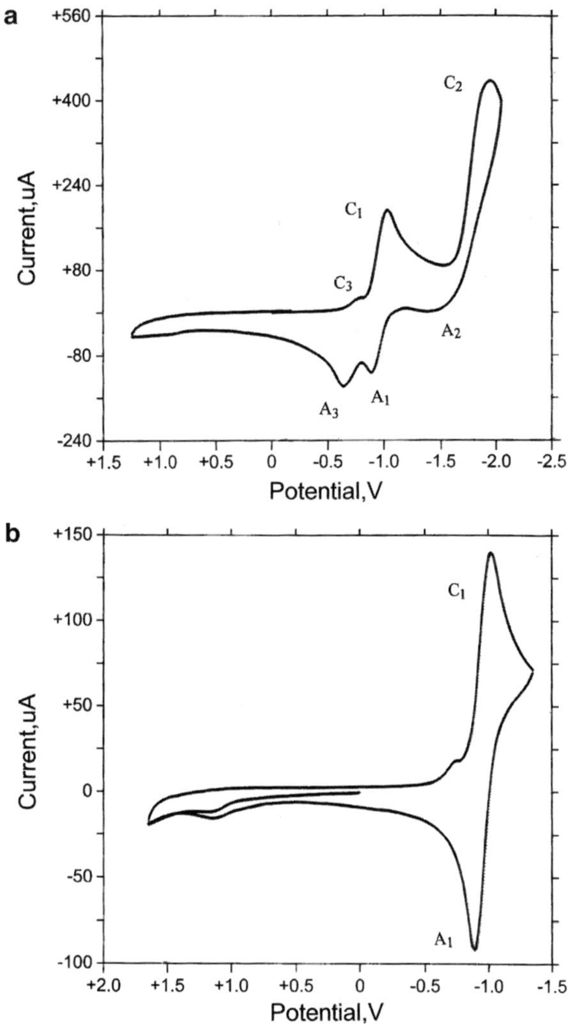

Fig. 17 CVs at the GCE of a 2.0 mM solution of **17** in MeCN (0.10 M Bu_4NPF_6). Potential scan rate of 200 mV/s. (*left*) Potential range: +1.25/−2.05 V; (*right*) potential range: +1.65/−1.35 V

metal ions to metal deposits. The results obtained in the presence of Zn^{2+} are similar to those described above for Cd^{2+}.

Complexation constants and complexes stoichiometry were determined for both Zn^{2+} and Cd^{2+}. The 2:1 stoichiometry obtained corresponds to that of the reduced complexes formed between M^{2+} and the receptor dianion resulting from the biphenyl-centered reduction process.

Reduction of the biphenyl moiety of **17** gives rise to the formation of a planar anion, that is, able to complex Zn^{2+} and Cd^{2+} with an LM_2 stoichiometry (Scheme 5). This behavior is directly related to the planarity of the dianion, which precludes the possibility of the "clamp-type" complexes observed under different experimental conditions [26].

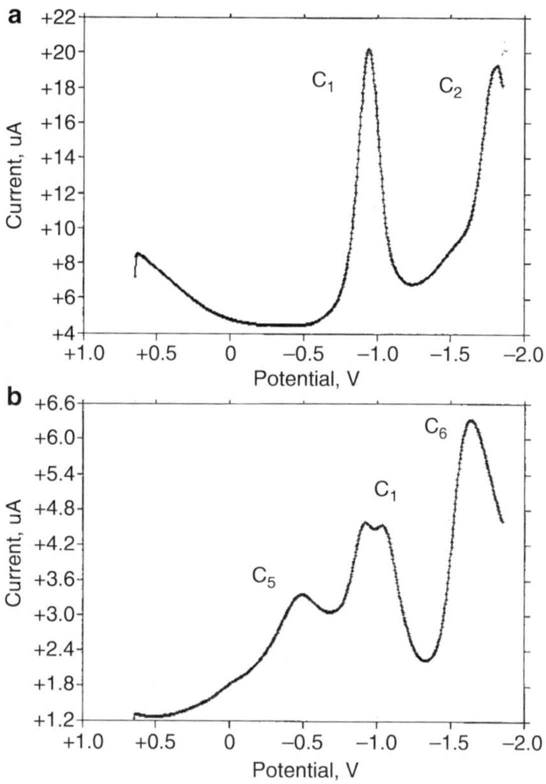

Fig. 18 SQWVs at a Pt electrode of (**a**) 0.48 mM **17**, (**b**) 2.16 mM Cd^{2+} plus 1.69 mM **17** solutions in MeCN (0.10 M Bu_4NPF_6). Potential step increment of 4 mV, square wave amplitude of 25 mV, frequency of 15 Hz

2.2 *Recognition and Sensing of Anions*

Demand for chemosensors which are selective for specific anions continuously increases [5, 8, 10, 27–35]. With this in mind, one especially important group is sensors that monitor small organic anions of biological interest. Inside this group, we are interested in α,ω-dicarboxylates as they play key roles in different biochemical process [36–38]. Major efforts have centered on the design and application of chemosensors based on the binding site-signaling unit approach for anion recognition using ureas, thioureas, or their amido derivatives, given their ability to act as H-bond donors. Moreover, they are highly suitable for carboxylate recognition with the appropriate geometry [39, 40].

As we indicate above, biphenyl moieties have demonstrated that the conformational changes induced by modifying the dihedral angle have a strong influence on the fluorescent properties of sensors. A similar behavior has been observed in bipyridines, and for this reason, ligands based on this system have been also considered.

As a first approach to colorimetric sensors of anions, the selective sensing of fluoride was studied. Thus, amide-based macrocyclic amine sensors **18–20** were

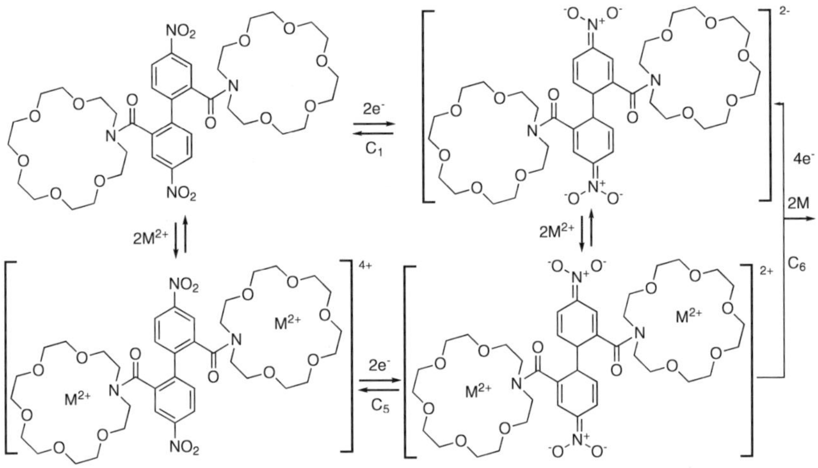

Scheme 5 Complexation of the planar reduction product of ligand **17** with Zn^{2+} and Cd^{2+} with a LM_2 stoichiometry

prepared and studied [33, 34]. The reactivity of compound **9** to anions was also studied for comparison purposes.

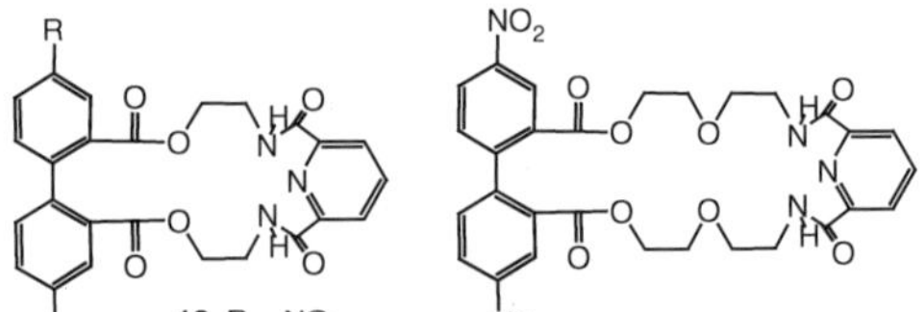

18 R = NO_2
20 R = $N(CH_3)_2$
19

Anion complexation experiments were carried out by adding 10 equiv. of the corresponding anion (F^-, Cl^-, Br^-, I^-, $H_2PO_4^-$), as tetrabutylammonium (TBA) salt, to the solutions (10^{-2} M) of the different receptors in acetonitrile.

The most remarkable effect was the selective color noted with **18** and **19** upon the addition of fluoride, either alone or in competitive assays (Fig. 19). The NMR studies of the complexation process reveal that two consecutive reactions take place in the presence of fluoride. Thus, after adding 1 equiv. of fluoride, a complex with a 1:1 stoichiometry and the geometry shown in Fig. 20 (left) formed.

However, color development was due to the deprotonation reaction taking place in the presence of at least 2 equiv. of fluoride. The color exhibited by ligands **18** and **19** upon the addition of fluoride ion is probably due to deprotonation, which leads to a charge-transfer complex [Fig. 20 (right)]. This suggestion agrees not only with the color extinction which occurs in the presence of water, but also with the results obtained in the additional experiments carried out with tetrabutylammonium *t*-butoxide [TBA(*t*-BO)] and tetrabutylammonium hydroxide (TBAOH). Finally, this proposal is reinforced by the behavior of ligand **20**; its substitution, with dimethylamino instead of nitro groups, precludes the formation of the charge transfer, and no color develops in the presence of fluoride [22]. From these results, it is possible to conclude that the electronic effect of substituents on the biphenyl moiety influences complex stability and sensing properties.

In order to determine whether the presence of the cavity controls the complexation and sensing capacity, a related open compound **21** was prepared [36].

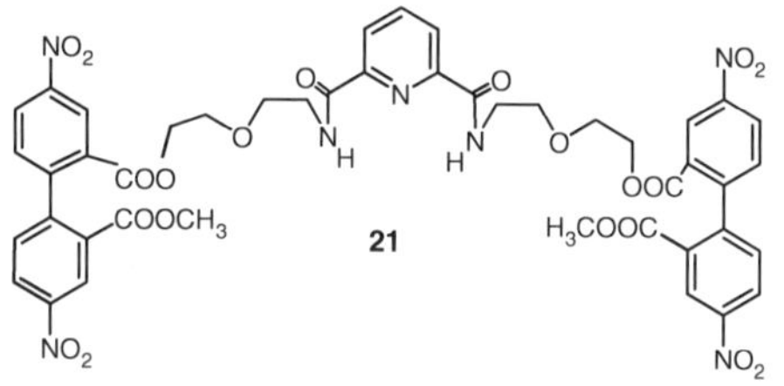

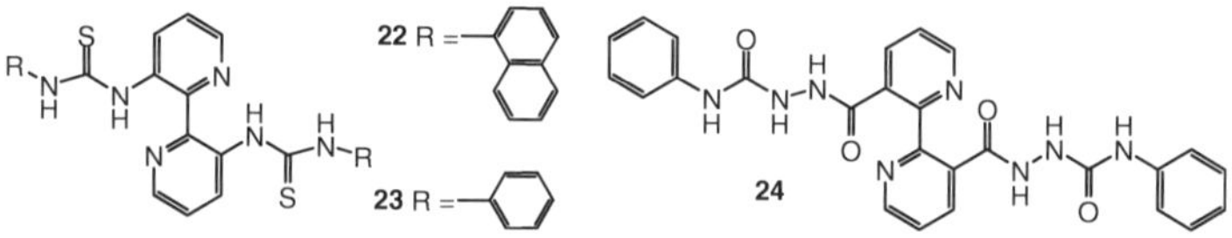

The complexation studies done with different halides, such as TBA salts, show a similar behavior to that previously described for ligands **18** and **19** but, in this case, the sensing response is faster. The solution changes in color from pale yellow to dark red in the presence of fluoride, immediately after addition. This change is due to a deprotonation reaction promoted by this anion's strong basicity.

The difference in kinetics is likely to be due to the greater flexibility of ligand **21**, which allows the anion easy access to the amide hydrogens. The sensitivity of **21** (0.001 M) is 1.5 mM for the naked eye and 0.1 mM for UV conditions. These detection limits are similar to those described in the literature for other fluoride chemosensors [37]. As observed with ligands **18** and **19**, even though coordination occurs also with bromide or chloride, no color changes are noted in these cases given the lower basicity of these anions, which are unable to produce ligand deprotonation.

Bipyridine-based ligands, compounds **22–24**, have been demonstrated to be suitable as anion sensors. Thioureas and amide derivatives were used as the binding site for dicarboxylate recognition [41, 42].

2,2′-Bipyridine is one of the most widely used samples of organic linkers. Furthermore, the chemical modification of the 2,2′-bipyridine ligands by introducing an additional functionality at the 4-, 5-, and 6- positions is well known, and the coordination chemistry of such compounds has been actively explored. However, fewer 3,3′-disubstituted ones are available and the coordination chemistry of only a low number of them has been studied. These ligands are very interesting because they present two coordination points; thus, they are able to

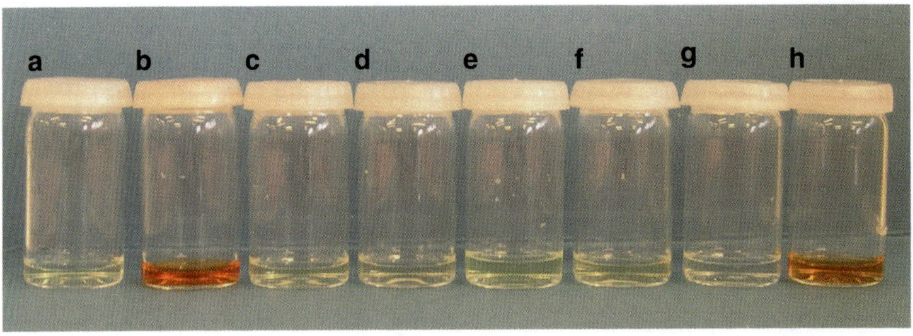

Fig. 19 From *left* to *right* (**a**): free ligand **18**, (**b**) **18** + F^-, (**c**) **18** + Cl^-, (**d**) **18** + Br^-, (**e**) **18** + I^-, (**f**) **18** + HSO_4^-, (**g**) **18** + $H_2PO_4^-$, (**h**) **18** + F^- + Cl^- + Br^- + I^- (10 equiv. of tetrabutylammonium salts for each anion were used)

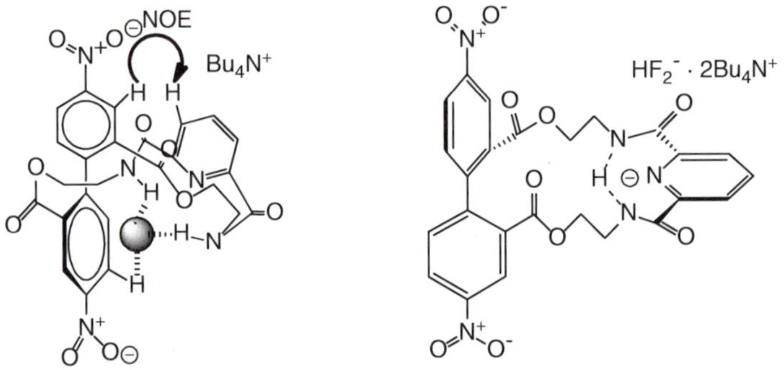

Fig. 20 1:1 complex formed between **18** and F^- (*left*); deprotonated specie responsible for the color (*right*)

complex anions at the thiourea moiety, as well as cations such as Ni^{2+} at the bipyridine moiety. This fact can confer allosteric properties to these ligands.

Optical properties of bipyridine systems are strongly dependent on the solvent [43, 44]. Thus, compound **24**, for example, shows a maximum of absorption at 236 nm in acetonitrile and at 258 nm in dimethyl sulfoxide (DMSO). In both solvents, a shoulder at 280 nm is observed, which corresponds to the phenylamido chromophore. In order to clarify this compound's solvatochromic behavior, a set of solvents was used. It was concluded that solvents with strong hydrogen-bond acceptor (HBA) properties [DMSO, tetrahydrofuran (THF) and ethyl acetate] induce a bathochromic shift, whereas those with strong hydrogen-bond donor (HBD) properties (methanol and ethanol) give rise to a clear hypsochromic effect of the UV absorption maxima. Differences in the absorption maximum can relate to the presence of different conformations in solution, depending on the solvent's characteristics. Thus, ligands **22** and **23** appear in DMSO solutions as *s-trans* conformers, which is stabilized through the formation of hydrogen bonds between the NH of the thiourea group attached to one ring with the pyridine nitrogen in the

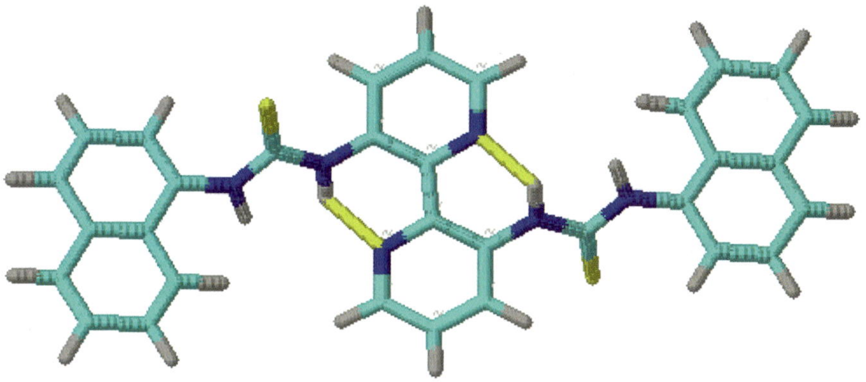

Fig. 21 Conformation calculated by PcModel for ligand **22**

other ring, and between the H (at the *para* position to one pyridine unit) and the sulfur atom of the thiourea in the same ring (Fig. 21).

Protonation experiments were carried out because the literature establishes that the absorption band of bipyridine systems shows a clear bathochromic shift after monoprotonation, which takes place giving a conformational change between the *s-trans* and the *s-cis* forms. However, ligands **22** and **23** display a different behavior; in both cases, the UV spectra exhibited no changes upon protonation. This suggests that protonation occurs on the sulfur atom and not on the nitrogen atom. This suggestion agrees with the evolution observed in solution where a hydrolysis of the ligands was observed. In contrast, complexation with $NiCl_2$ induces the expected conformational change from the *s-trans* to the *s-cis* conformation with the corresponding modification in the UV spectra. However, modifications in fluorescence proved more interesting; ligand **22** gives an emission band at 420 nm (λ_{exc} = 290 nm), which undergoes a slight bathochomic shift after protonation (426 nm), but with an enhancement of 130%, which is probably due to increased rigidity (Fig. 22).

These ligands are able to complex different α,ω-dicarboxylates, which leads to modifications in fluorescence emission. The ability of these new ligands to act as receptors for carboxylates in DMSO was studied by UV, fluorescence, and NMR spectroscopy. All three ligands form complexes with acetate and different dicarboxylates, but their stoichiometries depend on the nature of both the ligand and the carboxylate (see Table 1).

Ligands **22** and **24** give rise to 1:2 complexes with both dicarboxylates and acetate, whereas ligand **23** forms 1:2 complexes with acetate and 1:1 complexes with dicarboxylates. These observations may relate to the different substituents on thiourea moieties and their steric hindrance (Fig. 23).

The different stoichiometries of the complexes allow ligand **23** to act as a sensor which is able to discriminate between mono- and dicarboxylates. In fact, the complexes with a 1:2 stoichiometry strongly enhance fluorescence, whereas those with a 1:1 stoichiometry bring about minor changes (Fig. 24).

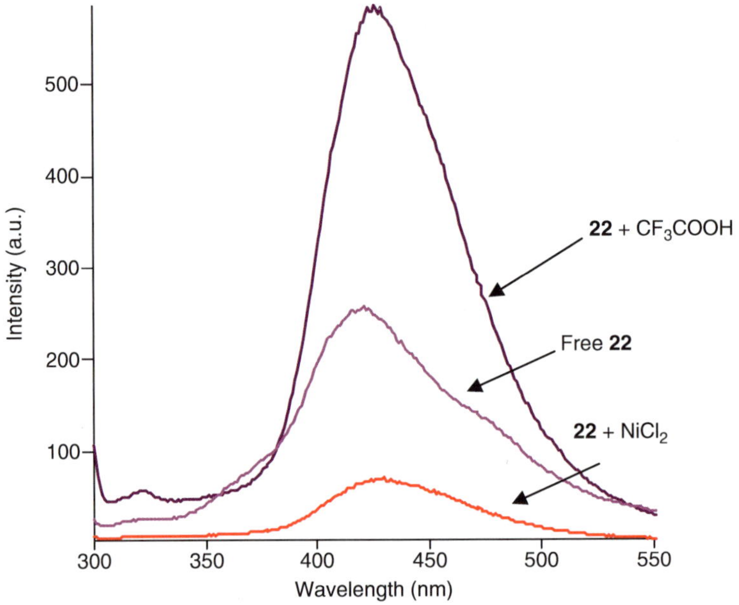

Fig. 22 Fluorescence spectra of free, protonated (immediately after protonation) and Ni^{2+} complexed ligand **22**

Table 1 Stoichiometry and the overall binding constants (log β) in DMSO for **22**–**24** with various anions measured by UV–Vis spectroscopy at 25°C

	22		**23**		**24**	
Anion[a]	Log β	L:A	Log β	L:A	Log β	L:A
Oxalate	7.2 ± 0.2	1:2	6.0 ± 0.7	1:1	7.7 ± 0.3	1:2
Malonate	7.7 ± 0.1	1:2	5.7 ± 0.6	1:1	8.9 ± 0.2	1:2
Succinate	7.0 ± 0.2	1:2	6.2 ± 0.7	1:1	7.9 ± 0.2	1:2
Acetate	7.4 ± 0.1	1:2	8.3 ± 0.1	1:2	–	–

[a]All the anions were used as their TBA salts

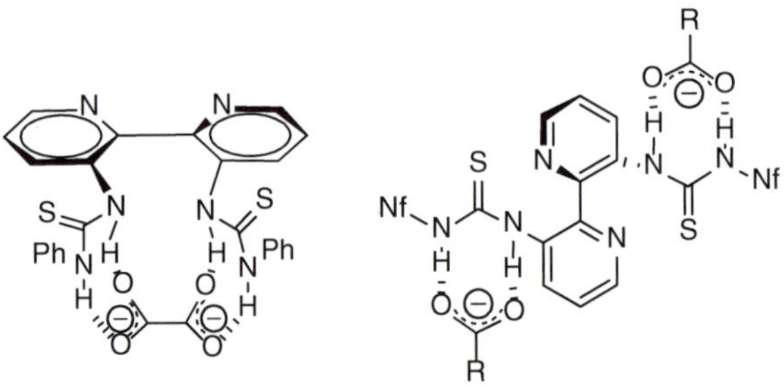

Fig. 23 Geometry of the 1:1 and 1:2 complexes

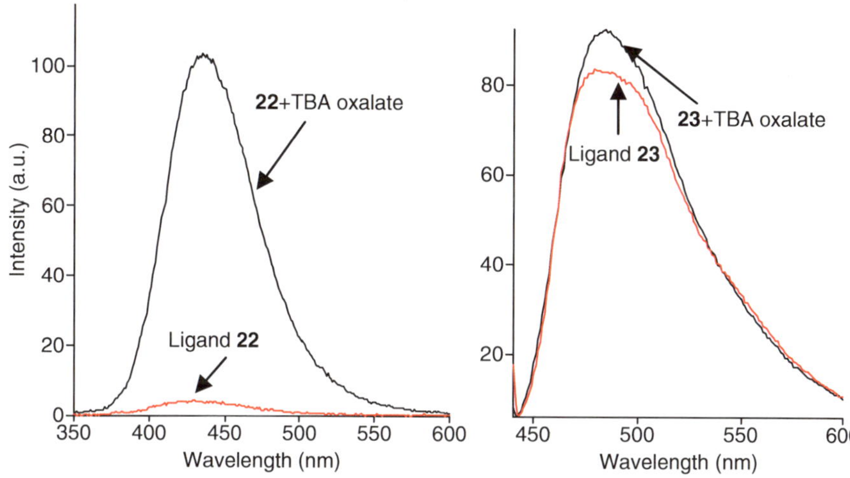

Fig. 24 Ligand **22** with TBA oxalate (in DMSO $\lambda_{exc} = 340$ nm) (*left*) and ligand **23** with TBA oxalate (in DMSO $\lambda_{exc} = 435$ nm) (*right*)

3 Cyclohexane Derivatives in Anion Recognition

A different approach to conformational-regulated fluorescent sensors for specific anions was explored by using cyclohexane derivatives with the appropriate configuration. Thus, *trans-transoid-trans* 1,2,4,5-tetrasubstituted cyclohexanes have proved useful in recognition processes, and the rigidity of this system has also been established to control the complex geometry [45].

The first studies refer to ligands **25** and **26** as racemic compounds (Fig. 25). The preferred conformation of these ligands in DMSO was unambiguously established by ^{1}H NMR. Thus, cyclohexane moiety is mainly in a chair conformation with the thioureas at equatorial sites and, therefore, axial ester groups. Theoretical studies have also confirmed that this conformation is a relative energy minimum.

The fluorescence studies carried out with these ligands in the presence of increasing amounts of fumarate and maleate anions show that ligand **25** is able to act as a selective sensor for maleate (MA) versus fumarate (FA) in DMSO solutions [46]. The fluorescence spectrum of compound **25** presents only an emission band at 410 nm ($\lambda_{exc} = 290$ nm) after adding increasing amounts of TMA fumarate only slight fluorescence quenching is observed (Fig. 26a). Conversely, addition of TMA maleate results in a new band at 495 nm (Fig. 26b). This fluorescent behavior appears to be related to the stoichiometry of the complexes being 1:2 for fumarate and 1:1 for maleate, which bring about different geometries. Thus, the maleic dianion with a *cis* configuration perfectly fits into the 1:1 complex inducing a conformational change in the ligand. This change places both naphthalene groups close and almost in parallel alignment with each other, which leads to excimer emission. In contrast, the fumaric dianion with a *trans* disposition of the carboxylate moieties is unable to form the 1:1

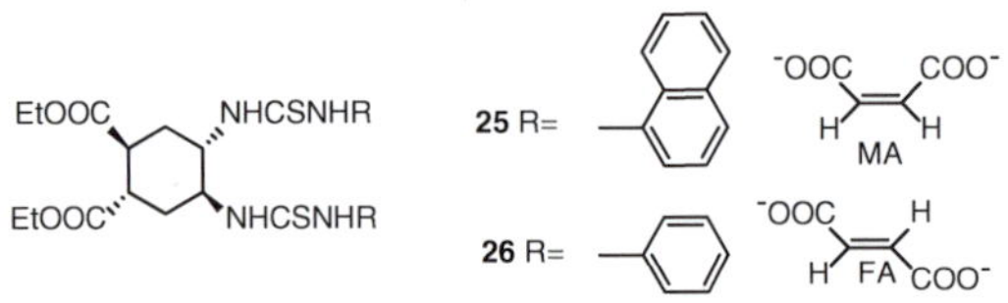

Fig. 25 Structure of racemic compounds **25** and **26**

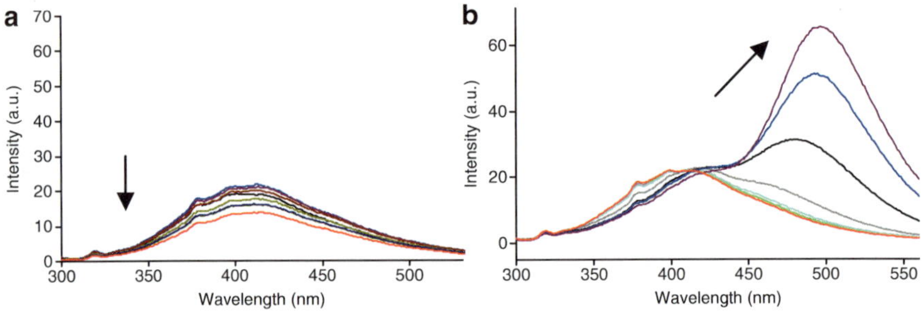

Fig. 26 The fluorescence spectra in DMSO (**a**) **25** + TMA fumarate and (**b**) **25** + TMA maleate ($\lambda_{exc} = 290$ nm)

complex, and two molecules bind to the ligand. This complex does not induce changes in conformation, and only minor modifications in fluorescence are noted. In conclusion, compound **25** is able to distinguish between diastereoisomeric dicarboxylates (Fig. 26), and sensing properties are also observed even in the presence of 5% water.

In order to verify this proposal, ^{1}H NMR experiments were carried out which clearly established that, in the complex formed between the host and TMA maleate, the cyclohexane moiety adopts a boat-like conformation which places both naphthalene groups close together in the space (Fig. 27).

Sensing experiments in DMSO were also performed toward aliphatic α,ω-dicarboxylates of different chain lengths [oxalate (DC2), malonate (DC3), succinate (DC4), glutarate (DC5), and adipate (DC6)], all in the form of their TMA salts [47]. Titration experiments, followed by ^{1}H NMR, UV and fluorescence spectroscopy, led to the stoichiometry and complexation constants values shown in Table 2.

The 1:1 complexes formed with the larger dicarboxylates can be explained by the complexation of each of the guest's carboxylate with one of the thiourea groups in the ligand (Fig. 28). Conversely, small dicarboxylates (i.e., oxalate) cannot be accommodated in one host as the complexation of the first carboxylate places the second one far away from the other thiourea group, and 1:2 complexes are a more stable arrangement for this guest.

The fluorescence study resulted in more interesting differences. Thus, the fluorescence spectra of ligand **25** resulting upon addition of increasing amounts of TMA succinate and malonate exhibited a new band (around $\lambda = 490$ nm), whose intensity can be related to the formation of an excimer species similarly induced by

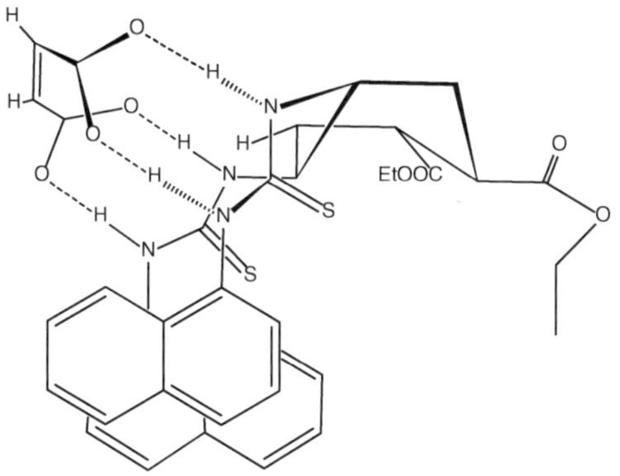

Fig. 27 Structural proposal for the complex formed between ligand **25** and TMA maleate

Table 2 Stoichiometry and log β for ligands **25** and **26** with TMA salts of oxalate (DC2), malonate (DC3), succinate (DC4), glutarate (DC5), and adipate (DC6) in DMSO by UV–Vis spectroscopy

Ligand		DC2	DC3	DC4	DC5	DC6
25	Log β	7.9 ± 0.2	2.8 ± 0.2	3.41 ± 0.01	3.3 ± 0.2	3.86 ± 0.03
	L:DC	1:2	1:1	1:1	1:1	1:1
26	Log β	3.66 ± 0.03	3.8 ± 0.4	–	–	3.1 ± 0.4
	L:DC	1:2	1:1	–	–	1:1

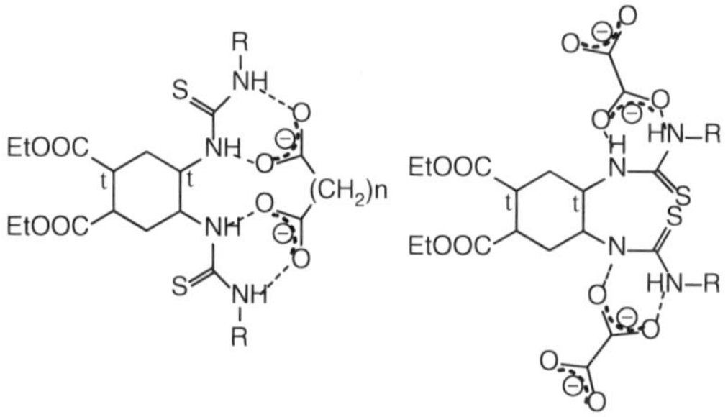

Fig. 28 Proposal structures for 1:1 and 1:2 complexes between ligand **25** and α,ω-dicarboxylates

the complexation observed for maleate. However, oxalate with a different complex stoichiometry does not induce marked changes in fluorescence. In conclusion, ligand **25** is able to distinguish the shortest dicarboxylate from longer dianions. These results are of much interest because discrimination of dicarboxylates according to their chain length is not usual [48, 49].

Ligands **25** and **26** can be used as sensors through a second channel: UV–Vis spectroscopy. Therefore, complexation with the studied dicarboxylates gives rise to modifications in the UV–Vis spectra, which depend on the complex stoichiometry.

In any case, the observed modifications in fluorescence are more intense than the corresponding changes in UV.

Related racemic ligands with only one thiourea group and a free hydroxyl group, **27** and without ethoxycarbonyl groups, **28** have been also evaluated. Compound **27** brings about a new band at 350 nm upon complexation with dicarboxylates whose intensity depends on chain length reaching a maximum for DC4. Regarding the fluorescence channel, CD3 and DC4 cause intermolecular excimer formation with the corresponding band at 295 nm. The complexes formed with DC2, DC5, and DC6 present no modifications in the fluorescence spectra.

On the other hand, the studies carried out with ligand **28** reveal that the presence of the two ethoxycarbonyl groups is essential for the ligand to act as a sensor. In the absence of these groups the higher flexibility of the ligand precludes both inter and intramolceular excimer formation [50].

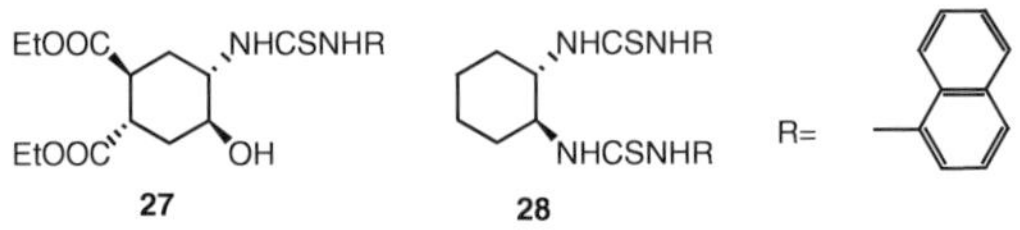

The good results obtained with cyclohexane-based naphthylthiourea ligands prompted us to synthesize the homochiral series to explore their application in chiral sensing [51]. Thus, ligands **25** and **27** were obtained as pure enantiomers, and their absolute configurations were determined from the circular dichroism spectra with the help of the exciton chirality rule [52].

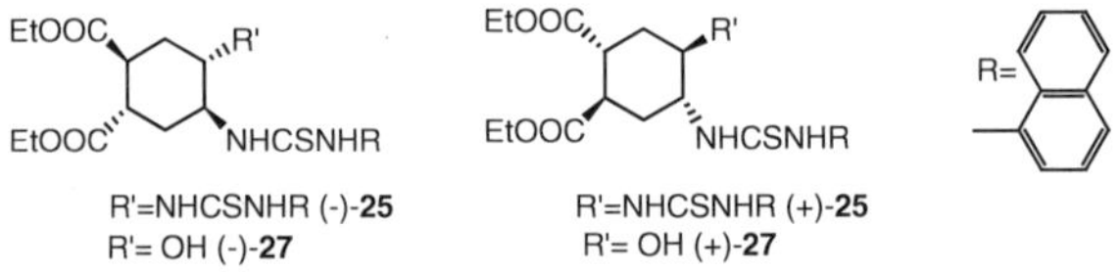

From the fluorescence titration studies performed with TMA salts of L-aspartate, D-aspartate, L-glutamate, and D-glutamate, the complexes' stoichiometry and equilibrium constants were determined. They follow a similar trend to that previously described, but the excimer band is observed only for ligands (+)-**25** with D-aspartate and (−)-**25** with L-aspartate, which leads to 1:1 complexes. The R configuration of D-aspartate perfectly fits the complex formed with (+)-**25** and induces a conformational change in the ligand, which places both naphthalene groups close and in an almost parallel arrangement to give rise to intense excimer emission. A less intense emission is observed upon the addition of (+)-**25** with L-aspartate and (−)-**25** with D-aspartate as its configuration precludes an effective approach of both naphthyl moieties.

Similar results were obtained upon the addition of L- and D-glutamate to the solutions of (+)-**25** and (−)-**25**. However, less intense excimer bands were observed in these cases. Longer chain length probably leads to a less tight arrangement compared with that for aspartate.

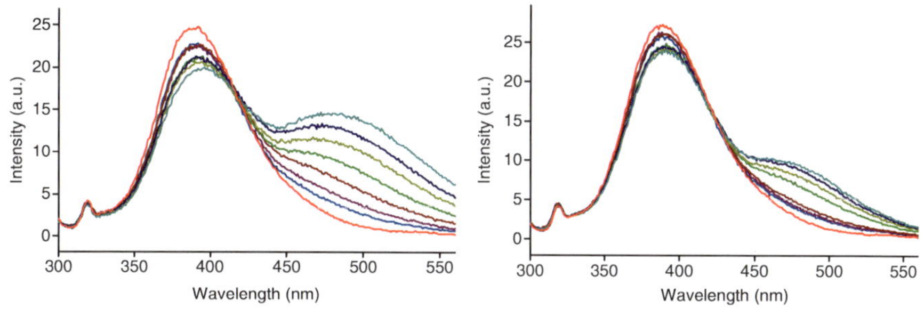

Fig. 29 Fluorescent spectrum of (+)-**27** (*left*) in the presence of TMA L-aspartate and (*right*) in the presence of TMA D-aspartate

Both (+)-**27** and (−)-**27** modify the fluorescence spectrum with L-aspartate showing a more marked effect (Fig. 29). With these ligands, the formed excimer should be intermolecular and in agreement with the 2:1 stoichiometry shown in the complexes. This stoichiometry may also be responsible for the similar effect that the same enantiomer has with both ligands because, in this case, the role played by the stereochemistry of the stereocentre can only relate to steric factors.

The above-described results clearly demonstrate that these cyclohexane derivatives have sensing properties when faced with different types of carboxylates, where the sensing mechanism relates directly to changes in ligand conformation. Changes in the different groups' spatial disposition determine the possible formation of the excimer responsible for the sensing response. However, it is important to point out that complex geometry is dependent on complexes stoichiometry, where 1:1 complexes give rise to more pronounced modifications in the ligand's cyclohexane moiety. It is also important to bear in mind that the sensing mechanism is possibly due to the rigidity of the cyclohexane ring and to the correct configuration of substituents.

4 Final Remarks

In summary, even though the literature reports a large number of transduction mechanisms used in designing chemical sensors, many of them are related to an electronic transmission of the changes induced by the binding event. We herein describe several examples relating to information transmission through conformational changes. The sensors designed by following this approach have to be based on partially rigid structures, which may undergo profound changes after complexation, and this change in conformation should induce modifications in the photophysical or electrochemical properties.

References

1. De Silva AP, Gunaratne HQG, Huxley CP, McCoy CP, Radamacher JT, Rice TE (1997) Signaling recognition events with fluorescent sensors and switches. Chem Rev 97:1515–1566
2. Full issue of (2000) Coord Chem Rev 205:1–232
3. Martínez-Máñez R, Sancenón F (2003) Fluorogenic and chromogenic chemosensors and reagents for anions. Chem Rev 103:4419–4476
4. Gale PA (2006) Structural and molecular recognition studies with acyclic anion receptors. Acc Chem Res 39:465–475
5. Katayev EA, Ustynyuk YA, Sessler JL (2006) Receptors for tetrahedral oxyanions. Coord Chem Rev 250:3004–3037
6. Davis P (2006) Anion binding and transport by steroid-based receptors. Coord Chem Rev 250:2939–3051
7. Schmidtchen FP (2006) Reflections on the construction of anion receptors: is there a sign to resign from design? Coord Chem Rev 250:2918–2928
8. Yoon J, Kim SK, Singh NJ, Kim KS (2006) Imidazolium receptors for the recognition of anions. Chem Soc Rev 35:355–360
9. Bowman-James K (2005) Alfred Werner revisited: the coordination chemistry of anions. Acc Chem Res 38:671–678
10. Suksai C, Tuntulani T (2003) Chromogenic anion sensors. Chem Soc Rev 32:192–202
11. Sessler JL, Camiolo S, Gale PA (2003) Pyrrolic and polypyrrolic anion binding agents. Coord Chem Rev 240:17–55
12. Gale PA (2001) Anion receptor chemistry: highlights from 1999. Coord Chem Rev 213:79–128
13. Schmidtchen FP, Berger M (1997) Artificial organic host molecules for anions. Chem Rev 97:1609–1646
14. McFarland SA, Finney NS (2001) Fluorescent chemosensors based on conformational restriction of a biaryl fluorophore. J Am Chem Soc 123:1260–1261
15. McFarland SA, Finney NS (2002) Fluorescent signaling based on control of excited state dynamics. Biarylacetylene fluorescent chemosensors. J Am Chem Soc 124:1178–1179
16. Mello JS, Finney NS (2001) Dual-signaling fluorescent chemosensors based on conformational restriction and induced charge transfer. Angew Chem Int Ed 40:1536–1538
17. Benniston AC, Harriman A, Patel PV, Sams CA (2005) A strategy for controlling the central torsion angle in biphenyl-based molecular-scale bridges. Eur J Org Chem 2005:4680–4686
18. Costero AM, Andreu R, Monrabal E, Martínez-Máñez R, Sancenón F, Soto J (2002) 4,4′-Bis (dimethylamino)biphenyl containing binding sites. A new fluorescent subunit for cation sensing. J Chem Soc Dalton Trans 2002:1769–1775
19. Ghosh P, Bharadwaj PK, Mandal S, Ghosh S (1996) Ni(II), Cu(II), and Zn(II) cryptate-enhanced fluorescence of a trianthrylcryptand: a potential molecular photonic OR operator. J Am Chem Soc 118:1553–1554
20. Costero AM, Bañuls MJ, Aurell MJ, Domenech A (2006) 4,4′-Subtituted biphenyl coronands. Preparation of a new selective fluorescent sensor for mercury salts. Tetrahedron 62:11972–11978
21. Costero AM, Andreu R, Martínez-Máñez R, Sancenón F, Soto J (2002) A fluorescent chemosensor able to distinguish between ionic and covalente mercury compounds. J Inclusion Phenom Mol Recognit Chem 46:121–124
22. Costero AM, Bañuls MJ, Aurell MJ, Ochando LE, Doménech A (2005) Cation and anion fluorescent and electrochemical sensors derived from 4,4′-substituted biphenyl. Tetrahedron 61:10309–10320
23. Costero AM, Sanchis J, Gil S, Sanz V, Ramirez de Arellano MC, Williams JAG (2004) Polyazapodands derived from biphenyl. Study of their behaviour as conformationally regulates fluorescent sensors. Supramol Chem 16:435–446

24. Costero AM, Sanchis J, Gil S, Sanz V, Williams JAG (2005) Poly(amine)biphenyl derivatives as fluorescent sensors for anions and cations. J Mater Chem 15:2848–2853
25. Costero AM, Gil S, Sanchis J, Peransí S, Sanz V, Williams JAG (2004) Conformationally regulated fluorescent sensors. Study of the selectivity in Zn^{2+} versus Cd^{2+} sensing. Tetrahedron 60:6327–6334
26. Costero AM, Sanchis J, Peransi S, Gil S, Sanz V, Domenech A (2004) Bis(crown ethers) derived from biphenyl: extraction and electrochemical properties. Tetrahedron 60:4683–4691
27. Gale PA (2003) Anion and ion-pair receptor chemistry: highlights from 2000 and 2001. Coord Chem Rev 240:191–221
28. Carvalho S, Delgado R, Fonseca N, Felix V (2006) Recognition of dicarboxylate anions by a ditopic hexaazamacrocycle containing bis-p-xylyl spacers. New J Chem 30:247–257
29. Liu S-Y, He Y-B, Wu J-L, Wei L-H, Qin H-J, Meng L-Z, Hu L (2004) Calix[4]arenes containing thiourea and amide moieties: neutral receptors towards α, ω-dicarboxylate anions. Org Biomol Chem 2:1582–1586
30. Singh N, Lee GW, Jang DO (2008) p-tert-Butylcalix[4]arene-based fluororeceptor for the recognition of dicarboxylates. Tetrahedron 64:1482–1486
31. Nie L, Lie Z, Han J, Zhang X, Yang R, Liu W-X, Wu F-Y, Xie J-W, Zhai Y-Z, Jiang Y-B (2004) Development of N-benzamidothioureas as a new generation of thiourea-based receptors for anion recognition and sensing. J Org Chem 69:6449–6454
32. Evans LS, Gale PA, Lighat ME, Quesada R (2006) Anion binding vs. deprotonation in colorimetric pyrrolylamidothiourea based anion sensors. Chem Commun 965–967
33. Costero AM, Bañuls MJ, Aurell MJ, Ward MD, Argent S (2004) Biphenyl macrolactames in anion complexation. Selective naked-eye fluoride recognition. Tetrahedron 60:9471–9478
34. Costero AM, Bañuls MJ, Aurell MJ, Ramirez de Arellano MC (2006) Biphenyl macrolactames as colorimetric sensors for anions through displacemente reactions. J Incl Phenom Macrocycl Chem 54:61–66
35. Whitlock BJ, Whitlock HW (1990) Concave functionality: design criteria for nonaqueous binding sites. J Am Chem Soc 112:3910–3915
36. Costero AM, Peransí S (2007) Colorimetric sensing of anions by a neutral biphenyl based amide receptor. Arkivoc IV:92–101
37. Amendola V, Esteban-Gómez D, Fabbrizzi L, Licchelli M (2006) What anions do to N–H-containing receptors. Acc Chem Res 39:343–353
38. Neumann T, Dienes Y, Baumgartner T (2006) Highly sensitive sensory materials for fluoride ions based on the dithieno[3,2-b:2′,3′-d]phosphole system. Org Lett 8:495–497
39. Badr IHA, Meyerhoff ME (2005) Highly selective optical fluoride ion sensor with submicromolar detection limit based on aluminum(III) octaethylporphyrin in thin polymeric film. J Am Chem Soc 127:5318–5319
40. Lin Z, Zhao Y, Duan C, Zhang B, Bai Z (2006) A highly selective chromo- and fluorogenic dual responding fluoride sensor: naked-eye detection of F^- ion in natural water via a test paper. J Chem Soc Dalton Trans 2006:3678–3684
41. Costero AM, Gil S, Parra M, Allouni Z, Lakhmiri R, Atlamsani A (2008) Complexation of α, ω-dicarboxylates by 3,3′-bis(5-phenyl-1,4-dioxo-2,3,5-triaza)-2,2′-bipyridine. J Inclusion Phenom Mol Recognit Chem 62:203–207
42. Costero AM, Gil S, Parra M, Huguet N, Allouni Z, Lakhmiri R, Atlamsani A (2008) New 3,3′-disubstitued 2,2′-bipyridines as carboxylate receptors. Conformational regulation of the bipyridine moiety. Eur J Org Chem 2008:1079–1084
43. Palmans ARA, Vekemans JAJM, Meijer EW (1995) Intramolecular hydrogen bonding in acylated 2,2′-bipyridine-3,3′-diamines. Recl Trav Chim 114:277–284
44. Mutai T, Araki K (2007) Fluorescent oligopyridines and their photo-functionality as tunable fluorophores. Curr Org Chem 11:195–211
45. Samoshin VV (2005) Cyclohexane-based conformationally controlled crowns and podands. MROC 2:225–235

46. Costero AM, Colera M, Gaviña P, Gil S (2006) Fluorescent sensing of maleate *versus* fumarate by a neutral cyclohexane based thiourea receptor. Chem Commun 761–763
47. Costero AM, Colera M, Gaviña P, Gil S, Llaosa U (2008) Fluorescent chemosensors based on cyclohexane: selective sensing of succinate and malonate versus their longer or shorter homologues. Tetrahedron 64:7252–7257
48. Ragusa A, Rossi S, Hayes JM, Stein M, Kilburn JD (2005) Novel enantioselective receptors for N-protected glutamate and aspartate. Chem Eur J 11:5674–5688
49. Pu L (2004) Fluorescence of organic molecules in chiral recognition. Chem Rev 104:1687–1716
50. Costero AM, Llaosa U, Gil S, Parra M, Colera M (2009) Enantioselective sensing of dicarboxylates. Influence of the stoichiometry of the complexes on the sensing mechanism. Tetrahedron Asymmetry 20:1468–1471
51. Costero AM, Colera M, Gaviña P, Gil S, Kubinyi M, Pál K (2008) Chiral chemosensors for enantiomeric recognition of aspartate. Tetrahedron 64:3217–3224
52. Berova N, Nakanishi N (2000) Circular dichroism: principles and applications, 2nd edn. Wiley-VCH, New York, pp 337–382

Learning from Proteins and Drugs: Receptors That Mimic Biomedically Important Binding Motifs

Fraser Hof and Thomas Pinter

Abstract Proteins are highly evolved, sophisticated machines which function together to maintain homeostasis in their hosts. While merely a collection of amino acids covalently bonded in a specific sequence, their wide variety of functions is truly remarkable. Of course these covalent sequences are essential for proper function, but equally important for proper function are *weak* interactions: protein folding, enzyme-substrate interaction, and protein–protein communication are all controlled by forces weaker than covalent bonds and understanding these forces is fundamental in medicinal chemistry and drug design. Many inhibitory drugs mimic natural substrates for protein binding sites but inhibition of the substrate by mimicking the binding site is also possible. This mimicry and the biological consequences are under investigation.

Keywords Aromatic interactions, Cation-pi interactions, Host-guest chemistry, Hydrogen bonds, Molecular Recognition

Contents

F. Hof (✉) and T. Pinter
Department of Chemistry, University of Victoria, Victoria, BC, Canada V8W 3V6
e-mail: fhof@uvic.ca

S.A. Piletsky and M.J. Whitcombe (eds.), *Designing Receptors for the Next Generation of Biosensors*, Springer Series on Chemical Sensors and Biosensors (2013) 12: 33–52
DOI 10.1007/5346_2012_15,
Published online: 27 June 2012

1 Introduction: Weak Interactions

Weak interactions determine protein size and shape and are therefore an essential part of normal protein function. Discreet binding pockets and motifs that have evolved to be highly selective for only a very particular class of substrates come about as a result of a myriad of these non-covalent interactions. The helical shape of DNA is so because of an intricate combination of H-bond donor and acceptor pairs, stacking between the bases and solvation effects. Weak hydrogen bonding, electrostatic, and hydrophobic interactions play roles in all protein–substrate and protein–protein interactions.

The most commonly observed weak interactions and arguably most important for normal protein function are hydrogen bonds. Ubiquitous in complex natural systems, almost all biological processes involve hydrogen bonding in some form or another and these interactions have been the topic of extensive study for decades [1–6]. Proteins which act on anionic substrates universally contain highly evolved hydrogen bond networks in their active sites. Figure 1a depicts the pore of a ClC chloride channel whose crystal structure was solved in 2002 [7]. The ion is coaxed into the pore by four key hydrogen bond donating residues. These attractive interactions pull the chloride in close proximity to an aspartic acid residue which is displaced, thus opening the ion channel. During drug design, medicinal chemists often seek to emulate the natural substrate of a biological target and attempt to preserve all attractive forces in the host–guest complex. Replacement of the phosphate linker in natural RNA with an acylsulfonamide in a simple dinucleoside mimic (**2**) preserved a key H-bond with His119 and resulted in inhibition of RNase A (Fig. 1b, c) [8].

Other essential, yet not as well understood non-covalent interactions present in biological systems are those involving aromatic residues [9–12] and the hydrophobic effect [13–16]. The former is often considered to be a result of the latter, as hydrophobic aromatic residues pack close together in the core of the protein while hydrophilic residues are more often observed near the protein surface [16]. Although disfavored on the basis of configurational entropy, as a folded protein loses huge numbers of degrees of freedom relative to its unfolded state, the favored enthalpic gains of water molecules able to hydrogen bond to each other causes the overall energy of the system to be favorable. As a direct result, hydrophobic aromatic residues are often seen stacked on one another. Doubly mutating two stacked tyrosine residues in bacterial ribonuclease (barnase) resulted in a 4.6 kcal/mol decrease in protein stability (Fig. 2a) [9]. Because many hydrophilic residues reside near protein surfaces, salt bridges and cation–pi interactions are often seen at protein–protein interfaces [3]. The latter is sometimes observed as a quaternary ammonium residue bound inside an electron-rich pocket heavily populated with aromatic residues. These binding sites, or "hotspots," contain highly preorganized tryptophan, phenylalanine, and tyrosine residues which stabilize the incoming cation through their electron-rich pi-clouds. An important interaction of this type is between methylated lysine and the CBX class of proteins, when methylation of lysine is misregulated, disease often follows [17–20]. Figure 2b depicts the co-

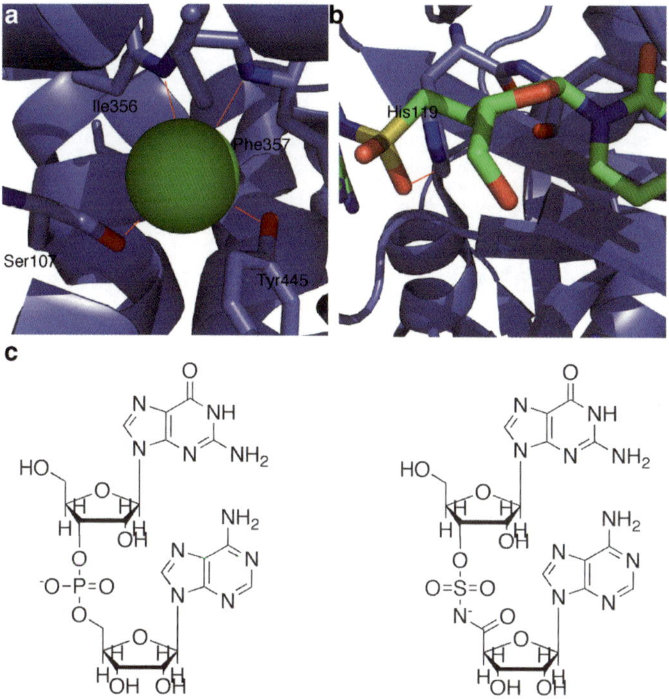

1 **2**

Fig. 1 (**a**) Crystal structure of chloride (*green sphere*) bound in the pore of a ClC chloride channel (PDB 1KPL). Key hydrogen bond contacts are observed with surrounding Ile, Ser, Tyr, and Phe residues. (**b**) *N*-acyl sulfonamide linked dinucleoside mimic bound to RNase A. A key H-bond between the sulfone of the inhibitor and a nearby histidine is observed in the crystal structure (PDB 2XOI). H-bonds are shown as *red lines*. (**c**) Natural dimeric RNA fragment (*left*) and *N*-acyl sulfonamide functionalized mimic (*right*). Both compounds are deprotonated at physiological pH and the mimic displays moderate inhibitory activity against RNase A

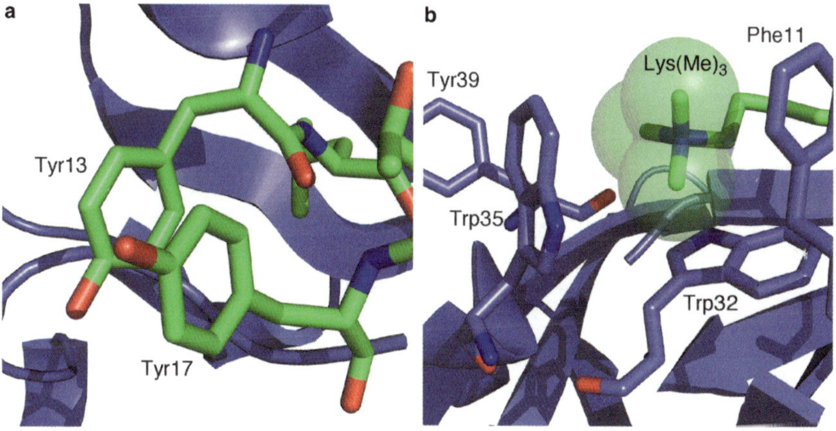

Fig. 2 (**a**) Aromatic–aromatic stacking in barnase stabilizes the protein (PDB 1A2P). (**b**) Trimethylated lysine reisdue is recognized by an "aromatic cage" binding motif in the protein CBX6 (PDB 3I90)

structure of a peptide fragment with trimethylated lysine binding in the aromatic cage pocket of CBX6 [21].

Development of compounds which emulate a natural substrate, be it for protein inhibition or detection if a fluorescent tag is attached, is commonplace. Is it possible to develop sensors which do the opposite—that is, mimic the binding domain of the protein itself and sequester the substrate before it enters? Our group and others are actively exploring this avenue of research.

2 Anion Recognition by Carboxylic Acid Bioisosteres

Since the advent of crown ethers in the 1960s, cation coordination chemistry has been thoroughly studied and rational design of synthetic cation receptors has matured a great deal. Anion recognition chemistry, however, had received less attention until the 1980s. The importance of anionic species to living systems is critical. Anions are ubiquitous in biological systems: careful regulation of intra- and extracellular charge gradients is necessary to maintain homeostasis, and the majority of enzyme substrates and cofactors carry a negative charge. DNA owes its helical shape to well-defined hydrogen bond networks between complimentary base pairs, phosphates provide the energy source crucial to all biochemical processes, transport channels for small anions such as chloride and sulfate regulate the flow of nutrients and osmotic pressure in and out the cell. Misregulation of these certain chloride channels have been proven to cause the debilitating respiratory illness cystic fibrosis [22], along with degenerative renal ailments Dent's disease [23] and Bartter's syndrome [24]. Many pollutants, be it from agricultural runoff (lake eutrophication from excess phosphate), nuclear waste (radioactive pertechnetate discarded into the ocean), are a cause of growing environmental concern and are anionic in nature. It is not surprising then that much attention has been focused on creating potent receptors that are selective for anionic species of interest [25–27].

In biological systems, the most prominent anion present is carboxylate. Negatively charged amino acids are often present at cation-binding hotspots to offer favorable electrostatic attractions. They are also one half of the common salt-bridge binding motif and are present in myriad enzyme substrates and cofactors. Development of drugs and sensors that mimic these substrates often requires the creation of esterified analogs that upon passage through a cell membrane are hydrolyzed to the corresponding acid derivative by native esterases. An alternate strategy in drug development is replacement of the carboxylic acid group with a functionally similar moiety, one for which the body lacks the evolved metabolic pathways to affect its degradation. These are known as bioisosteres of carboxylic acid. Common acid bioisosteres include aryl sulfonamides, acyl sulfonamides, and tetrazoles (Fig. 3). Utilizing these functional groups in drug development often leads to improved oral availability, metabolic stability, and potency relative to carboxylate bearing analogs [28–30]. It follows that a vast number of small molecule therapeutic targets contain tetrazole [31, 32], aryl sulfonamide [33, 34], and acyl sulfonamide functionality [35–37].

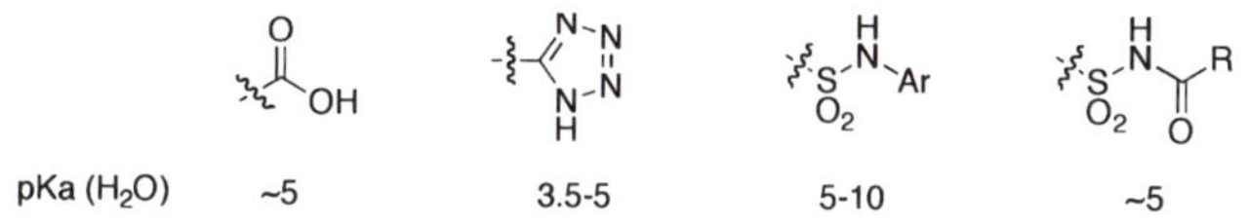

Fig. 3 Some common carboxylic acid bioisosteres

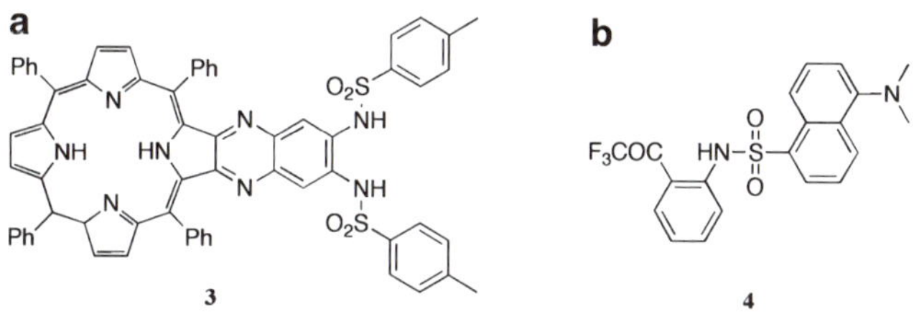

3 4

Fig. 4 (**a**) Fluoride selective sensor developed by Starnes et al. (**b**) Cyanide selective receptor developed by Ahn et al.

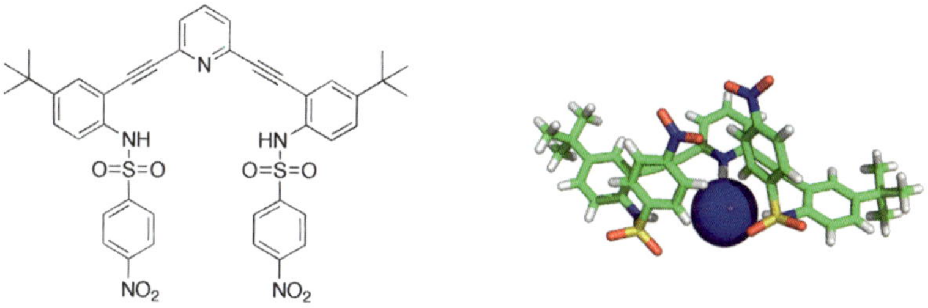

5

Fig. 5 Chloride selective fluorescent sensor developed by Johnson et al. (*left*) and its crystal structure complexed with chloride (*right*)

These groups have two possible modes of molecular recognition: when protonated, the acidic hydrogen atoms should make excellent donors useful for anion binding. When deprotonated, the anionic conjugate bases of these functional groups are potential cation binders. Our group and others have investigated the use of deprotonated tetrazolates for cation binding [38–43]. While aryl sulfonamides have been used as anion binders, the affinities of tetrazoles and acyl sulfonamides in neutral, protonated form have been largely unexplored.

Starnes and co-workers appended bis-aryl sulfonamide functionality to a well-known porphyrin scaffold (**3**) and observed changes in the UV–vis absorption spectra when titrating with various anions (Fig. 4a) [44], while Ahn et al. observed fluorescence enhancement and selectivity towards cyanide with their naphthalene-based aryl sulfonamide receptor (**4**) (Fig. 4b) [45]. Recently, Johnson and co-workers have been developing aryl ethynyl scaffolds as anion sensors and have shown remarkable fluorescence imaging of chloride in vitro [46]. They have also fabricated a scaffold decorated with aryl sulfonamide functionality (**5**) that is also a successful anion binder [47] (Fig. 5). Against this backdrop of favorable results, it is

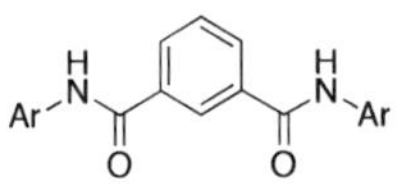

Ar = p-(n-Bu)C_6H_4

6

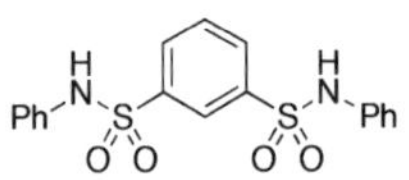

7

Fig. 6 Amide and sulfonamide functionalized anion receptors exhibit different selectivities

Table 1 Anion affinities of aryl sulfonamide functionalized hosts **8** and **9**, acyl sulfonamide functionalized hosts **10** and **11**, and tetrazole functionalized hosts **12** and **13**

$K_{assoc}(M^{-1})$ in CD_3CN					
Host	$Bu_4N^+Cl^-$	$Bu_4N^+NBr^-$	$Bu_4N^+OTs^-$	$Bu_4N^+NO_3^-$	$Bu_4N^+HSO_4^-$
10	600(10)	110(10)	60(5)	40(40)	50(5)
11	1,030(5)	320(5)	250(5)	100(5)	180(10)
12	100(50)	20(10)	100(100)	80(5)	70(20)
13	390(10)	90(5)	120(5)	60(5)	100(5)
14	8,450(10)	720(10)	1,400(20)	5,800(30)	660(50)
15	3,560(40)	800(10)	520(10)	330(5)	340(5)

Errors in brackets are standard deviations of 2–3 replicate titrations

interesting to note that the original report of isophthalamides as anion binders by Crabtree also reports aryl sulfonamide analogs that have varying affinities and selectivities relative to amides (Fig. 6) [48]. Compound **6** exhibits threefold more potent binding for chloride relative to **7** while **7** binds fluoride approximately twice as strongly as **6**.

We recently explored anion binding by all three bioisosteres—aryl sulfonamides, acyl sulfonamides, and tetrazoles—affixed to a common calixarene scaffold. Hosts **8–13** were prepared and their binding with several biologically important halides and oxyanions was determined. The results are summarized in Table 1. These studies revealed that although the N–H proton acidities of these three classes of compounds are similar (representative N–H pK_a values are 4.6, 8.5, and 5.2 for tetrazole [49], aryl sulfonamide [50], and acyl sulfonamide [36] moieties, respectively), tetrazoles proved to be superior anion-binding elements relative to their sulfonamide analogs within this structural context. The expected trends based on N–H proton acidities were in fact observed with aryl sulfonamides, as **8** bound all anions studied less strongly than did electron poor **9**. Notably, acyl sulfonamide functionalized hosts **10** and **11**, although known to be more acidic than simple aryl sulfonamides were less competent hosts. These findings implied that there were more forces at work in these systems than simply hydrogen bond donation (Fig. 7).

We hypothesized that their varying conformational preferences, largely ignored in their simple classification as interchangeable replacements for carboxylic acids, might play a role in determining their anion-binding affinities. Molecular modeling studies were carried out to investigate the host–guest complexes. We identified geometries for each host–guest complex in which the calix[4]arene is in a perfect "cone" conformation and all four hydrogen bond donors are engaging the central

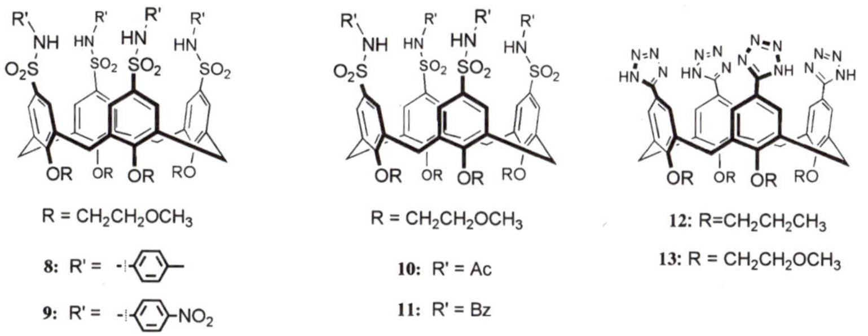

Fig. 7 New class of carboxylic acid bioisostere functionalized calix[4]arene recognition elements

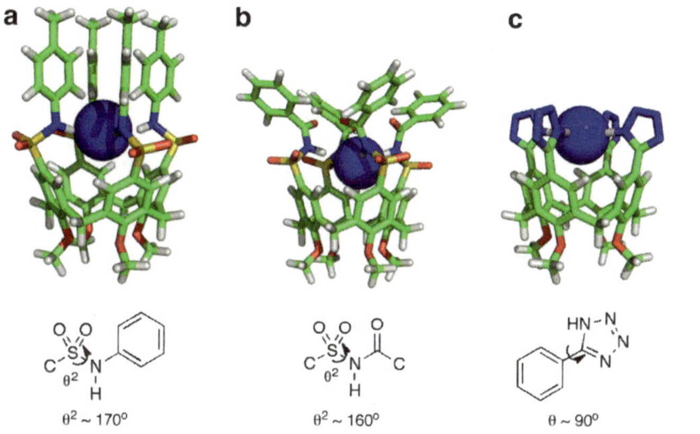

Fig. 8 Local minima that involve the maximum four host–guest hydrogen bonds for representative host–guest complex with key dihedral angles calculated (HF/6-31+G*). Lower rim substituents have been omitted. (**a**) Aryl sulfonamide functionalized host complexed with Cl^-. (**b**) Acyl sulfonamide functionalized host complexed with Cl^-. (**c**) Tetrazole functionalized host complexed with Cl^-

anion symmetrically. These provided values for the key torsion angles of each functional group that would be optimal for binding (Fig. 8). The actual inherent conformational preferences for each functional group were then determined by dihedral driving calculations and CSD surveys (not shown) and compared to the values required for anion binding in this structural context.

The combined computational analyses for the key dihedral angles that define the inherent shapes of these moieties are presented in Fig. 9.

These comparisons showed that the tetrazole hosts **12** and **13** must adopt a nonideal dihedral that takes the tetrazole out of conjugation with its neighboring ring to bind an anion, and that it paid ~13 kJ/mol in order to do so (Fig. 9c). The shapes of the aryl and acyl sulfonamides are defined by three important dihedral angles. The conformations of the rotatable carbon–sulfur bonds for aryl sulfone type functionalities (θ_1, Fig. 9d) and are known to be similar (θ_1 = ~90°) [51].

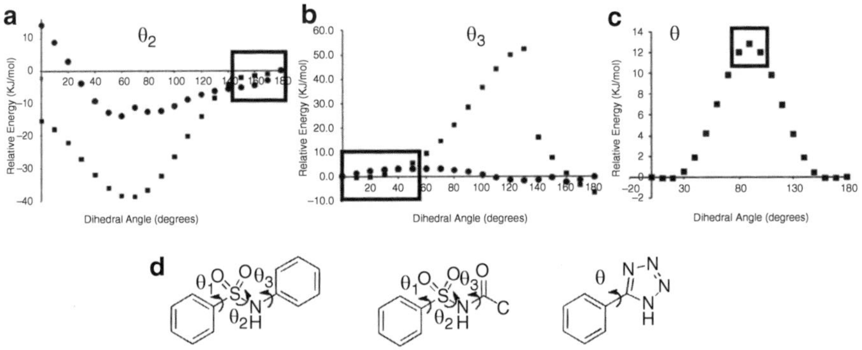

Fig. 9 Energy diagrams calculated at the HF/6-31+G* level of theory when driving key dihedral angles from 0° to 180° for simplified fragments of aryl (*square*) and acyl (*circle*) sulfonamides (**a**, **b**) and for a simplified tetrazole fragment (**c**). (**d**) Labeling of key dihedral angles studied computationally. The approximate angles required to fully engage a central guest in a fourfold symmetrical manner are highlighted with *boxes*

Focusing on the dihedral for rotation about the sulfonamide S–N bond, (θ_2; Fig. 9a) and the amide/aniline dihedrals that define rotation about the neighboring N–C bonds (θ_3; Fig. 9b) provides a different picture. Computational analysis revealed that these functional groups also have preference for co-planarity with their aryl neighbors. An analysis similar to that done for the tetrazole shows that both compounds pay little energy in order for θ_3 to adopt a good binding geometry. When examining θ_2 however, the acyl sulfonamide fragment pays a much higher penalty to orient itself toward the guest than the aryl sulfonamide. These lessons inform on sulfonamide recognition in general, as they can also explain the failure of the Crabtree bis-sulfonamide **7**. Clearly, the chosen scaffold was not ideal for maximum host–guest binding and within the right structural context, the rigid acyl sulfonamides have potential to bind anions with more potency.

Tetrazoles are the most highly acidic of these isosteres, and should be the best hydrogen bonders of anions. Their utility as anion recognition elements has been demonstrated in a variety of contexts that are informed on by considerations of host shape and electronics. A tripodal receptor bearing three tetrazoles (**15**) was compared to a carboxylic acid analog (**14**) [52]. Binding studies showed that the tetrazole-functionalized host bound halides up to six orders of magnitude more strongly than did its carboxyl functionalized counterpart despite their nearly identical acidities. Again, the apparent similarity covers up major differences in shape and stereoelectronics that drive molecular recognition. While the acidic OH and acidic NH groups in **14** and **15** are arrayed in nearly identical positions in three-dimensional space, the OH groups are directed outward and away from the guest because of a carboxylic acid's strong preference for a *syn* OH conformation. On the contrary, the tetrazole's NH (which prefers strongly to exist as the 1H-tetrazole tautomer in polar solutions) is oriented such that all three host NHs can bind the guest simultaneously. The impact of these stereoelectronic effects all cooperating in

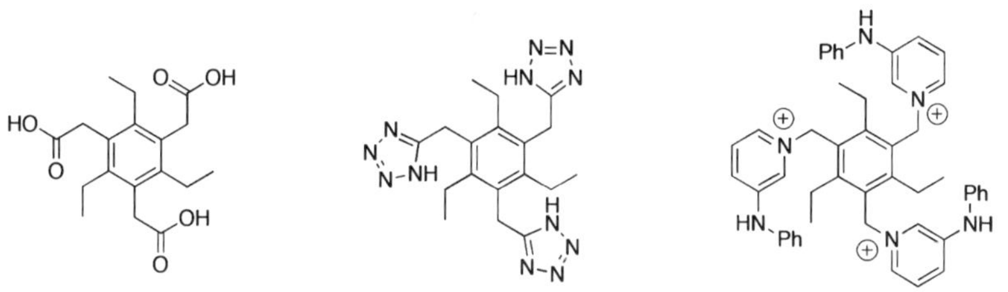

14 $K_{assoc}(Cl^-) \sim 2 \times 10^2$

15 $K_{assoc}(Cl^-) \sim 1 \times 10^7$

16 $K_{assoc}(Cl^-) > 10^6$

Fig. 10 Tripodal anion receptors with association constants measured in acetonitrile

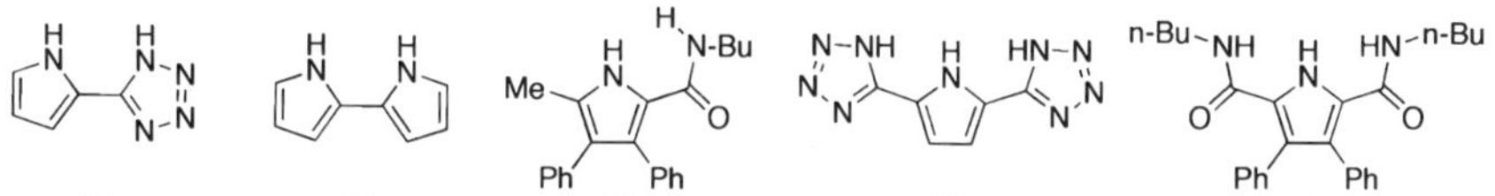

17 **18** **19** **20** **21**

Fig. 11 Previously studied amidopyrroles and new tetrazole containing anion binders

the binding of the guest is dramatic. Comparison with other recognition elements appended to the same scaffold reveals that this host is among the strongest neutral binders of anions ever reported—even comparable to triply cationic hosts such as **16** [53] (Fig. 10).

Further inspiration from medicinal chemistry was found in the natural product prodigiosin and its synthetic analogs (prodigiosenes), which are being heavily investigated as novel therapeutics in a variety of contexts and *whose highly potent binding of anions are linked to their biological activities* [54]. The core anion-binding motif of prodigiosenes is a 2,2′-bipyrrole; prior lessons, the shape-preferences, and anion-binding potencies of tetrazoles suggested the utility of a scaffold in which tetrazoles are installed next to a neighboring, conjugated pyrrole. 5-(Pyrrolyl)-tetrazole (**17**) is a small fragment that satisfies these design criteria (Fig. 11) [55]. The recognition properties of the tetrazole are revealed by comparisons to two key relatives, 2,2′-bipyrrole (**18**), which reproduces the geometry of **17** with a nearly isosteric tetrazole-for-pyrrole swap and a simple amidopyrrole (**19**), which is a representative member of a larger family of amidopyrroles that have been shown to bind and transport anions. Compound **16** binds almost any anion with tenfold greater affinity than the structurally related bipyrrole (**18**) (Table 2). The similar selectivity trends observed for **17** and **18** suggest that their structures and directional hydrogen bonding are similar and that the acidity of the tetrazole is the key to higher affinities. Compound **17** also binds anions at least 47-fold stronger than amidopyrroles. The tridentate 2,5-(bis-tetrazolyl)-pyrrole (**20**) was also produced and compared to a forerunner that used bis(amido)pyrroles for anion binding (**21**) [56]. Affinities improved again—this time by two orders of magnitude. These amide-to-tetrazole comparisons revealed

Table 2 Association constants for selected anions with compounds **17–21**

$K_{assoc}(M^{-1})$ in CD_3CN				
Host	$Bu_4N^+Cl^-$	$Bu_4N^+NBr^-$	$Bu_4N^+OTs^-$	$Bu_4N^+NO_3^-$
17	3,300(1,200)	450(50)	900(50)	160(20)
18	310(10)	50(3)	40(4)	20(1)
19	28	<10	nd	nd
20	26,000(2,300)	1,500(430)	34,000(3,500)	1,600(300)
21	138	<10	nd	nd

Errors in brackets are standard deviations of 2–3 replicate titrations. For compounds **19** and **21**, errors are estimated to be <15% [56]

that the tetrazole-functionalized hosts encoded a non-Hofmeister preference for sulfonate/sulfate type anions relative to chloride, showing that even a subtle structural change (in this case, the angle of the N–H donors relative to each other) can have large effects on binding selectivities.

3 Receptors That Mimic Natural Aromatic Cage Motifs

As with receptors inspired by common drug motifs, inspection of natural binding motifs in proteins provides a multitude of lessons on molecular recognition. One that has found particular resonance and utility in supramolecular chemistry is the aromatic cage motif that is used throughout nature to bind tertiary and quaternary ammonium cations. This motif is typically defined as a rigid cluster of 2–4 aromatic amino acid side chains (Trp, Phe, Tyr) describing a central binding site [57]. Notable examples include the choline-binding proteins, which include acetylcholinesterase and the nicotinic acetylcholine receptor, and several gene regulation proteins that recognize and bind to the "histone code" marks of dimethyllysine and trimethyllysine side chains (see Fig. 2) [58]. The latter are especially interesting, because they have evolved in a competitive environment where they must reject binding of proteins that are identical to their targets but are unmethylated at the critical lysine side chain. Thus, their aromatic cages are the sole and unique hot spots that are responsible for this biologically driven selectivity.

It is notable that, for tertiary and especially quaternary ammonium ions, nature rarely employs a negatively charged molecular recognition element such as carboxylate or phosphate. This leads one to the idea that the cation–pi interaction and possibly the hydrophobic effect are the key operators. A large set of synthetic receptors have been used both to demonstrate that the cation–pi interaction is adept at encoding the strong and selective binding of quaternary ammonium ions in water. The variety of such structures is large, and this area has been extensively reviewed [59]. Exemplary evidence is provided by data from a single family of macrocyclic hosts invented by Dougherty (Fig. 12). Host **22**, a synthetic aromatic cage functionalized with polar solubilizing groups, demonstrated the ability to bind ammonium ion guests such as **23** and **24** with high affinities in pure water.

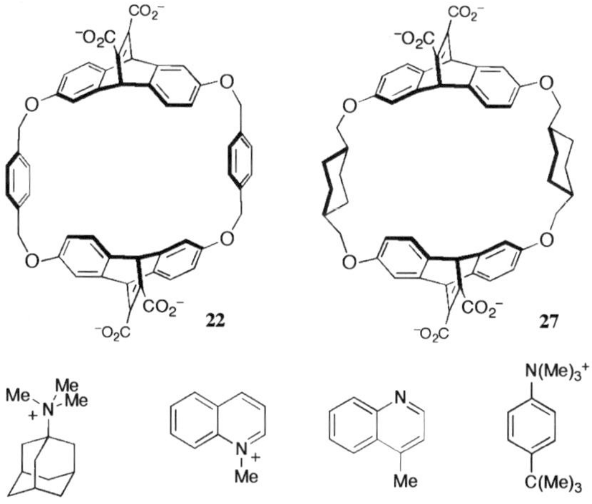

22 27

23 24 25 26

Fig. 12 Macrocyclic hosts that mimic aromatic cage motifs, and their guests that have been used to understand the nature of their interactions

It has always been a confounding element of these motifs (natural and synthetic) that a hydrophobic component (and not the cation–pi interaction) might be sufficient to drive the observed binding of quaternary ammonium ions (which are much more hydrophobic than primary ammonium ions). But comparisons in water of nearly isostructural guests such as **24** and **25** that bear different charges revealed a significant difference in affinities that can only be explained by the multiple cation–pi interactions that exist between the guest **24** and the cyclophane. Further, the guest **26**, with isosteric aliphatic $–C(Me)_3$ and ammonium $–N(Me)_3^+$ ends, shows a strong preference for binding with the charged (and slightly less hydrophobic) $–N(Me)_3^+$ end inside the aromatic cage of **22**; if hydrophobicity were the prime driver for guest binding, then the more hydrophobic $–C(Me)_3$ end would win out. Finally, the host analog **27**, with two of the benzene rings of **22** replaced with cyclohexane walls that are more hydrophobic and more polarizable than the benzenes in the parent host, also shows decreased affinities for ammonium ion guests that support the key role of cation–pi interactions in the parent host **22**.

Other informative comparisons have been conducted using natural, protein-based aromatic cage motifs as "receptors" themselves for physical organic studies. In one kind of study, strong-binding quaternary ammonium ion ligands of the type $–N(Me)_3^+$ (**31**) are compared to ligand analogs that are isostructural except for –C $(Me)_3$ substitutions (**32**, Fig. 13) [60, 61]. The affinities in both studies are higher for the charged species that can form multiple cation–pi interactions with their protein binding partners. In a different kind of study, the protein itself is altered to reduce the strength of the proposed cation–pi interaction by installation of unnatural aromatic amino acids with decreased electron density relative to the native Trp residues (e.g., F-Trp, F_2-Trp, F_3-Trp) [62]. Again, the dominant role of certain cation–pi contacts in these biologically important recognition events is supported

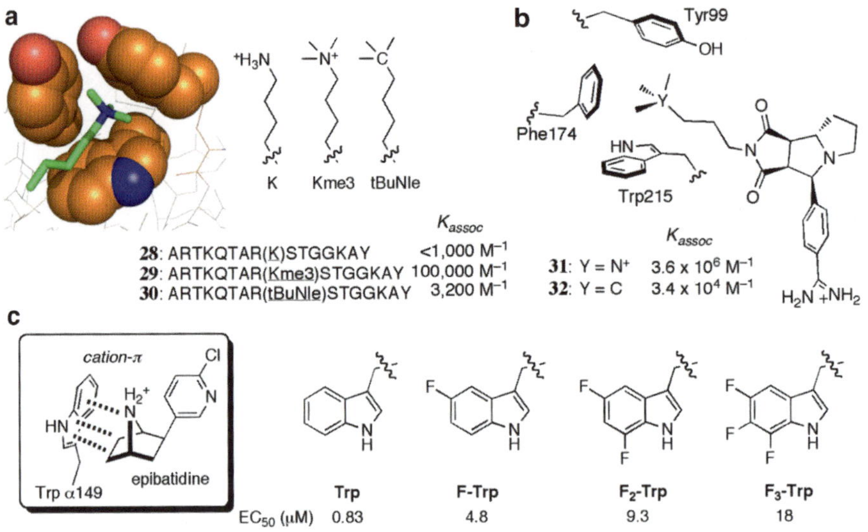

Fig. 13 Perturbation of protein–ligand interactions by modulating the cation–pi interaction. (**a**) Peptides bearing trimethyllysine $-N(Me_3)^+$ and neutral $-C(Me_3)^+$ side chains, as well as the parent peptide with an unmodified lysine ($-NH_3^+$) side chain, and their affinities for the aromatic cage protein HP1. The HP1 binding pocket containing bound trimethyllysine is shown at *left* (1KNE). (**b**) Two inhibitors of the enzyme Factor Xa that show 100-fold increased potency for the $-N(Me_3)^+$ type inhibitor. (**c**) Electron-deficient Trp analogs are introduced into nicotinic acetylcholine receptor (nAChR) at position α149 by mutagenesis. The resulting receptors show reduced activation by the agonist epibatidine with increasing fluorination, demonstrating the importance of the Trp149-epibatidine cation–pi interaction

by the dramatic weakening of protein–ligand interactions for the proteins with more highly fluorinated, and therefore more electron-poor aromatics.

Trp is the most electron-rich of the aromatic amino acids, and as such is the aromatic residue most frequently identified as participating in strong cation–pi interactions in nature [63]. In recent work, our group has been creating new aromatic cage mimics that involve the incorporation of Trp into receptor frameworks of different types. One variation on this theme involves the construction of small Trp-rich peptides, with augmentation of their aromatic character by incorporation of *N*-benzyltryptophan as a building block [64]. This artificial amino acid presents both the electron-rich indole and an appended benzyl substituent to cationic binding partners. The peptide-derived receptor Trp(Bn)–Trp(Bn) (**33**) shows the ability to bind to quaternary ammonium ions in water, with selectivity over unmethylated primary quaternary ammonium ions in this highly competitive medium. More telling, the construction and comparison of receptors based on Trp–Trp peptides (**33–35**) bearing zero, one, and two appended benzyl substituents showed an increasing ability to bind to acetylcholine in water with increasing benzylation (Fig. 14).

In another approach, we have created a variety of hosts based on indole carboxylic acids, including receptor **37** made from three copies of the de-aminated Trp building block indole-3-propionic acid (**36**, Fig. 15) [65]. Despite its extreme

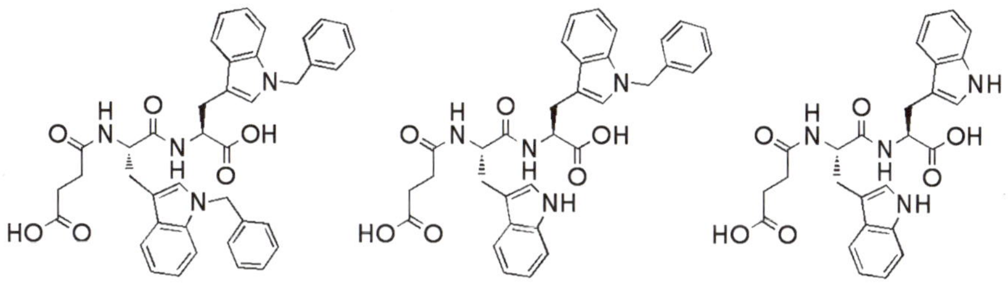

33 Trp(Bn)-Trp(Bn)

34 Trp-Trp(Bn)

35 Trp-Trp

Fig. 14 Hosts for quaternary ammonium ions based on *N*-benzylated Trp residues

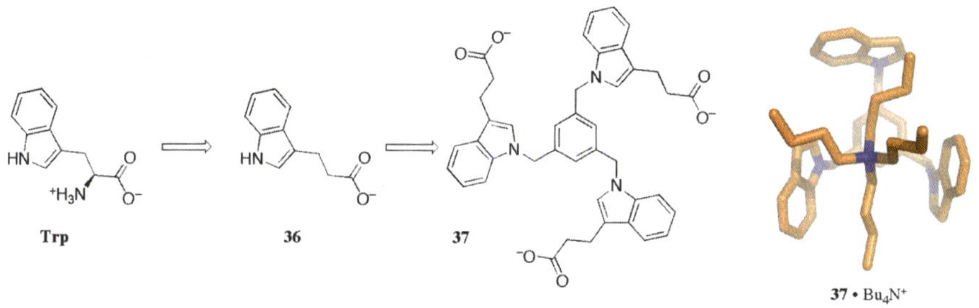

Fig. 15 A host based on Trp derivative indole-3-propionic acid binds quaternary ammonium ions in buffered water

simplicity and lack of pre-organized binding pocket, host **37** binds quaternary ammonium ions like Me_4N^+, AcCh, and Kme3 in buffered water, with two- to fivefold selectivity over some analogous primary and secondary ammonium ions. In this case, however, studies that extended to examination of more hydrophobic cations showed more dramatic increases in affinity, including a >800-fold higher affinity for Bu_4N^+ than Me_4N^+. This difference is completely absent in $CDCl_3$ (where in fact, Bu_4N^+ is not bound at all), demonstrating that it must be largely driven by the hydrophobic effect and not by specific weak interactions between host **37** and Bu_4N^+. This lesson—that hydrophobicity can cooperate with cation–pi interactions in aqueous medium—likely extends to naturally occurring aromatic cages that can recognize peptides bearing quaternary trimethyllysine side chains over their unmethylated congeners. As previously stated, the observation that a substrate bearing neutral $–C(Me)_3$ side chains (**32**) binds more weakly to an aromatic cage protein than does the natural partner (**31**) bearing $–N(Me)_3^+$ side chains demonstrates the importance of the cation–pi interaction. But the comparison of the methylated peptide **29** to its unmethylated analog **28** raises a question: all other things being equal, the strength of the cation–pi between the aromatic cage protein and unmethylated **28** should be stronger, because the cation is more compact (and charge dense) and can form shorter cation–pi contacts. But of course,

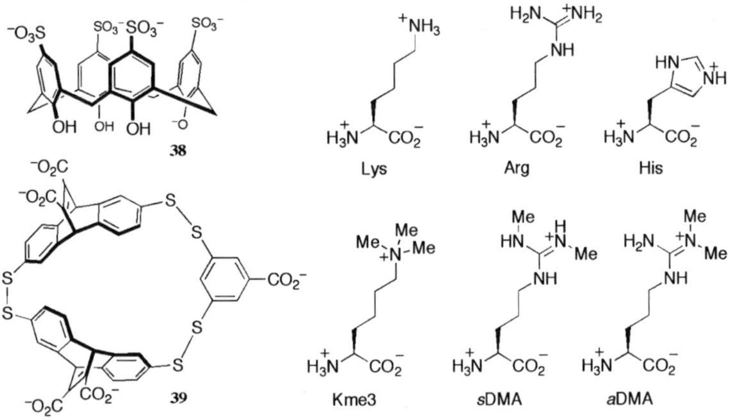

Fig. 16 *p*-Sulfonatocalix[4]arene (**38**) and macrocycle **39** mimic the aromatic-rich pockets of methyllysine-binding proteins in different ways. Both bind peptides bearing trimethyllysine side chains in buffered water with protein-like affinities and selectivities

the unmethylated peptide **28** has a handicap, in that the primary $-NH_3^+$ cation of its side chain is strongly engaged with a shell of hydrating water molecules. The affinities for these peptides with their partner protein HP1 are provided in Fig. 13 [61]. The hydrophobic effect that is induced upon methylation is an important player that works alongside cation–pi interactions and dispersive interactions in generating the natural selectivities of aromatic cage proteins for their trimethyllysine targets over unmethylated counterparts.

The calixarenes, a family of hosts that present multiple phenol rings in a macrocyclic arrangement, have been widely explored as biomimetic hosts. A particularly large body of work exists on sulfonated calixarenes, which are soluble in water and capable of binding a wide variety of natural and unnatural guests presenting ammonium ions. Hosts like *p*-sulfonatocalix[4]arene (**38**) have long been known to bind biologically important quaternary ammonium ions like acetylcholine [66, 67]. More recent work has broadly explored their abilities to bind cationic amino acids, peptides, and proteins [68, 69]. Host **38** binds lysine, arginine, and histidine in buffered water with affinities of 520, 330, and 20 M^{-1}, respectively (Fig. 16) [70–72]. Affinities rise upon methylation, with asymmetric dimethylarginine (*a*DMA*)* and symmetric dimethylarginine (*s*DMA*)* binding threefold stronger than unmethylated arginine [72]. The lysine series displays even stronger dependence on methylation state, with affinities ranging up to 37,000 M^{-1} for the 1:1 complexation of trimethyllysine (Kme3) and 96,600 M^{-1} for the short trimethyllysine-containing peptide R-(Kme3)-S-T [72]. Such high selectivities for the quaternary ammonium ions over primary counterparts are dramatic and consistently displayed for this host. The difference probably arises from a change in binding mode: unmethylated lysine side chains bury their most hydrophobic elements, the aliphatic CH_2 groups, inside of the aromatic cavity of host **38** and leave the polar $-NH_3^+$ group exposed to the polar sulfonate functional groups and external solvent [73]. Trimethyllysine, instead, buries its $-NMe_3^+$ group snugly inside the cavity of the host, and in doing so forms multiple strong cation–pi, CH–pi, and dispersive interactions assisted by the hydrophobic effect [72].

Other synthetic hosts have recently been explored as receptors for trimethyllysine, including macrocycle **39**, which contains disulfide bridges that enable its participation in and selection from a dynamically equilibrating host library. This aromatic-rich host also binds trimethyllysine-containing peptides with excellent selectivities, displaying an association constant of 40,000 M^{-1} for a trimethylated peptide and >50-fold weaker binding to the unmethylated analog [74]. The affinities of both calixarene **38** and cyclophane **39** for their trimethyllysine targets are of the same order as the naturally evolved aromatic cage proteins, which typically range from $K_{assoc} =$ 50,000–200,000 M^{-1} ($K_d = 5$–20 μM). In both of these cases, protein-like affinities and selectivities, which are rarely displayed by supramolecular hosts, are achieved. Both receptors profit from the approach of "teaching old dogs new tricks," i.e., identifying existing host scaffolds that mimic naturally evolved protein binding partners, and using them to engage biological targets that had previously been unconsidered by supramolecular chemistry. The creation of biomimetic receptors for post-translationally methylated protein residues is a promising area for future developments in biotechnology and biomedical research.

4 Conclusions

The technological promise of biomimetic receptor-type compounds as both sensors and disruptors of biological pathways is only now beginning to be realized [67, 75] but there remain many challenges to converting this type of biologically inspired receptor into advances that are biomedically important. The most fundamental is that strong and specific molecular recognition in the medium of life—pure, warm, salty water—remains difficult to achieve using the simple scaffolds that are familiar in the world of supramolecular chemistry. Examples of success of the types described here are relatively rare. As we continue to seek simple molecules that can achieve complex recognition tasks, we find an almost inexhaustible source of inspiration for these studies in the huge diversity of proteins and drugs that are known encode strong and selective binding in water.

Acknowledgements We thank NSERC for funding. FH is a Career Scholar of the Michael Smith Foundation for Health Research and Canada Research Chair.

References

1. Perrin CL, Nielson JB (1997) "Strong" hydrogen bonds in chemistry and biology. Annu Rev Phys Chem 48(1):511–544
2. Xu D, Tsai CJ, Nussinov R (1997) Hydrogen bonds and salt bridges across protein–protein interfaces. Protein Eng 10(9):999–1012

3. Wahl MC, Sundaralingam M (1997) C–H. . .O hydrogen bonding in biology. Trends Biochem Sci 22(3):97–102
4. Resnati G (2002) Book review: the weak hydrogen bond—in structural chemistry and biology. Gautam Desiraju and Thomas steiner. ChemPhysChem 3(2):225–226
5. Gao J, Bosco DA, Powers ET, Kelly JW (2009) Localized thermodynamic coupling between hydrogen bonding and microenvironment polarity substantially stabilizes proteins. Nat Struct Mol Biol 16(7):684–690
6. Alkorta I, Elguero J (1998) Non-conventional hydrogen bonds. Chem Soc Rev 27(2):163–170
7. Dutzler R, Campbell EB, Cadene M, Chait BT, MacKinnon R (2002) X-ray structure of a ClC chloride channel at 3.0 Å reveals the molecular basis of anion selectivity. Nature 415 (6869):287–294
8. Thiyagarajan N, Smith BD, Raines RT, Acharya KR (2011) Functional and structural analyses of *N*-acylsulfonamide-linked dinucleoside inhibitors of rnase a. FEBS J 278(3):541–549
9. Serrano L, Bycroft M, Fersht AR (1991) Aromatic-aromatic interactions and protein stability: investigation by double-mutant cycles. J Mol Biol 218(2):465–475
10. Bhattacharyya R, Samanta U, Chakrabarti P (2002) Aromatic-aromatic interactions in and around α-helices. Protein Eng 15(2):91–100
11. Hunter CA, Lawson KR, Perkins J, Urch CJ (2001) Aromatic interactions. J Chem Soc Perkin Trans 2 (5):651–669
12. Ma M, Kuang Y, Gao Y, Zhang Y, Gao P, Xu B (2010) Aromatic-aromatic interactions induce the self-assembly of pentapeptidic derivatives in water to form nanofibers and supramolecular hydrogels. J Am Chem Soc 132(8):2719–2728
13. Spolar RS, Ha JH, Record MT (1989) Hydrophobic effect in protein folding and other noncovalent processes involving proteins. Proc Natl Acad Sci USA 86(21):8382–8385
14. Lins L, Brasseur R (1995) The hydrophobic effect in protein folding. FASEB J 9(7):535–540
15. Southall NT, Dill KA, Haymet ADJ (2001) A view of the hydrophobic effect. J Phys Chem B 106(3):521–533
16. Tsai C-J, Lin SL, Wolfson HJ, Nussinov R (1997) Studies of protein–protein interfaces: a statistical analysis of the hydrophobic effect. Protein Sci 6(1):53–64
17. Bachmann IM, Halvorsen OJ, Collett K, Stefansson IM, Or S, Haukaas SA, Salvesen HB, Otte AP, Akslen LA (2006) EZH2 expression is associated with high proliferation rate and aggressive tumor subgroups in cutaneous melanoma and cancers of the endometrium, prostate, and breast. J Clin Oncol 24(2):268–273
18. Eckerle S, Brune V, Doring C, Tiacci E, Bohle V, Sundstrom C, Kodet R, Paulli M, Falini B, Klapper W, Chaubert AB, Willenbrock K, Metzler D, Brauninger A, Kuppers R, Hansmann ML (2009) Gene expression profiling of isolated tumour cells from anaplastic large cell lymphomas: insights into its cellular origin, pathogenesis and relation to Hodgkin lymphoma. Leukemia 23(11):2129–2138
19. Fiskus W, Wang Y, Sreekumar A, Buckley KM, Shi H, Jillella A, Ustun C, Rao R, Fernandez P, Chen J, Balusu R, Koul S, Atadja P, Marquez VE, Bhalla KN (2009) Combined epigenetic therapy with the histone methyltransferase EZH2 inhibitor 3-deazaneplanocin A and the histone deacetylase inhibitor panobinostat against human AML cells. Blood 114(13): 2733–2743
20. Basak SK, Veena MS, Oh S, Huang G, Srivatsan E, Huang M, Sharma S, Batra RK (2009) The malignant pleural effusion as a model to investigate intratumoral heterogeneity in lung cancer. PLoS One 4(6):e5884
21. Kaustov L, Ouyang H, Amaya M, Lemak A, Nady N, Duan S, Wasney GA, Li Z, Vedadi M, Schapira M, Min J, Arrowsmith CH (2011) Recognition and specificity determinants of the human CBX chromodomains. J Biol Chem 286(1):521–529
22. Gribkoff VK, Champigny G, Barbry P, Dworetzky SI, Meanwell NA, Lazdunski M (1994) The substituted benzimidazolone NS004 is an opener of the cystic fibrosis chloride channel. J Biol Chem 269(15):10983–10986

23. Cox JPD, Yamamoto K, Christie PT, Wooding C, Feest T, Flinter FA, Goodyer PR, Leumann E, Neuhaus T, Reid C, Williams PF, Wrong O, Thakker RV (1999) Renal chloride channel, CLCN5, mutations in Dent's disease. J Bone Miner Res 14(9):1536–1542
24. Schlingmann KP, Konrad M, Jeck N, Waldegger P, Reinalter SC, Holder M, Seyberth HrW, Waldegger S (2004) Salt wasting and deafness resulting from mutations in two chloride channels. N Engl J Med 350(13):1314–1319
25. Beer PD, Gale PA (2001) Anion recognition and sensing: the state of the art and future perspectives. Angew Chem Int Ed 40(3):486–516
26. Gale PA, Quesada R (2006) Anion coordination and anion-templated assembly: highlights from 2002 to 2004. Coord Chem Rev 250(23–24):3219–3244
27. Lavigne JJ, Anslyn EV (2001) Sensing a paradigm shift in the field of molecular recognition: from selective to differential receptors. Angew Chem Int Ed 40(17):3118–3130
28. Herr RJ (2002) 5-Substituted-1H-tetrazoles as carboxylic acid isosteres: medicinal chemistry and synthetic methods. Bioorg Med Chem 10(11):3379–3393
29. Patani GA, LaVoie EJ (1996) Bioisosterism: a rational approach in drug design. Chem Rev 96(8):3147–3176
30. Olesen PH (2001) The use of bioisosteric groups in lead optimization. Curr Opin Drug Discov Devel 4:471–478
31. Myznikov LV, Hrabalek A, Koldobskii GI (2007) Drugs in the tetrazole series. Chem Heterocycl Compd 43(1):1–9
32. Nattel S (2005) Can losartan prevent new-onset atrial fibrillation in hypertensive patients more effectively than atenolol? Nat Clin Pract Cardiovasc Med 2(7):332–333
33. Skulnick HI, Johnson PD, Aristoff PA, Morris JK, Lovasz KD, Howe WJ, Watenpaugh KD, Janakiraman MN, Anderson DJ, Reischer RJ, Schwartz TM, Banitt LS, Tomich PK, Lynn JC, Horng M-M, Chong K-T, Hinshaw RR, Dolak LA, Seest EP, Schwende FJ, Rush BD, Howard GM, Toth LN, Wilkinson KR, Kakuk TJ, Johnson CW, Cole SL, Zaya RM, Zipp GL, Possert PL, Dalga RJ, Zhong W-Z, Williams MG, Romines KR (1997) Structure-based design of nonpeptidic HIV protease inhibitors: the sulfonamide-substituted cyclooctylpyranones. J Med Chem 40(7):1149–1164
34. Reddy NS, Mallireddigari MR, Cosenza S, Gumireddy K, Bell SC, Reddy EP, Reddy MVR (2004) Synthesis of new coumarin 3-(*N*-aryl) sulfonamides and their anticancer activity. Bioorg Med Chem Lett 14(15):4093–4097
35. Lobb KL, Hipskind PA, Aikins JA, Alvarez E, Cheung Y-Y, Considine EL, De Dios A, Durst GL, Ferritto R, Grossman CS, Giera DD, Hollister BA, Huang Z, Iversen PW, Law KL, Li T, Lin H-S, Lopez B, Lopez JE, Cabrejas LMM, McCann DJ, Molero V, Reilly JE, Richett ME, Shih C, Teicher B, Wikel JH, White WT, Mader MM (2004) Acyl sulfonamide anti-proliferatives: benzene substituent structure, activity relationships for a novel class of antitumor agents. J Med Chem 47(22):5367–5380
36. Stansfield I, Pompei M, Conte I, Ercolani C, Migliaccio G, Jairaj M, Giuliano C, Rowley M, Narjes F (2007) Development of carboxylic acid replacements in indole-*N*-acetamide inhibitors of hepatitis C virus NS5B polymerase. Bioorg Med Chem Lett 17(18):5143–5149
37. Hu X, Sun J, Wang H-G, Manetsch R (2008) Bcl-xl-templated assembly of its own protein-protein interaction modulator from fragments decorated with thio acids and sulfonyl azides. J Am Chem Soc 130(42):13820–13821
38. Mahnke DJ, McDonald R, Hof F (2007) A shape-dependent hydrophobic effect for tetrazoles. Chem Commun (36):3738–3740
39. Tominey A, Andrew D, Oliphant L, Rosair GM, Dupre J, Kraft A (2006) Supramolecular binding of protonated amines to a receptor microgel in aqueous medium. Chem Commun (23):2492–2494
40. Peters L, Froehlich R, Boyd ASF, Kraft A (2001) Noncovalent interactions between tetrazole and an N, N-diethyl-substituted benzamidine. J Org Chem 66(10):3291–3298
41. Khazaei A, Manesh AA (2006) Selective oxidation of oximes to carbonyl compounds using n-bromo-n-benzoyl-4-toluenesulfonamide. Mendeleev Commun 16(2):109–111

42. Chen Y-J, Chung W-S (2009) Tetrazoles and para-substituted phenylazo-coupled calix[4] arenes as highly sensitive chromogenic sensors for ca2+. Eur J Org Chem 2009 (28):4770–4776
43. Boyko V, Rodik R, Danylyuk O, Tsymbal L, Lampeka Y, Suwinska K, Lipkowski J, Kalchenko V (2005) Tetrazolecalix[4]arenes as new ligands for palladium(ii). Tetrahedron 61(52):12282–12287
44. Starnes SD, Arungundram S, Saunders CH (2002) Anion sensors based on β, β′-disubstituted porphyrin derivatives. Tetrahedron Lett 43(43):7785–7788
45. Chung YM, Raman B, Kim D-S, Ahn KH (2006) Fluorescence modulation in anion sensing by introducing intramolecular H-bonding interactions in host-guest adducts. Chem Commun (2):186–188
46. Carroll CN, Naleway JJ, Haley MM, Johnson DW (2010) Arylethynyl receptors for neutral molecules and anions: emerging applications in cellular imaging. Chem Soc Rev 39 (10):3875–3888
47. Berryman OB, Johnson CA, Zakharov LN, Haley MM, Johnson DW (2008) Water and hydrogen halides serve the same structural role in a series of 2+2 hydrogen-bonded dimers based on 2,6-bis(2-anilinoethynyl)pyridine sulfonamide receptors. Angew Chem Int Ed 47(1):117–120
48. Kavallieratos K, Bertao CM, Crabtree RH (1999) Hydrogen bonding in anion recognition: a family of versatile, nonpreorganized neutral and acyclic receptors. J Org Chem 64(5): 1675–1683
49. Shchipanov VP, Krashina KI, Skachilova AA (1973) Tetrazole derivatives. Chem Heterocycl Compd 9(11):1423–1426
50. Seydel JK (1968) Sulfonamides, structure-activity relationship, and mode of action. Structural problems of the antibacterial action of 4-aminobenzoic acid (paba) antagonists. J Pharm Sci 57(9):1455–1478
51. Hof F, Schütz A, Fäh C, Meyer S, Bur D, Liu J, Goldberg DE, Diederich F (2006) Starving the malaria parasite: inhibitors active against the aspartic proteases plasmepsins I, II, and IV. Angew Chem Int Ed 45(13):2138–2141
52. McKie AH, Friedland S, Hof F (2008) Tetrazoles are potent anion recognition elements that emulate the disfavored anti conformations of carboxylic acids. Org Lett 10(20):4653–4655
53. Wallace KJ, Belcher WJ, Turner DR, Syed KF, Steed JW (2003) Slow anion exchange, conformational equilibria, and fluorescent sensing in venus flytrap aminopyridinium-based anion hosts. J Am Chem Soc 125(32):9699–9715
54. Davis J (2010) Anion binding and transport by prodigiosin and its analogs anion recognition in supramolecular chemistry. In: Gale PA, Dehaen W (eds) Topics in heterocyclic chemistry, vol 24. Springer, Berlin, pp 145–176
55. Courtemanche RJM, Pinter T, Hof F (2011) Just add tetrazole: 5-(2-Pyrrolo)tetrazoles are simple, highly potent anion recognition elements. Chem Commun 47 (47):12688–12690
56. Gale PA, Camiolo S, Tizzard GJ, Chapman CP, Light ME, Coles SJ, Hursthouse MB (2001) 2-Amidopyrroles and 2,5-diamidopyrroles as simple anion binding agents. J Org Chem 66(23):7849–7853
57. Campagna-Slater V, Schapira M (2010) Finding inspiration in the Protein Data Bank to chemically antagonize readers of the histone code. Mol Inform 29(4):322–331
58. Taverna SD, Li H, Ruthenburg AJ, Allis CD, Patel DJ (2007) How chromatin-binding modules interpret histone modifications: lessons from professional pocket pickers. Nat Struct Mol Biol 14(11):1025–1040
59. Ma JC, Dougherty DA (1997) The cation-π interaction. Chem Rev 97(5):1303–1324
60. Schärer K, Morgenthaler M, Paulini R, Obst-Sander U, Banner DW, Schlatter D, Benz J, Stihle M, Diederich F (2005) Quantification of cation–π interactions in protein–ligand complexes: crystal-structure analysis of Factor Xa bound to a quaternary ammonium ion ligand. Angew Chem Int Ed 44(28):4400–4404

61. Hughes RM, Wiggins KR, Khorasanizadeh S, Waters ML (2007) Recognition of trimethyllysine by a chromodomain is not driven by the hydrophobic effect. Proc Natl Acad Sci USA 104(27):11184–11188
62. Cashin AL, Petersson EJ, Lester HA, Dougherty DA (2004) Using physical chemistry to differentiate nicotinic from cholinergic agonists at the nicotinic acetylcholine receptor. J Am Chem Soc 127(1):350–356
63. Gallivan JP, Dougherty DA (1999) Cation-π interactions in structural biology. Proc Natl Acad Sci USA 96(17):9459–9464
64. Beshara CS, Hof F (2010) Modular incorporation of 1-benzyltryptophan into dipeptide hosts that bind acetylcholine in pure water. Can J Chem 88(10):1009–1016
65. Whiting AL, Neufeld NM, Hof F (2009) A tryptophan-analog host whose interactions with ammonium ions in water are dominated by the hydrophobic effect. Tetrahedron Lett 50(50):7035–7037
66. Koh KN, Araki K, Ikeda A, Otsuka H, Shinkai S (1996) Reinvestigation of calixarene-based artificial-signaling acetylcholine receptors useful in neutral aqueous (water/methanol) solution. J Am Chem Soc 118(4):755–758
67. Guo D-S, Uzunova VD, Su X, Liu Y, Nau WM (2011) Operational calixarene-based fluorescent sensing systems for choline and acetylcholine and their application to enzymatic reactions. Chem Sci 2(9):1722–1734
68. Perret F, Lazar AN, Coleman AW (2006) Biochemistry of the *para*-sulfonato-calix[n]arenes. Chem Commun (23):2425–2438
69. Coleman A, Perret F, Moussa A, Dupin M, Guo Y, Perron H (2007) Calix[n]arenes as protein sensors. Creative chemical sensor systems. In: Schrader T (ed) Topics in current chemistry, vol 277. Springer, Berlin, pp 31–88
70. Arena G, Casnati A, Contino A, Magri A, Sansone F, Sciotto D, Ungaro R (2006) Inclusion of naturally occurring amino acids in water soluble calix[4]arenes: a microcalorimetric and ^{1}H NMR investigation supported by molecular modeling. Org Biomol Chem 4(2):243–249
71. Douteau-Guevel N, Coleman AW, Morel J-P, Morel-Desrosiers N (1999) Complexation of the basic amino acids lysine and arginine by three sulfonatocalix[n]arenes (n=4, 6 and 8) in water: microcalorimetric determination of the Gibbs energies, enthalpies and entropies of complexation. J Chem Soc Perkin Trans 2 (3):629–634
72. Beshara CS, Jones CE, Daze KD, Lilgert BJ, Hof F (2010) A simple calixarene recognizes post-translationally methylated lysine. Chembiochem 11(1):63–66
73. Selkti M, Coleman AW, Nicolis I, Douteau-Guevel N, Villain F, Tomas A, de Rango C (2000) The first example of a substrate spanning the calix[4]arene bilayer: the solid state complex of *p*-sulfonatocalix[4]arene with -lysine. Chem Commun (2):161–162
74. Ingerman LA, Cuellar ME, Waters ML (2010) A small molecule receptor that selectively recognizes trimethyl lysine in a histone peptide with native protein-like affinity. Chem Commun 46(11):1839–1841
75. Wang K, Guo D-S, Zhang H-Q, Li D, Zheng X-L, Liu Y (2009) Highly effective binding of viologens by *p*-sulfonatocalixarenes for the treatment of viologen poisoning. J Med Chem 52 (20):6402–6412

Molecular Recognition of Nucleotides

Hannes Y. Kuchelmeister and Carsten Schmuck

Abstract Nucleotides are amongst the most targeted anionic species for artificial host systems, because they are ubiquitously present in biological systems in which they exercise key roles in many cellular functions and thus offer interesting structural features for the benchmarking of designed host systems in supramolecular chemistry (Voet et al., Fundamentals of biochemistry. Wiley-VCH, Weinheim, 2002). Nucleotides are, for example, involved in DNA synthesis, energy, and electron transfer events (Nath, Bioenerg Biomembr 42:301–309, 2010), cell signaling (Riedl and Salvesen, Nat Rev Mol Cell Biol 8:405–413, 2007), or membrane transport (Tojima et al., Nat Rev Neurosci 12:191–203, 2011). These are complex events which require the molecular recognition of a specific nucleotide. In order to increase our insight into these processes the development of model systems for nucleotide recognition is desirable.

Keywords Artificial receptors • Host-guest chemistry • Supramolecular chemistry • Nucleotides

Contents

H.Y. Kuchelmeister and C. Schmuck (✉)
Institute for Organic Chemistry, University of Duisburg-Essen, Universitätsstr. 7, 45117 Essen, Germany
e-mail: carsten.schmuck@uni-due.de

S.A. Piletsky and M.J. Whitcombe (eds.), *Designing Receptors for the Next Generation of Biosensors*, Springer Series on Chemical Sensors and Biosensors (2013) 12: 53–66
DOI 10.1007/5346_2012_16,
Published online: 27 June 2012

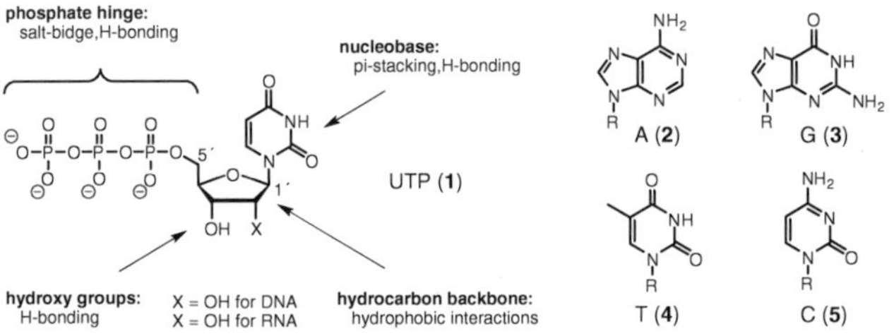

Fig. 1 Uridine triphosphate (**1**), adenine (**2**), guanine (**3**), thymine (**4**), and cytosine (**5**)

1 Nucleotide Recognition in Water

Due to their structural features, nucleotides are quite interesting targets from a chemist's point of view. As depicted in Fig. 1 they consist of three main parts—nucleobase, phosphate hinge, and sugar backbone. The sugar scaffold is a five-membered ribose for RNA and a 2′-deoxyribose ring for DNA building blocks. It connects the phosphate, which is attached to the 5′-hydroxyl group with the nucleobase at position 1′, this way clearly defining the nucleotide's geometry. The ribose ring itself is rather difficult to target. The hydroxyl groups offer potential for hydrogen bonding. However, they are hard to differentiate from the surrounding water molecules. The hydrocarbon chain can be targeted by means of hydrophobic interactions. These are, however, rather weak and nonspecific. The best artificial carbohydrate receptors known today complex their targets with binding constants of merely $5 \times 10^3\ M^{-1}$ (pH 7.4, buffered water) [1]. The phosphate hinge, consisting of one to three phosphates, is an easier target. Due to its multiple negative charges and the presence of several hydrogen bond acceptors, it may be addressed via hydrogen bonds and salt bridges. In comparison with the aforementioned carboxylates, phosphate anions are of tetrahedral shape and therefore require a different kind of binding motif. The nucleobases are either purine- (A, G) or pyrimidine based (C, U, T). Their aromatic nature allows for π-stacking, while their heterocyclic nature and the exocyclic carbonyl and amine functions offer potential hydrogen bonding acceptor and donor sites.

Due to the huge number of nucleotide receptors reported in the literature, this chapter will only focus on systems which have been used in aqueous media.

1.1 Polyamine-Based Receptors

Nakai has shown that the naturally occurring polyamines shown in Fig. 2 (top), putrescine (6), spermidine (7), and spermine (8), are able to bind to nucleotides [2]. Binding constants were determined by means of an anion-exchange resin method in buffered water at pH 7.5. While putrescine only forms a weak 1:1 complex with AMP ($K = 82\ M^{-1}$), spermidine is already able to form somewhat stronger 1:1

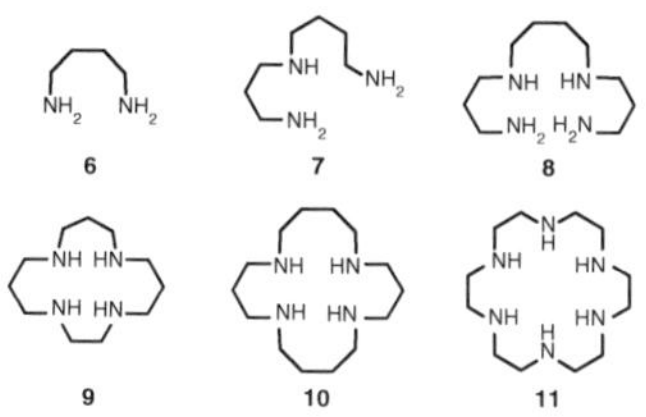

Fig. 2 Linear polyamines putrescine **6**, spermidine **7** and spermine **8** (*top*) and exemplary macrocyclic polyamine receptors **9–11** (*bottom*)

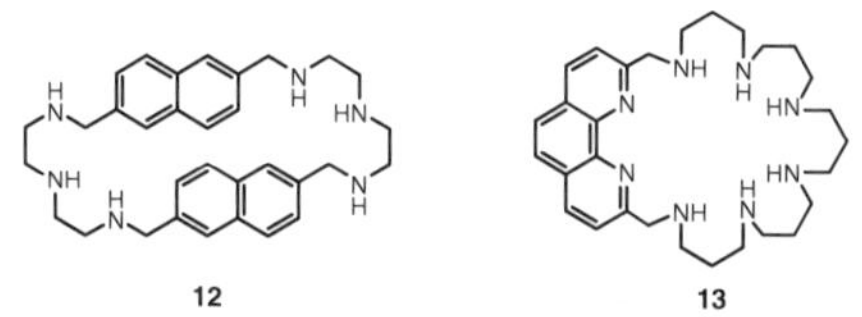

Fig. 3 Macrocyclic polyamine receptor **12** with two naphthalene bridges (*left*) and phenanthroline linked receptor **13** (*right*)

complexes with ADP ($K = 10^3\ M^{-1}$). Finally, spermine forms even stronger 1:1 complexes with ATP, with a binding constant of $10^4\ M^{-1}$. Obviously complex formation becomes stronger the greater the number of charge–charge interactions it is possible to form.

Not surprisingly, the first artificial nucleotide receptors, developed by Kimura, were 15- to 18-membered macrocyclic polyamines—some exemplary representatives are shown in Fig. 2 (**9–11**) [3]. By studying their binding properties towards phosphate, AMP, ADP, and ATP via polarographic and NMR measurements in buffered water at neutral pH, binding constants of up to $4 \times 10^6\ M^{-1}$ could be observed. Again, ATP was bound strongest followed by ADP and AMP. Phosphate on its own was bound more weakly than AMP by a factor of 10–100, although their charge is the same. This effect was attributed to additional hydrogen bonds between the nucleobase and the receptor as indicated by NMR. By comparing spermine with its cyclic analog **10** it could also be shown that the macrocycle binds more strongly to nucleotides by approximately two orders of magnitude. Furthermore, only polyamines which are capable of incorporating at least three charges at neutral pH formed stable complexes.

The next generation of receptors incorporated aromatic rings into the macrocycle. Lehn developed the nucleotide host **12** depicted in Fig. 3, which is able to bind to nucleotides in buffered water at pH 6 with binding constants of up to $2 \times 10^5\ M^{-1}$ (ATP, NMR titration) [4]. The less charged AMP is bound more weakly by one order of magnitude. Interestingly the purine-derived nucleotides AMP and GMP are bound more strongly than the smaller pyrimidine analogs CMP and UMP by a factor of 10. Lacking the phosphate hinge, the corresponding nucleosides are more weakly bound by 2–3 orders of magnitude. Thus it can be concluded that both the stacking between nucleobase and aromatic scaffold as well

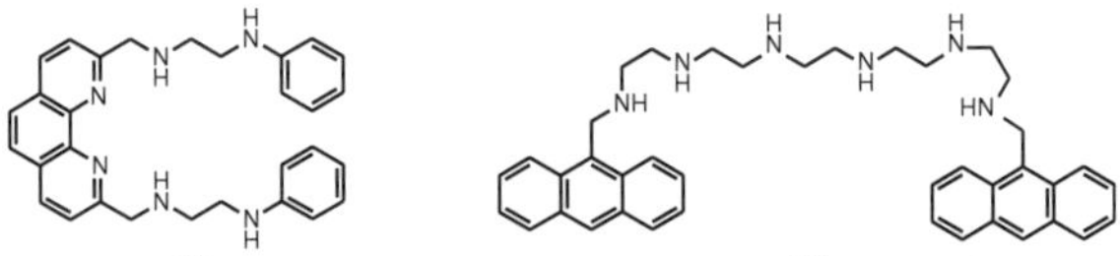

14 15

Fig. 4 Tweezer receptor **14** with two polyamine side chains linked via a phenanthroline template (*left*) and receptor **15** with two anthracene units bridged by a polyamine chain

as charge–charge interactions between the polyammonium chains and the phosphate part of the nucleotide are essential for efficient binding.

Bencini developed the macrocyclic receptor **13** (Fig. 3) with a phenanthroline unit inside the polyamine macrocycle [5]. With the help of potentiometric measurements, the highest affinity was found for ATP with an extremely high binding constant of $3 \times 10^9\ M^{-1}$. The corresponding nucleotides TTP, CTP, and GTP were bound less well by 1–2 orders of magnitude. Furthermore, the artificial host could selectively sense ATP thanks to fluorescence quenching caused by a photoinduced electron transfer from an amine group of the receptor to the excited phenanthroline. Concerning the complex geometry, it could be shown by means of NMR experiments, molecular modeling and a crystal structure of one of the complexes that the polyamine chain forms a cavity for the phosphate, enforced by ion pairs and hydrogen bonds, while the nucleobase stacks with the phenanthroline moiety. A macrocycle analog with reduced ring size and one fewer amine group lost a huge part of its affinity, i.e., the binding constant to ATP was only $10^6\ M^{-1}$—lower by a factor of 3,000. These results demonstrate the importance of an appropriate mix of attractive interactions with the correct geometric prerequisites needed for efficient recognition of a given substrate molecule.

Lin designed the tweezer receptor **14** (Fig. 4) with two polyamine side chains linked via a phenanthroline scaffold; each arm incorporates a phenyl headgroup [6]. The binding constant for ATP was measured via potentiometry to be an excellent $7.9 \times 10^{10}\ M^{-1}$. As expected, NMR and molecular modeling revealed charge–charge interactions between phosphate and polyammonium side chains and π-stacking between phenanthroline and phenyl groups with the nucleobase. These results prove that in order to achieve high binding affinity it is not necessary to make use of cyclic systems: Essentially the combination of several attractive non-covalent interactions in the correct molecular topology for binding is more important for the strength of the molecular recognition event.

A similar system (**15**), also depicted in Fig. 4, was developed by Garcia-España. Two anthracene moieties are linked by a polyamine chain [7]. As determined by potentiometry, the binding constant for ATP was also similarly high, being measured as $10^9\ M^{-1}$. ADP and ATP were bound consecutively worse by two and, respectively, three orders of magnitude. Furthermore, in this work also the effective binding constant for ATP at neutral pH was determined to be approximately $5 \times 10^5\ M^{-1}$. This value was confirmed with the help of a fluorescence titration. It is noteworthy that the effective binding constant is lower by more than three orders of magnitude compared to the highest possible value reported, based on potentiometric measurements.

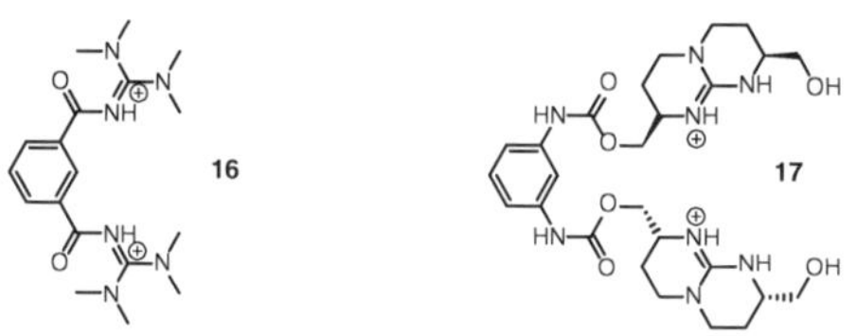

Fig. 5 Tweezer receptor **16** for AMP with two tetramethylguanidinio groups (*left*) and ditopic receptor **17** for TMP with two bicyclic, chiral guanidinium moieties (*right*)

At this point it is important to state that great care has to be taken when comparing different binding constants. Potentiometric measurements represent the binding constant of host and guest in a clearly defined protonation state. However, at a given pH a mixture of different protonation states always exists. Consequently these K values only refer to single theoretical protonation states and are therefore not valid for the macroscopic ensemble but represent rather the maximum possible binding constant between two single species within the mixture. Without the calculation of an effective binding constant these data cannot be used to compare binding constants of different host systems. The above-mentioned examples made use of potentiometry. The effective binding constant was never calculated except in the work conducted by Garcia-España. This is also why many of the extremely high binding constants must be put into perspective—the effective binding constants of the macroscopic sample are likely to be lower by several orders of magnitude. A more appropriate way to obtain comparable binding constants is to utilize methods such as NMR, UV/Vis, fluorescence, or ITC. Whatever the choice may be, it is always strongly advisable to use at least two different methods to verify one's results.

1.2 Guanidinium-Based Receptors

A statistical evaluation of 3,003 X-ray crystal structures of phosphate binding proteins revealed that two third of them do not make use of metal ions for phosphate binding [8]. More than half of the metal-free proteins use lysine or arginine instead, especially those with phosphate binding sites located on their surface. Consequently, ammonium and guanidinium moieties seem to be well suited to bind to phosphate in natural systems, even when directly opposed to the competitive influence of the surrounding water molecules and counter ions. While most of the receptors reported in the literature so far make use of ammonium groups in analogy to lysine, only very few host systems have been designed yet which utilize the corresponding arginine analogus—guanidinium moieties—for phosphate recognition.

Wang reported the ditopic receptor **16** with two tetramethylguanidinio moieties linked via an aromatic scaffold (Fig. 5) [9]. With the help of an NMR titration in D_2O he was able to show that this rather simple molecule alone was able to bind AMP with $K = 7 \times 10^4\ M^{-1}$.

Fig. 6 The amphiphilic receptor **18** combines a long hydrophobic chain with a positively charged guanidinium headgroup. At the air–water interface the positively charged guanidinium moieties point towards the aqueous phase while the hydrophobic part points towards the air

In a similar approach Schmidtchen developed the tweezer **17**, depicted in Fig. 5 [10]. Two chiral, bicyclic guanidinium moieties were connected via an aromatic template. With the help of NMR titrations in water a very high binding constant of 10^6 M^{-1} to TMP could be observed, which was explained by a perfect preorganization of the two guanidine units perpendicular to each other. This way their geometry is perfectly matched to bind to the tetrahedral phosphate anion via two hydrogen bonded salt bridges. Once more, preorganization and structural complementarity proved to be the key to obtaining high affinity. For phosphate binding, especially ditopic receptors with two anion binding groups seem to be perfectly suitable.

The most successful approach towards guanidinium-based nucleotide recognition so far was Kunitake's receptor **18**, shown in Fig. 6 [11]. By attaching a long, lipophilic chain to a positively charged guanidinium head group, amphiphilic host systems were obtained which were able to form micelles or bilayers in aqueous systems. The binding properties towards adenine-based nucleotides were characterized via ultrafiltration. With 2×10^7 M^{-1} being the highest binding constant that was observed at an air–water interface in buffered water at pH 7 for ATP, AMP was bound more weakly by one order of magnitude. The results of this approach stress the importance of the microenvironment on binding events. Kunitake's system is actually a simple guanidinium moiety which is set into a defined microenvironment with the help of a long amphiphilic side chain, yet very good binding constants could be achieved this way.

1.3 Calixarene- and Cyclodextrin-Based Receptors

The focus of most of the above-mentioned nucleotide receptors clearly lies within charge–charge interactions to their substrate, sometimes accompanied by stacking between the nucleobase and one or more aromatic moieties of the host systems. While π-stacking is a powerful tool for molecular recognition, especially in aqueous media, it is not the only possible way to utilize non-covalent interactions in water. The following examples put more focus on hydrophobic interactions, which may contribute substantial amounts to the Gibbs free energy in water.

Diederich, for example, made use of a bowl-shaped calixarene with a resorcin[4] arene scaffold and four phenylamidinium groups at the upper rim for his nucleotide receptor **19** (Fig. 7) [12]. Calixarenes form hydrophobic cavities in aqueous media

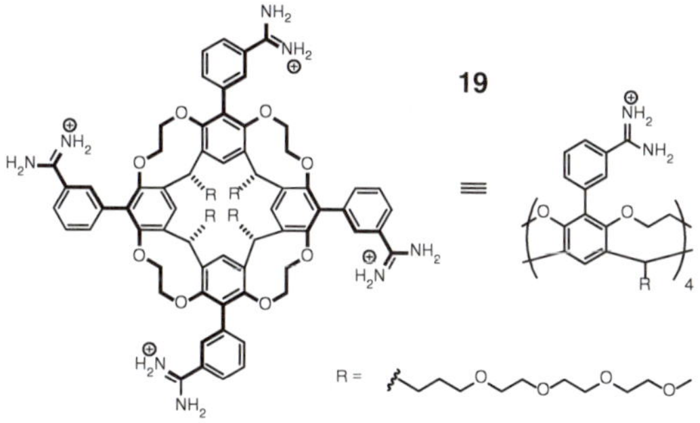

Fig. 7 Calixarene-based receptor **19** with four phenylamidinium groups

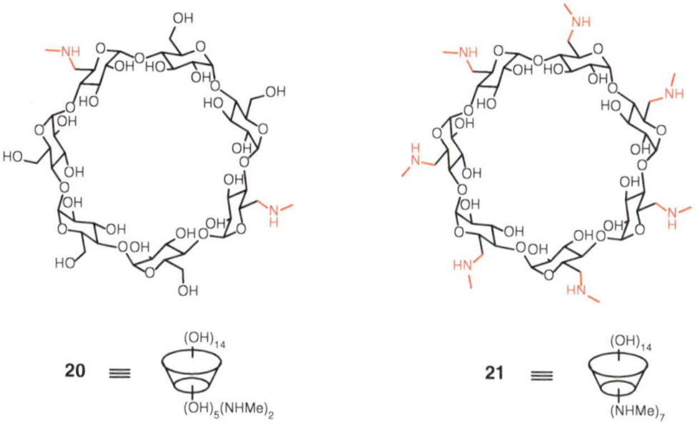

Fig. 8 Cyclodextrin-based receptors **20** and **21** with two and, respectively, three methylammonium groups

for the accommodation of nonpolar guest molecules. NMR titrations in buffered D_2O at pH 8.3 revealed affinities corresponding to the charge on the nucleotide guest (ATP > ADP > AMP > cAMP) and binding constants for the 1:1 host–guest complexes of up to $7 \times 10^5\ M^{-1}$. AMP was slightly preferred over the G, C, T, and U derivatives by a factor of 2–3. As derived from NMR and molecular modeling the nucleobase is bound within the hydrophobic bowl and the phosphate interacts with two phenylamidinium groups via hydrogen bonded salt bridges. The preference for A-nucleotides can be ascribed to a superior stereoelectronic complementarity of this nucleobase with regard to its inclusion into the bowl-type cavity.

Another sort of cavity-forming molecules was used as a scaffold by Schneider [13]. By exchanging two or, respectively, all seven 6-hydroxy groups in β-cyclodextrin (CyD) for aminomethyl moieties, the two nucleotide receptors **20** and **21**, shown in Fig. 8, were obtained. CyDs are highly water soluble, cyclic, 1,4-linked oligo-α-D-glucopyranoside units. Their topology is toroidal, with a larger and a smaller opening which present their hydroxyl groups to the solvent. Although the

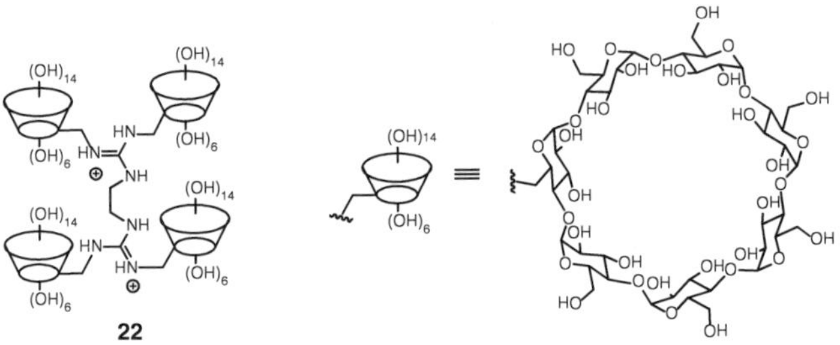

Fig. 9 Receptor **22** combines four cyclodextrin units with two guanidinium scaffolds

interior of the cavity is not hydrophobic, it is less hydrophilic than water. Thus **21** forms strong nucleotide-complexes, with binding constants of up to 3×10^6 M^{-1} with ATP (effective binding constant obtained by potentiometry, verified via NMR titration). This is stronger by more than one order of magnitude compared to its complex with AMP. However, this host system is not able to differentiate between nucleobases or between ribose and deoxyribose sugar moieties. The receptor **20**, on the other hand, forms weaker complexes (e.g., 10^5 M^{-1} for ATP) but is able to differentiate both between nucleobases and sugars.

With the help of knockout analogs (e.g., ribose, phosphoribose, phosphate) the different contributions to the overall binding constant could be assigned. The main driving force stems from ionic interactions between phosphate and methylammonium groups. Surprisingly, the nucleobase does not contribute in an advantageous way to the binding. Instead, the second most important contribution to the binding strength stems from interactions with the sugar moiety. With the help of NMR and molecular modeling it could be shown that the ribose resides inside the CyD cavity and interacts with it via hydrogen bonds. This is also why the deoxyribose is bound more weakly than ribose, despite the fact that it is more hydrophobic. The nucleobase-selectivity of **20** was explained by secondary, rather weak hydrogen bonds between nucleobase and upper rim hydroxyl groups. The higher charged **21** pulls the nucleotide deeper into the cavity and thus prevents these secondary interactions from forming. Again, these experiments are excellent examples of the difficult task of designing receptors which have high selectivity as well as high affinity to a given substrate. All too often these two desirable properties are antipodal. Furthermore Schneider's work is an instructive example on how model systems help to improve our understanding of anion recognition.

Marsura made use of the same principle by bridging four β-cyclodextrin cavities via two guanidinium groups to receptor **22**, as depicted in Fig. 9 [14]. The cationic scaffold allows for hydrogen bonded ion pairing to nucleotide phosphates while the CyDs display pockets for hydrophobic inclusion. NMR experiments in D_2O at pH 6.5 revealed that 1:2 host–guest complexes are formed to AMP, ADP, and ATP with similar binding constants of 2×10^6 M^{-2}. Two substrate molecules are bound with similar binding constants, representing a noncooperative binding mode with two independent complexation steps. NMR also revealed that the phosphate hinge

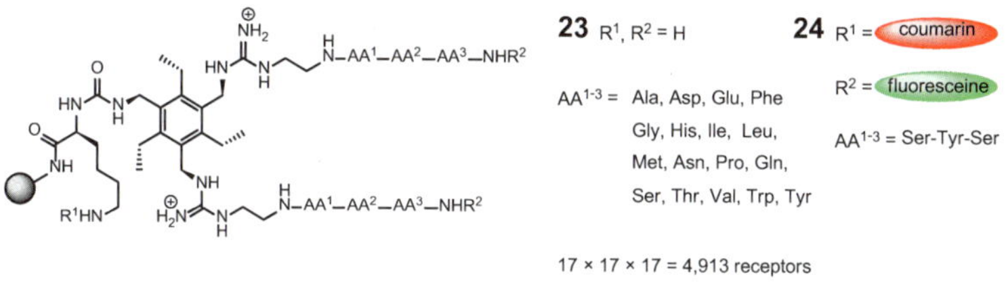

Fig. 10 Receptor library **23** with 4,913 members and fluorescence labeled hit structure **24** for ATP binding with amino acid sequence AA^{1-3} = Ser-Tyr-Ser

interacts with the guanidinium moieties via hydrogen bond enforced salt bridges. For ATP and ADP the nucleobase and the sugar moiety are inside the CyD cavity; for AMP, on the other hand, only the nucleobase is included. This is probably due to the distance between the cavity and the cationic template, which is too large to be bridged by AMP's monophosphate hinge when sugar and nucleobase are inside the CyD.

1.4 Peptide-Based Receptors

As shown for carboxylate and oligopeptide recognition (vide supra), peptides have proven to be excellent building blocks for the synthesis of artificial receptors. Furthermore, they are excellent model systems for proteins to investigate, e.g., ATP binding sites, which are commonly present in many enzymes.

Anslyn used the combinatorial library **23** (R_1, R_2 = H), shown in Fig. 10, with 4,913 members (17^3) to obtain ATP selective receptors [15]. Two combinatorial tripeptide side chains were linked by an aromatic scaffold which preorganizes adjacent groups alternately "up" and "down" with respect to the benzene ring. As a result, a cavity is formed for substrate binding. Each arm was attached to the resin via a guanidinium group in order to act as a phosphate binding site. The third arm was used as the connection to the solid phase and features a lysine group. The library was screened in buffered water at pH 7.1, with fluorescence labeled ATP, revealing that approximately 15% of the library members bind to ATP. Several hit structures were resynthesized and labeled with two different dyes: fluoresceine at the N-termini of the two arms (R^2) and coumarin in the third arm (R^1, as internal standard). Binding studies in solution verified the binding properties in all but one example. This result again stresses the importance of confirming on-bead results with those obtained in solution. The best receptor, with the amino acid sequence AA^{1-3} = Ser-Tyr-Ser (**24**), binds to ATP with $K = 3 \times 10^3\ M^{-1}$ (pH 7.1, buffered water). The competing analytes AMP and GTP were bound significantly less effectively. The presence of Tyr and Ser in the best receptor hints at hydrogen bonding and π-stacking between receptor and adenine and/or ribose. By attaching the fluorescence labeled receptors onto a solid support they could be used as selective sensors in arrays [16].

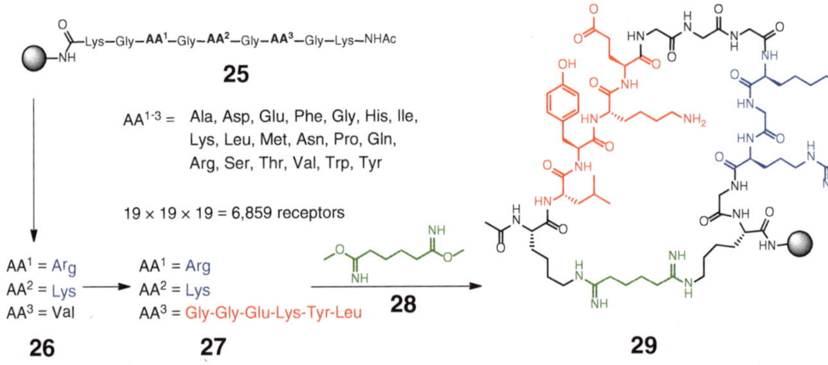

Fig. 11 Combinatorial library of 6,859 (19^3) ATP receptors **25** and hit structure **26**. Rationally improved compound **27** and crosslinked ATP selective receptor **29** with phosphate (*blue*) and nucleobase (*red*) binding site. Cyclization was achieved by crosslinking the first and last Lys residue with dimethyl adipimidate (*green*)

Matsui derived his ATP selective receptor **29** (Fig. 11) in a three-step procedure [17]. Firstly, the nonapeptide library **25**, with three combinatorial amino acid positions and comprising 6,859 (19^3) members was screened with the help of fluorescently labeled ATP. The binding constant of the best receptor **26**, with AA^{1-3} = Arg-Lys-Val, was determined on-bead in buffered water at pH 7 to be 6×10^3 M^{-1}. Molecular modeling revealed that Arg and Lys bound to the phosphate hinge, while there is no interaction with the nucleobase. Furthermore, Val does not seem to take part in the recognition process. Thus Val was replaced by Gly-Gly-Glu-Lys-Tyr-Leu, a sequence derived from the adenine binding site of biotin carboxylase. The corresponding receptor **27** had an increased affinity to ATP ($K = 10^4$ M^{-1}). In the last step, the thus-derived peptide was crosslinked with dimethyl adipimidate (**28**) in the presence of ATP.

The cyclic peptide **29** now had a binding constant of 5×10^4 M^{-1} to ATP. Additionally, in contrary to its precursors, **29** was now able to distinguish between ATP and ADP, AMP, GDP, GTP which are all bound significantly weaker by one order of magnitude. When the cyclization was carried out without the presence of ATP the affinity of the corresponding receptors was decreased, which suggests the formation of alternative cyclization products. This effect was attributed to the two inner Lys moieties being involved with binding to the phosphate hinge. Only the first and the last lysine side chains are then freely available for crosslinking. Unfortunately, the authors did not provide additional data obtained in solution to support this hypothesis. However, they were able to prove the general concept of combining combinatorial chemistry with rational design and molecular imprinting in order to increase the affinity of a library-derived receptor—in this case by one order of magnitude.

Schmuck developed the tweezer receptor **30** (Fig. 12) with two symmetric peptidic arms which are connected via an aromatic template and contain phenylalanine, lysine and a guanidiniocarbonyl pyrrole (GCP) oxoanion binding site as the

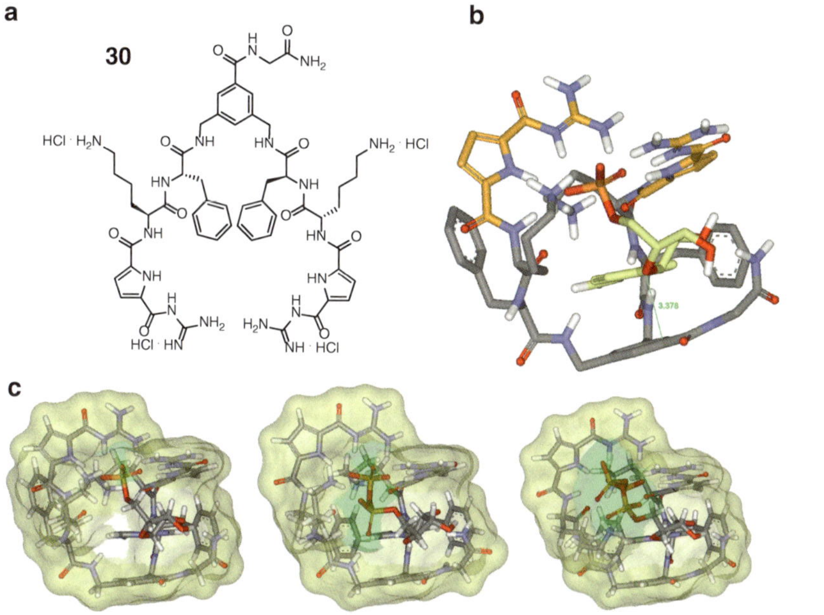

Fig. 12 (**a**) Symmetric tweezer receptor **30**; (**b**) side view of the calculated energy-minimized structure of the complex between **30** and AMP as obtained from a Monte Carlo conformational search. Nonpolar hydrogen atoms have been removed for clarity. Color code: substrate (*yellow*), GCP (*orange*), and receptor backbone (*gray*); (**c**) solvent-accessible surfaces of **30** (*yellow*) and the phosphate residue of the nucleotides (*green*). From *left* to *right*: AMP, ADP, and ATP. The molecular modeling images are reprinted with permission from [18]. Copyright 2011 John Wiley and Sons

headgroup [18]. With the help of UV/Vis and fluorescence experiments it could be shown that **30** forms 1:1 complexes with nucleotides in buffered water (pH 7) with binding constants of up to $K \approx 10^5\ M^{-1}$. Simple phosphates are bound less well, by one order of magnitude. An even more striking characteristic of this host is that it prefers to bind to AMP ($8 \times 10^4\ M^{-1}$) over ADP ($3 \times 10^4\ M^{-1}$) and ATP ($8 \times 10^3\ M^{-1}$)—the monophosphate is bound stronger than the triphosphate by one order of magnitude. This binding behavior is unusual and so far is unprecedented for artificial nucleotide receptors. As described above, the more highly charged nucleotides are normally preferred due to the formation of additional charge–charge interactions between receptor and substrate.

According to NMR spectroscopic measurements and molecular modeling studies, the efficient binding is due to strong electrostatic interactions between the GCP and lysine groups of the host with the highest charged α-phosphate of the substrates, combined with π–π interactions within the cavity-shaped receptor. The host and guest are thus locked in a well-defined complex with specific interactions. The larger the phosphate hinge on the guest molecule, the more it is located outside of the cavity, because the receptor cannot wrap around it so effectively (see the solvent accessible surface in Fig. 12c). This way, the competitive influence of the surrounding water becomes more significant with the consequence that the binding strength is reduced. Hence, the complex formation between **30** and the phosphate-based substrates is a good example for the importance of combining different types of non-covalent interactions within a defined structure in order to achieve selective molecular recognition of biologically relevant targets in aqueous solution.

In conclusion, nucleotide recognition in water has received considerable attention. Many different artificial host systems have been developed and high binding constants of up to $10^7\ M^{-1}$ could be achieved in aqueous media. The majority of nucleotide receptors are based on polyamines—either cyclic or linear in form. Another, less frequently used binding motif for ion pairing to the phosphate hinge is the guanidinium group. Very often aromatic moieties are implemented into the hosts in order to allow for π-stacking. Following a similar approach, calixarenes or cyclodextrins are used in order to present hydrophobic pockets for nucleobase and/or sugar. Another successful principle is the utilization of tweezer receptors which form a preorganized cavity for accommodation of the substrate. Very often two anion binding motifs are incorporated—one into each arm. This way the receptors may adopt a geometry that matches the tetrahedral shape of the phosphate anion. Hydrogen bonding patterns play a crucial role for the selective recognition of nucleotides. While strong binding is mainly achieved by increasing charge–charge interactions, introducing selectivity is more subtle, because purine and, respectively, pyrimidine nucleobases are quite similar in shape and hydrophobicity. Another aspect of selectivity in almost all artificial host systems is the preference for tri- over di- and monophosphates, which can easily be explained by the presence of additional ionic interactions. Peptide-based systems have also been used for the recognition of nucleotides. Although the results are not yet as good as those seen with polycationic host systems, the use of amino acids as building blocks offers the intriguing potential to generate artificial receptors by means of rational approaches, combinatorial chemistry or combinations thereof.

References

1. Rauschenberg M, Bomke S, Karst U, Ravoo BJ (2010) Dynamic peptides as biomimetic carbohydrate receptors. Angew Chem Int Ed 49:7340–7345
2. Nakai C, Glinsmann W (1977) Interactions between polyamines and nucleotides. Biochemistry 16:5636–5641
3. Kimura E, Kodama M, Yatsunami T (1982) Macromonocyclic polyamines as biological polyanion complexons. 2. Ion-pair association with phosphate and nucleotides. J Am Chem Soc 104:3182–3187
4. Dhaenens M, Lehn J-M, Vigneron J-P (1993) Molecular recognition of nucleosides, nucleotides and anionic planar substrates by a water-soluble bis-intercaland-type receptor molecule. J Chem Soc Perkin Trans 2 1379–1381
5. Bazzicalupi C, Bencini A, Biagini S, Faggi E, Meini S, Giorgi C, Spepi A, Valtancoli B (2009) Exploring the binding ability of phenanthroline-based polyammonium receptors for anions: hints for design of selective chemosensors for nucleotides. J Org Chem 74:7349–7363

6. Guo Y, Ge Q, Lin H, Lin H, Zhu S, Zhou C (2003) Recognition promoted by Zn^{2+} between phenanthroline bridging polyaza ligands and nucleotides – Zn^{2+} acts as 'messenger' between the receptor and substrate. J Mol Recognit 16:102–111
7. Albelda MT, Bernardo MA, Garcia-España E, Godino-Salido ML, Luis SV, Melo MJ, Pina F, Soriano C (1999) Thermodynamics and fluorescence emission studies on potential molecular chemosensors for ATP recognition in aqueous solution. J Chem Soc Perkin Trans 2 2545–2549
8. Hirsch AKH, Fischer FR, Diederich F (2007) Molekulare Erkennung von Phosphaten in der Strukturbiologie. Angew Chem 119:342–357
9. Gao H, Cai L, Qi Y, Wang H (2003) Synthesis of 1,3-{Di-[N-bis(dimethylamino)methane]}-benzyl-diamide and its molecular recognition of nucleotides in aqueous solution. Supramol Chem 15:323–325
10. Schmidtchen FP (1989) A non-macrocyclic host for binding organic phosphates in protic solvents. Tetrahedron Lett 30:4493–4496
11. Onda M, Yoshihara K, Koyano H, Ariga K, Kunitake T (1996) Molecular recognition of nucleotides by the guanidinium unit at the surface of aqueous micelles and bilayers. A comparison of microscopic and macroscopic interfaces. J Am Chem Soc 118:8524–8530
12. Sebo L, Diederich F (2000) Tetrakis(phenylamidinium)-substituted resorcin[4]arene receptors for the complexation of dicarboxylates and phosphates in protic solvents. Helv Chim Acta 83:93–113
13. Eliseev AV, Schneider HJ (1994) Molecular recognition of nucleotides, nucleosides, and sugars by aminocyclodextrins. J Am Chem Soc 116:6081–6088
14. Menuel S, Duval RE, Cuc D, Mutzenhardt P, Marsura A (2007) Molecular recognition of nucleotides by a new bis(guanidinium)tetrakis(β-cyclodextrin) tetrapod. New J Chem 31:995–1000
15. Schneider SE, O'Nei SN, Anslyn EV (2000) Coupling rational design with libraries leads to the production of an ATP selective chemosensor. J Am Chem Soc 122:542–543
16. McCleskey SC, Griffin MJ, Schneider SE, McDevitt JT, Anslyn EV (2003) Differential receptors create patterns diagnostic for ATP and GTP. J Am Chem Soc 125:1114–1115
17. Matsui J, Nagano J, Miyoshi D, Tamaki K, Sugimoto N (2009) An approach to peptide-based ATP receptors by a combination of random selection, rational design, and molecular imprinting. Biosens Bioelectron 25:563–567
18. Kuchelmeister HY, Schmuck C (2011) Nucleotide recognition in water by a guanidinium-based artificial tweezer receptor. Chem Eur J 17:5311–5318

Molecular Recognition of Oligopeptides and Protein Surfaces

Hannes Y. Kuchelmeister and Carsten Schmuck

Abstract Many physiological functions of life are controlled by the interplay between natural or synthetic agents with their corresponding receptors in the human body (Highlights in bioorganic chemistry: methods and applications. Wiley-VCH, Weinheim, 2000). Such molecular recognition events are based on the combination of many weak attractive non-covalent interactions between receptor and substrate, such as electrostatic, dipole, and dispersion interactions, π-stacking, and hydrogen bonding, together with entropic contributions, as, for example, the liberation of solvent molecules (Core concepts in supramolecular chemistry and nanochemistry. Wiley-VCH, West Sussex, 2007; Angew Chem Int Ed 46:2366–2393, 2007). To gain a more detailed insight into these complex processes, which are currently not entirely understood in all their complexity, the development of synthetic model systems is worthwhile. Within this chapter, selected illustrative examples of artificial receptors for biologically relevant targets such as oligopeptides or protein surfaces will be presented. A directed molecular recognition of these building blocks of life will lead to an increase in knowledge concerning the complex recognition processes taking place within the human body and in the best case will allow for biological processes to be directly targeted and can thus be used for future applications in analytical, biological, or medicinal chemistry.

Keywords Artificial receptors, Host-guest chemistry, Supramolecular chemistry, Peptides

---

H.Y. Kuchelmeister and C. Schmuck (✉)
Institute for Organic Chemistry, University of Duisburg-Essen, Universitätsstr. 7, 45117 Essen, Germany
e-mail: carsten.schmuck@uni-due.de

S.A. Piletsky and M.J. Whitcombe (eds.), *Designing Receptors for the Next Generation of Biosensors*, Springer Series on Chemical Sensors and Biosensors (2013) 12: 67–84
DOI 10.1007/5346_2012_17,
Published online: 27 June 2012

Contents

1 Oligopeptide and Protein Surface Recognition

1.1 Recognition of Short Peptides with Linear Receptors

Following Lehn's description, the term "receptor" will be used throughout this chapter as a chemical host which binds to a given guest [1]. According to this definition, peptide receptors are synthetic organic molecules which selectively bind to a peptide sequence via multiple non-covalent interactions. The molecular recognition of short peptide sequences is a promising goal due to their importance in biochemical and medicinal processes [2]. Artificial model systems may not only contribute to the acquisition of a more substantial understanding of the yet not fully understood molecular recognition process itself, but they may also serve as new starting points for drug development [3] or as sensors for diagnostics [4]. A prerequisite for potential applications is always a strong and selective complexation of the target peptide sequence. In the following chapter, selected instructive examples of oligopeptide receptors will be presented.

A well-studied example concerning a short peptide sequence of biological importance is the amino acid sequence D-Ala-D-Ala-OH (1) and its depsipeptide analog D-Ala-D-Lac-OH (2), which are both shown in Fig. 1. The first one is crucial for the mode of action of the antibiotic Vancomycin (3) against Gram-positive bacteria: Vancomycin binds to this sequence with high affinity ($K = 2 \times 10^5\ M^{-1}$), thereby sterically blocking a transpeptidase enzyme from crosslinking the peptide side chains of peptidoglycan strands, which act as precursors during bacterial cell wall biosynthesis. This leads to a decrease of mechanical cell wall stability and ultimately to lysis upon osmotic pressure changes. The depsipeptide 2 can be found in Vancomycin-resistant bacteria: by exchanging an amide for an ester bond, the complex stability is reduced by a factor of 1,000. This dramatic decrease is due to the loss of one hydrogen bond, which is instead replaced by an electrostatic repulsion between the oxygen lone pairs of the ester and the corresponding carbonyl group in the antibiotic [5].

With the help of the combinatorial library 4 shown in Fig. 2, comprising 39,304 members (343 proteinogenic and non-natural amino acids), Ellman was able to identify receptors which bind to the depsipeptide sequence N,N′-Ac_2-L-Lys-D-Ala-D-Lac-OH (5) in aqueous solution [6]. The design of Ellman's receptor mimics that

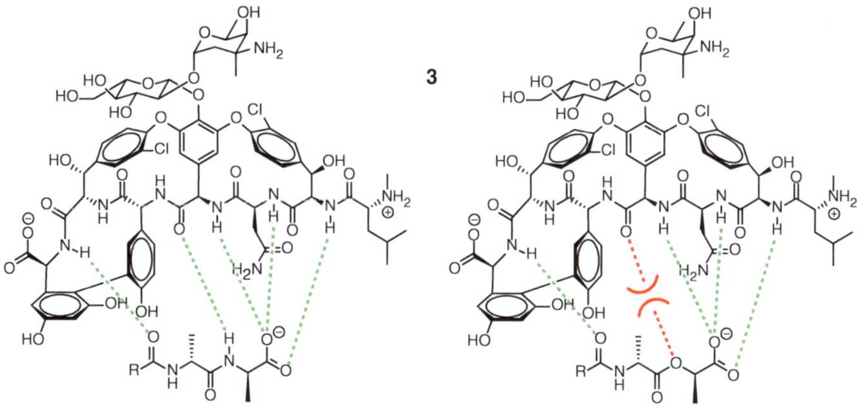

3

1 2

Fig. 1 Comparison of interactions between Vancomycin (**3**) with normal (**1**) and mutated (**2**) bacterial cell wall-forming peptide sequence. An attractive hydrogen bond is replaced for a repulsive electrostatic interaction, thus decreasing the affinity of the antibiotic by a factor of 1,000

4 6

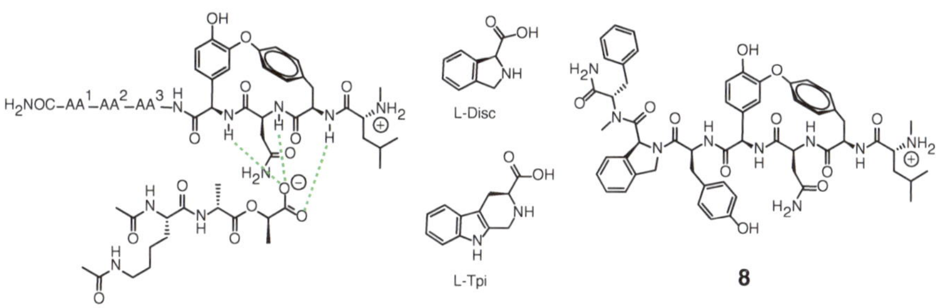

5 7 8

$AA^{1\text{-}3}$ = L/D: Arg, Asn, His, Leu, N-Me-Leu, Lys, Phe, N-Me-Phe, Pro, Thr, Trp, Tyr, Tic

L: Aib, Chg, Atc, Dapa, Disc, Isn, Tpi, N-Me-Gly

34 x 34 x 34 = 39,304 receptors

Fig. 2 Schematic representation of receptor library **4** for the substrate N,N′-Ac_2-L-Lys-D-Ala-D-Lac-OH (5) (*left*) and receptor of highest affinity **8** with the amino acid sequence L-N-Me-Phe-L-Disc-L-Tyr (*right*)

of Vancomycin: the right part corresponding to a simplified carboxylate binding pocket, which should be able to bind to both D-Ala-OH and D-Lac-OH via three hydrogen bonds. The left side of the receptor, on the other hand, consists of a variable tripeptide, which is free to rotate and should therefore be able to avoid electrostatic repulsion between the antibiotic's carbonyl group and the lactate oxygen. Indeed, an on-bead screening with a fluorescently labeled substrate revealed binding constants five times as high as for Vancomycin. The best receptors mainly contained the amino acid sequence L-Tpi-L-His (29%) and L-N-Me-Phe-L-Disc (20%) in position AA^1 and

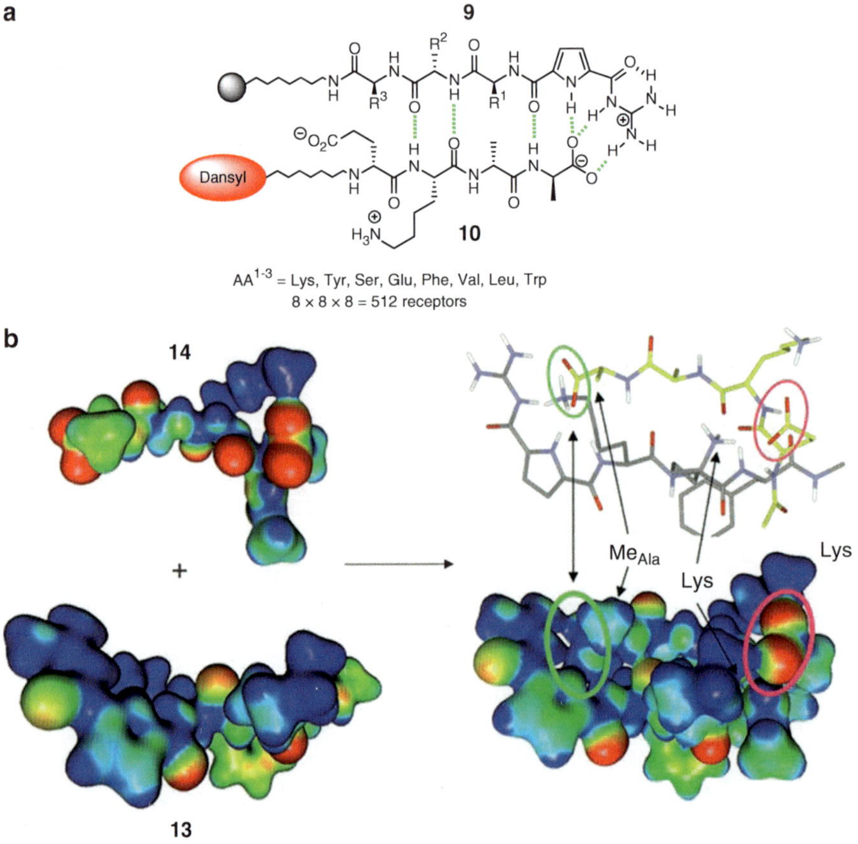

Fig. 3 (**a**) General scheme of the tripeptide receptor library **9** and the dansyl-labeled substrate EKAA (**10**); (**b**) calculated electrostatic surface potential, which shows the charge complementarity between receptor **11** (GCP-KKF) and tetrapeptide **12** (EKAA). The molecular modeling images are reprinted with permission from [7]. Copyright 2006 John Wiley and Sons

AA^2 (L-Disc 6 and L-Tpi 7 are shown in Fig. 2). For AA^3 no clear tendency could be observed. To determine the binding constants and to validate the results from the on-bead screening, the best performing sequences were synthesized and tested in solution for their binding affinity towards N,N′-Ac_2-L-Lys-D-Ala-D-Lac-OH (5) with the help of microcalorimetry. The overall best receptor **8**, with the sequence L-N-Me-Phe-L-Disc-L-Tyr, has a binding constant of $3 \times 10^4\ M^{-1}$ in water. For the recognition of peptides in aqueous solution this is a very good result. However, the substrate is fully acetylated and therefore rather nonpolar. This facilitates binding based on van der Waals interactions, but also is significantly different from the actual biological relevant substrate.

As depicted in Fig. 3, Schmuck was able to identify receptors for the dansyl-labeled polar, anionic peptide sequence N-Ac-D-Glu-L-Lys-D-Ala-D-Ala-OH

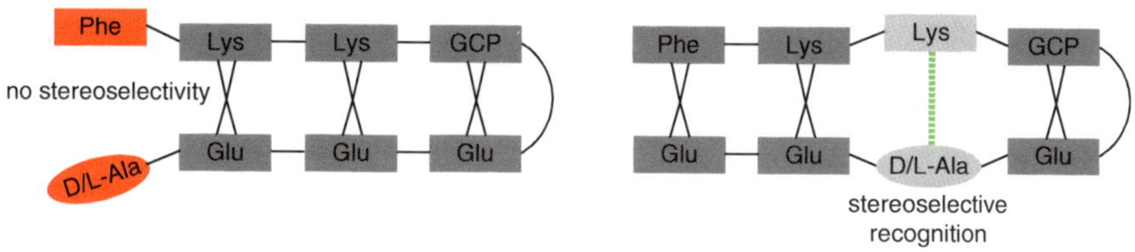

Fig. 4 Screening of a tetrapeptide library revealed that stereoselectivity is observed at positions where receptor and substrate are locked by strong electrostatic interactions at both sides of the stereogenic center: D-Ala: $K = 5{,}000\ M^{-1}$; L-Ala: $K = 900\ M^{-1}$ (*right*). When this is not the case no stereoselectivity can be witnessed: D-Ala: $K = 7{,}900\ M^{-1}$; L-Ala: $K = 8{,}500\ M^{-1}$ (*left*)

(EKAA, **10**) with binding constants of up to $K \approx 10^4\ M^{-1}$ in buffered water at pH 6 by means of screening a combinatorial library with 512 peptide-based artificial receptors (9) containing a guanidiniocarbonyl pyrrole unit (GCP) as the headgroup [9]. The rigid, planar GCP group allows for efficient binding of carboxylates in aqueous media by forming a hydrogen bond-enforced salt bridge with the C-terminus of the oligopeptide. The guanidinium group in 9 is connected to the pyrrole core by an acetyl group, thus decreasing the pK_a value of free guanidine from 13.5 to approximately 6–7 [10]. As a consequence, the potency of its action as a hydrogen donor was increased dramatically. Furthermore, the intramolecular hydrogen bond between the carboxyl and guanidinium group preorganizes the binding motif in the correct conformation for oxoanion binding. By combinatorially varying the amino acid hinge, which is attached to the GCP moiety, the receptor could be fine-tuned for the substrate. The best artificial receptor corresponds to the sequence $R^1 = R^2$ = lysine and R^3 = phenylalanine side chains (**11**). The interaction between the C-terminus of the peptidic substrate **10** with the GCP unit is the main driving force for complex formation. Additionally, the attached linear tripeptide chain interacts with the substrate by forming a hydrogen bond-mediated β-sheet with the peptidic substrate. An additional salt bridge is formed between the glutamate and the lysine ammonium group. In order to verify the substrate selectivity, the library was also screened against the inverse substrate sequence N-Ac-D-Ala-D-Ala-L-Lys-D-Glu-OH (AAKE), which resulted in binding constants that were lower by approximately one order of magnitude [7]. Thus, despite the fact that the complex stability is mainly based on a directed ion pair, the binding constants are strongly dependent on the amino acid sequence of the substrate. For binding of a polar substrate in water the binding constants achieved are very good.

In another set of experiments, the best receptor for the EKAA sequence (GCP-KKF) was tested against a rather small combinatorial tetrapeptide library consisting of 320 members [11]. With only three amino acids present in the library (Ala, Lys, and Glu) all substrates were closely related. Two amino acids just differed in their absolute configuration (D/L-Ala). Binding constants were determined via fluorescence spectroscopy in buffered water at pH 6.1 and ranged from $K < 50\ M^{-1}$ to $3 \times 10^4\ M^{-1}$. More importantly, a sequence-dependent stereoselectivity of the receptor could be observed, as depicted in Fig. 4, when the D/L-Ala amino acid of the substrate was fixed by strong electrostatic interactions between host and guest at both sides.

These experiments clearly demonstrate the potential of small but focused libraries, which allow the binding properties of all members to be determined. A large and random library would not have elucidated the subtle differences of the binding properties, as shown in Fig. 4. By characterizing all possible host–guest pairs it was possible to gain a more detailed insight into the binding event than would have been the case for a large and random library, in which only a few selected hit structures are isolated and characterized. Of course, due to the limitations of small libraries with regard to their structural and functional diversity, the library has to be carefully designed to give the correct answer to the problems addressed [12].

1.2 Di- and Multivalent Receptors for Peptide Recognition

Unlike linear receptors (vide supra) so-called tweezer receptors contain two arms which are bridged via a template. Depending on the linker structure this kind of molecules is able to adopt a preorganized, yet flexible cavity in which the substrate is held, as if by a pair of molecular tweezers [13]. When correctly designed, the second side chain is able to form additional attractive interactions to the substrate, thus increasing the complex stability and the substrate selectivity. The second arm may either be identical to the first, forming symmetrical tweezer receptors, or completely different in nonsymmetrical tweezers. This scheme may of course be further expanded to receptors with three or more arms. In general the aim of such a multivalent approach is to strengthen the complex stability by means of simultaneous interactions between multiple complementary functionalities between host and guest. Optimally the binding affinity is then higher than the sum of the corresponding monovalent interactions. This approach is also widely made use of in nature, be it in bacterium cell [8], antibody–antigen interactions [14], or the interplay of transcription factors with multiple sites of DNA [15]. In the following sections this chapter will be limited to the discussion of di- and trivalent receptors.

Pioneering work performed in the early 1990s by Still involved the design of a library comprising 10,000 (10^4) peptidosteroidal receptors for enkephalin-like peptides (**13**, Fig. 5) [16]. The two arms were linked via a steroidal cheno-12-deoxycholic acid template and differ in their amino acid sequence. The nonsymmetric substitution was achieved by making use of the different chemical reactivity of the two hydroxyl groups at positions C3 and C7. With the help of a two-color two-substrate assay, receptors could be identified which were able to selectively bind to one of the two differently labeled substrates: blue-dye-linker-L-Tyr-Gly-Gly-L-Phe-L-Leu-OH and red-dye-linker-L-Tyr-D-Ala-Gly-L-Phe-L-Leu-OH in chloroform, which only differ by one amino acid. Thus, by selecting beads which are stained by merely one color only those receptors which are selective for one of the two substrates could be isolated. When utilizing a flexible backbone, instead of

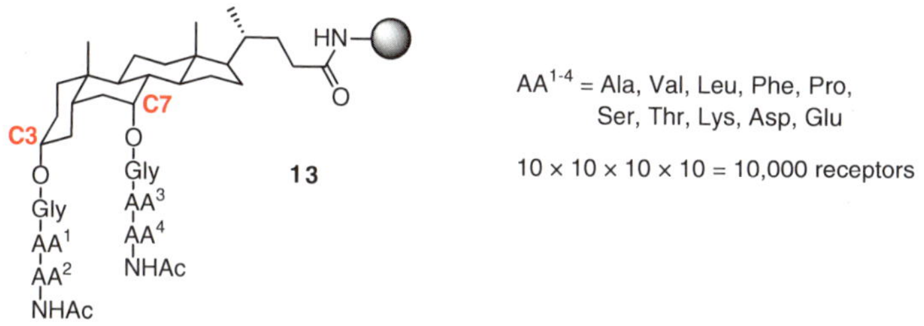

Fig. 5 The peptidosteroidal tweezer receptor library **13** with two differently substituted side chains was simultaneously screened for two substrates labeled with either a *blue* or a *red dye*

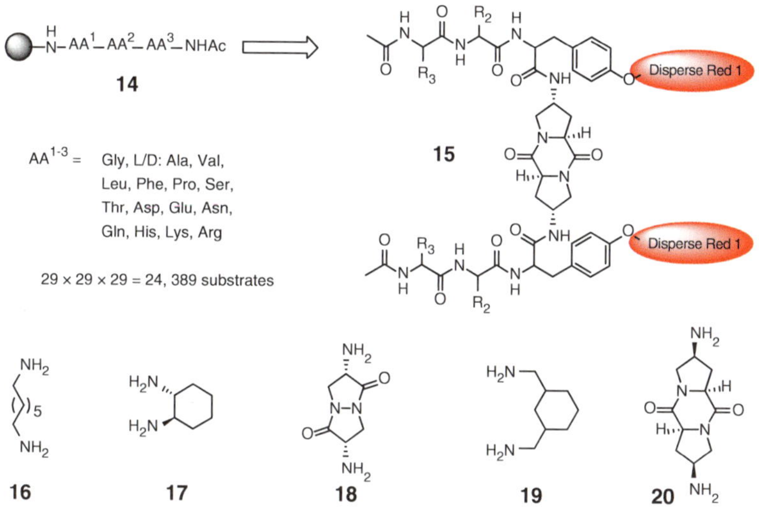

Fig. 6 Schematic design of tweezer receptors **15** containing two identical tripeptidic arms labeled with a *red dye* (*left*) and alternative templates **16–20** (*right*)

the rigid steroidal skeleton, binding strength was weakened by a factor of five and complete loss of selectivity was observed. In terms of the ability to recognize the pentapeptidic substrates, highly flexible receptors seem to be disadvantageous, probably due to a loss of pre-organization.

Wennemers synthesized five closely related, yet distinct, tweezer receptors **15**, which were labeled with a red dye, as depicted in Fig. 6. The two identical arms were linked via a rigid diamino diketopiperazine template [17]. Screening for substrate selectivity in chloroform was carried out with the help of a substrate library comprising $29^3 = 24{,}389$ tripeptides (**14**). Binding constants of up to $10^3\ M^{-1}$ (chloroform) and high selectivity were observed: only 1 out of 5,000 sequences was bound, e.g., the receptor with the amino acid sequence L-Tyr-L-Asn (Trt)-L-Phe exclusively selects peptides containing D-His followed by two hydrophobic D-amino acids. In order to verify whether a simplified receptor design is still

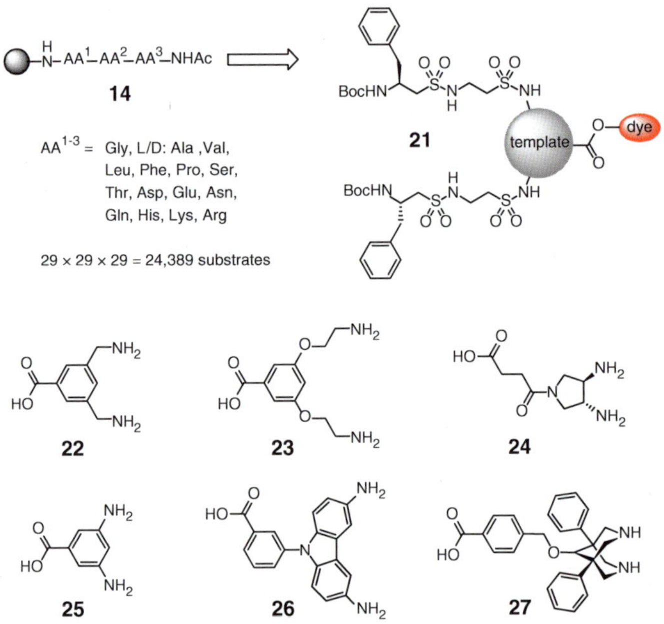

Fig. 7 Schematic representation of tweezer receptors **21** and templates **22–27**

able to achieve effective binding, the corresponding host systems with only two amino acids per side chain and one-armed analogs were screened with the same substrate library. However, as a result of truncation the host system completely lost its binding affinity.

Under the conditions used for the experiments **15** seems to be the minimum structure which is necessary for the recognition of the tripeptides under investigation. In order to evaluate the role of the template on the binding properties, similar receptors with the same side chains, but different diamine templates, **16–20**, were synthesized as well [18]. A screening against the same substrate library revealed that with only a few exceptions these analogs do not bind to the substrates at all. The U-shaped geometry of the original *trans–trans*-diketopiperazine template in **15** highly preorganizes the two arms in a fashion which is suitable for tripeptide binding. On the other hand, side chains which are connected via the *cis–cis*-diastereomer **20** adopt an almost linear geometry. Hence, a defined angle between template and side chains as well as a distance of approximately 8 Å in between (as is the case for the original diketopiperazine scaffold) seems to be superior to all other templates that were tested in this work. Accordingly, the template plays a decisive role in determining the affinity between receptor and substrate and thus has to be carefully chosen. However, with the help of these host systems only rather moderate affinities could be achieved. Although the selectivity is quite good, it has to be noted that the binding studies were only carried out in organic solvent and not in water.

Further studies concerning the role of the template for the performance of tweezer receptors were conducted by Liskamp [19]. As shown in Fig. 7, the

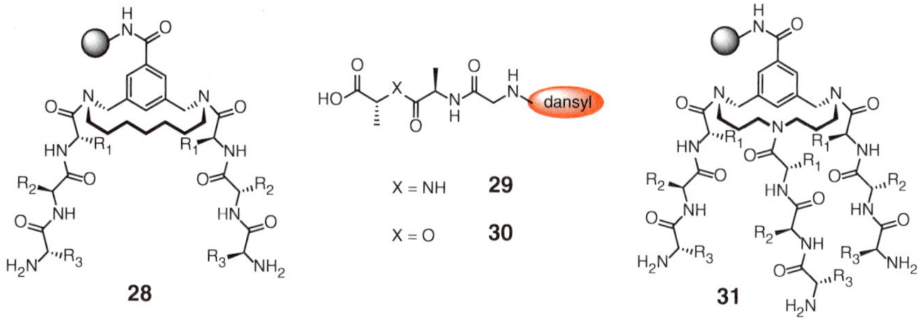

$AA^{1\text{-}3}$ = Gly, Ala, Val, Leu, Ser, Pro, Asp, Tyr, His, Lys, Gln, Phe

12 × 12 × 12 = 1,728 receptors

Fig. 8 Schematic illustration of receptor library **28** with an additional hydrocarbon chain. Side chains consist of three variable amino acids AA^{1-3}. The substrates D-Ala-D-Ala **29** and D-Ala-D-Lac **30** were labeled with a dansyl fluorescence marker. An additional arm attached to the hydrocarbon chain gave library **31**

receptors, **21**, consisting of two identical peptidosulfonamide side chains, were bridged by several templates **22–27**. These scaffolds differed in both their flexibility and the distance between the amine groups. A dye-label enabled screening of a tripeptide library containing $29^3 = 24{,}389$ members (**14**). The highest overall binding affinity in chloroform was found for the scaffold **22**, with a binding constant for the peptide D-Ala-L-Asp-D-Ser of $4 \times 10^3\ M^{-1}$. Since no binding affinity could be observed for a one-armed analog, it was concluded that the second arm is necessary for efficient tripeptide binding. The second best receptors, with binding constants of up to $8 \times 10^2\ M^{-1}$, were those with template **23**, which is more rigid and offers less space between the two arms. Third ranked are the two scaffolds **24** and **25**. The distance between side chains seems to be too large in these two cases for stable complexation of the peptidic substrates.

When implementing **26** as linker, the corresponding receptor showed high selectivity (>95%) for the amino acid sequence Glu-His-X. However, the binding affinity was rather poor—this was again probably due to the large distance between the two arms. No affinity at all could be observed for **27**. The cause for this effect can probably be found in the repulsion between the free electron pairs of the sulfonamide oxygen groups, which would be in too close proximity in the depicted chair–chair conformation of the template. Instead, it adopts a boat–chair conformation which results in the loss of tweezer structure. A common feature of all examined host systems in this work is the high guest selectivity: the amino acid AA^1 is almost always an acid; AA^2 is a polar amino acid such as asparagine or histidine. For the residue at position AA^3, which is bound directly to the resin, no tendency could be observed.

The binding studies of library **21** were again only conducted in organic solvents. For further studies in aqueous solution, two more combinatorial tweezer libraries (**28**, Fig. 8) each consisting of 1,728 members were synthesized, based on the best scaffold **22** [20]. As before, the two arms were identical but instead of sulfonamides, they now consisted of tripeptides, to facilitate the combinatorial

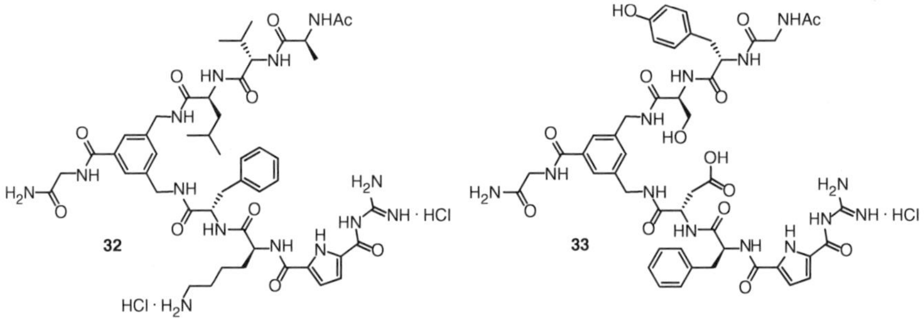

Fig. 9 Tweezer receptors **32** and **33** featuring two nonidentical peptidic arms—one of them is equipped with a guanidiniocarbonyl pyrrole oxoanion binding motif

approach. One of the two libraries contained the template without change, for the other the template was derivatized with a hydrocarbon chain, as illustrated in Fig. 8, to reduce its flexibility. Both libraries were then tested for their binding affinity towards the dansyl-labeled biologically relevant substrates Gly-D-Ala-D-Ala-OH (**29**) and Gly-D-Ala-D-Lac-OH (**30**) in aqueous phosphate buffer. Both substrates were linked to the fluorescent dansyl group via a glycine spacer. Characterization of hit structures revealed a highly consistent distribution of amino acids at position 3. For the dipeptide **29**, lysine was reported in 40% of cases and proline and alanine each accounted for a further 20%, which shows that these amino acids are thus clearly preferred with respect to a merely statistical distribution (ca. 3%). The results from the screening for the corresponding depsipeptide, **30**, showed that position 3 was mainly occupied by proline (50%), followed by lysine (25%) and alanine (20%). The conformity was less well pronounced for positions 1 and 2. However, in both cases mainly aromatic or aliphatic amino acids were present, which might be due to interaction of the nonpolar amino acid side chains with the dansyl label. The glycine linker is probably too short to prevent this unwanted hydrophobic interaction with the fluorescence dye. With a binding constant of $5 \times 10^2\ M^{-1}$ the highest affinity for the depsipeptide in aqueous phosphate buffer was found for the amino acid sequence AA^{1-3} = L-Leu-L-Phe-L-Lys. In chloroform binding constants were significantly higher with values of up to $10^4\ M^{-1}$. Due to the competitive influence of water, the achieved binding constants are rather low. The comparison between both libraries reveals that the hydrocarbon chain has no significant influence on the complex stability. The addition of a third arm, which was indirectly attached to the scaffold via the hydrocarbon chain (**31**), did neither improve binding affinity nor selectivity.

Schmuck developed the two tweezer receptors **32** and **33**, depicted in Fig. 9 [21, 22]. Based on the work of Liskamp, the aromatic scaffold **22** (Fig. 7) was used to bridge two peptidic arms. In order to synthesize receptors which feature two nonidentical arms, such as **32** and **33**, the scaffold was equipped with two orthogonal protecting groups (Fmoc, Boc). One of the receptor's arms consisted of three amino acids while the other featured two amino acids plus a GCP group as an

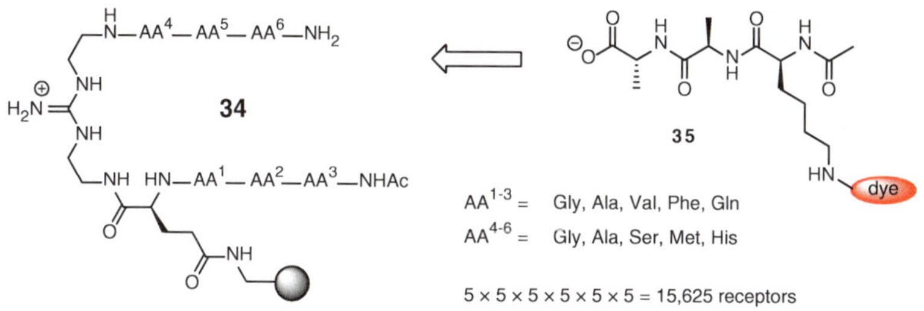

Fig. 10 Nonsymmetric tweezer receptor library **34** with a guanidinium template for carboxylate recognition and dye labeled substrate N-Ac-Lys(dye)-D-Ala-D-Ala-OH **35**

oxoanion binding motif. By means of UV/Vis experiments, a binding constant of $3 \times 10^3\ M^{-1}$ was determined for the complexation between **33** and the polar tripeptide N-Ac-Lys-D-Ala-D-Ala-OH in buffered water at neutral pH. Weaker binding, by a factor of 5, ($K = 6 \times 10^2\ M^{-1}$) was observed for **32**. While the main driving force for complex formation stems from the interaction between the free C-terminus of the tripeptide with the GCP moiety of the host systems, the difference between the two tweezer receptors could be attributed to additional non-covalent interactions between the substrate and **33**, such as an additional salt bridge between the aspartic acid, which is negatively charged at this pH, and the ammonium group of the substrate's lysine. Furthermore, in contrast to **32**, the hydroxyl groups in the serine and tyrosine side chains of the first arm of **33** are potential hydrogen bond donor and acceptor sites and might thus further stabilize the complex via additional hydrogen bonds. In conclusion, these two artificial receptors are illustrative examples of the crucial role of the building blocks in the interaction between host and guest, which can only form a stable complex if the building blocks are appropriate for the formation of non-covalent interactions between the two molecules.

Kilburn prepared the combinatorial tweezer receptor library **34**, depicted in Fig. 10, with 15,625 (5^6) members [23]. The two nonsymmetrical side chains consisted of varying tripeptide sequences. A guanidinium headgroup served as both the scaffold and as a carboxylate binding site in order to increase the binding affinity to the C-terminus of the substrate via a hydrogen bond-enforced salt bridge. The positive influence of the cationic headgroup was confirmed by comparing receptors that were bound to the solid support directly via the guanidine [24]. These model systems lost all of their affinity to the substrate because the guanidine could not be protonated under these circumstances. An on-bead screening of **34** with the dye labeled peptide N-Ac-Lys(dye)-D-Ala-D-Ala-OH (**35**) in buffered water at pH 8.75 revealed that less than 2% of the library members bound the substrate. The selected hit sequences were highly consistent, as shown in Table 1. Additional on-bead binding studies with the best receptor of the sequence AA^{1-6} = Gly-Val-Val-Met-His-Ser showed a binding constant of $10^3\ M^{-1}$. The corresponding diastereomeric substrate N-Ac-Lys(dye)-Ala-Ala-OH was bound less efficiently with a binding constant of $3 \times 10^2\ M^{-1}$. Binding studies in solution did not lead to

Table 1 Occurrence of amino acids of selected hit structures at each position of the receptor

Position	AA^1	AA^2	AA^3	AA^4	AA^5	AA^6
Amino acid	Gly (30%)	Val (40%)	Val (40%)	Met (50%)	Met (40%)	Ser (70%)
	Ala (30%)	Ala (40%)	Ala (40%)		His (30%)	

reliable results. This indicates clearly that the resin influenced the binding events occurring on the solid support. It is therefore crucial that results from measurements involving resin-bound substances be verified in solution.

In conclusion, multivalent receptors have been proven to possess the potential to increase both the selectivity and affinity of their binding to a defined substrate in comparison with their one-armed analogs. By increasing the functional and structural diversity and "summing up" multiple non-covalent interactions, the competing influence of the surrounding medium can be overcome. This is of special importance when working in water, which is necessary when biological targets are involved. When designing such molecules several things have to be taken into account. First, the template that bridges the arms plays a decisive role for the complexation. Optimally it preorganizes the arms in a suitable fashion for binding of the substrate and is still flexible enough to allow for its accommodation. For peptide recognition by tweezer receptors, the distance and angle between the side chains seems to be of utmost importance. Concerning the building blocks of the side chains, amino acids offer valuable resources especially, when preparing combinatorial receptor libraries, due to their diversity and commercial availability. Commonly the two arms are identical, as this facilitates the synthesis. Less well studied are host systems with two different arms, although this approach offers the possibility to add additional functionality. The introduction of a carboxylate binding site, such as a guanidinium group, into the receptors has proven to be worthwhile for improving binding properties. Other things that should be kept in mind are that linkers and labels may also take part in the binding event, as can the solid support. For this reason, it is advisable to always validate on-bead results with additional measurements performed in solution.

2 Protein Surface Recognition

Proteins play a central role in many biological processes including metabolism, regulatory activities, information transmission, transport, or as structural building blocks for cells [25]. Their *primary structure* is determined by the sequence of their building blocks, the 20 proteinogenic amino acids. Their size varies between 40 and 26,000 amino acids—with the majority having between 100 and 1,000 amino acids. The local orientation of the polypeptide chain is called the *secondary structure*: the α-helix has a right-handed twist, contains 3.6 building blocks per turn, and thus has a pitch of 0.54 nm. It is stabilized by an intra-strand hydrogen bond network, which is formed between the *n*th C=O and the (n + 4)th NH groups of the polypeptide chain. The β-sheet, on the other hand, is stabilized by a hydrogen bond network between two neighboring strands, which are either oriented in the same (parallel β-sheet) or in opposite directions (antiparallel β-sheet). Those parts of the polypeptide chain which do not feature any special secondary structure are called random coil. The folding of the secondary structure elements and the exact spatial positioning of each atom of the protein is called its *tertiary structure*. It can be determined by X-ray crystal structure analysis or by NMR spectroscopy. If a functional peptide unit consists of more than one polypeptide strand, the orientation of the different strands relative to each other is called the *quaternary structure*.

Due to the structural diversity of proteins it is possible to interact with them in a manifold of different ways [26]. While the overwhelming majority of known enzyme inhibitors are small molecules which competitively bind to the active center of the target protein (either reversibly or irreversibly) and thus prevent the binding of the actual substrate, in recent years other ways to influence proteins and especially enzymes are coming more and more to the fore. These novel approaches offer great potential for the development of new therapeutic strategies [27]. Within this chapter we will focus on selected examples of protein surface recognition by chemical ligands. Such interactions also allow the biological functions of a given protein to be influenced [28–30]. For example, protein–protein interactions can be prevented by a chemical ligand which binds to the area of the protein surface responsible for its interaction with another protein. Enzymes can be inhibited by surface binding ligands too if the conformation of the protein is altered upon binding or if the ligand blocks access of the substrate to the active site (non-competitive inhibition).

2.1 Anionic Ligands

In recent years more and more multivalent chemical peptide receptors were developed for the recognition of peptide surfaces. In particular Hamilton has synthesized and applied various tetravalent ligands for the recognition of protein surfaces [31–33]. Templates such as porphyrin, calix[4]arene, or rigid aromatic scaffolds were used to combine linear or cyclic peptidic side chains. Equipped with anionic side chains such as tetravalent ligands are then able to bind to positively charged protein surfaces with high affinity. Tetravalent **36** (Fig. 11), for example, binds to Cytochrome c, which features a strongly cationic surface ($pI = 10$), with a binding constant of $K \approx 10^6\ M^{-1}$.

Hayashida developed the cyclophane-based anionic, trivalent resorcinol-derivative **37**, depicted in Fig. 12, which is able to form stable complexes with histones in aqueous solution [34, 35]. As determined by changes of the fluorescence properties of the integrated dansyl-label upon binding, the affinity is as high as $K \approx 10^6\ M^{-1}$. Acetylated, neutral histones, on the other hand, are not bound by **37**, which clearly stresses the importance of electrostatic interactions for the recognition of the positively charged histone surface.

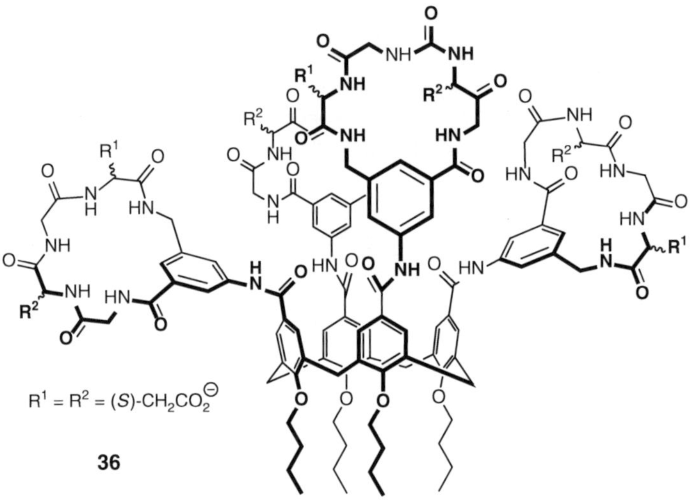

Fig. 11 Calix[4]arene-based, negatively charged tetravalent **36** protein surface ligand

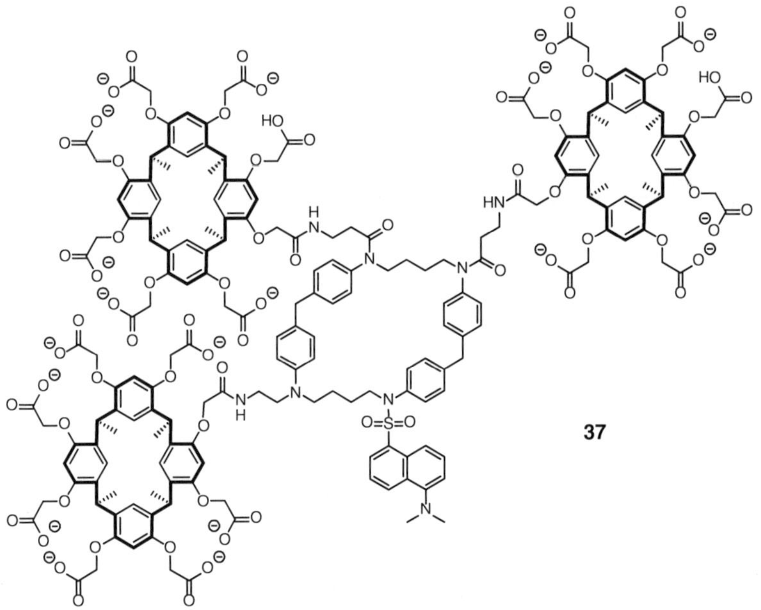

Fig. 12 Cyclophane-based, negatively charged trivalent resorcinol-derivative **37** for the recognition of positively charged protein surfaces

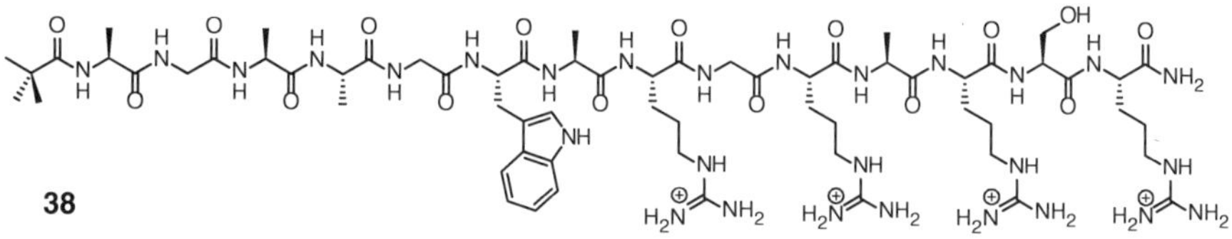

38

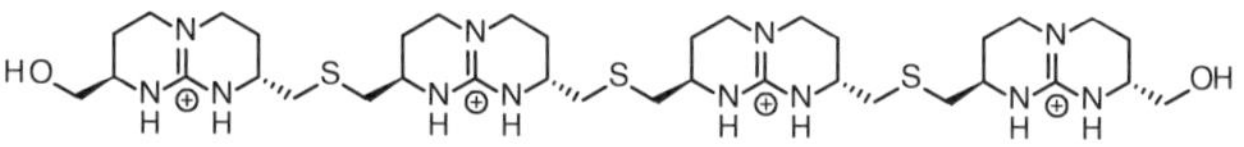

39

Fig. 13 The cationic peptide **38** and the positively charged tetraguanidinium moiety **39** recognize p53 via four carboxylates at the protein surface

2.2 *Cationic Ligands*

Giralt synthesized protein surface-binding molecules, including the cationic peptide **38** or the non-peptidic tetraguanidinium ligand **39**, which are illustrated in Fig. 13, for the recognition of the tetramerization domain of p53 via four glutamate moieties on the protein surface [36]. This protein is an important transcription factor for the regulation of apoptosis and cell division and as such is an interesting target for the development of novel drugs against cancer. The binding constant for **38** and similar peptides towards p53 was determined to be $K = 10^5\ M^{-1}$, and the tetraguanidinium derivative **39** binds with $K = 2 \times 10^4\ M^{-1}$.

Schmuck identified highly efficient tetravalent inhibitors of β-tryptase via an on-bead fluorescence screening of a ligand library, **40**, containing 216 (6^3) members (Fig. 14) [37]. The tryptase is a human serine protease, which plays a decisive role in the pathogenesis of asthma and other allergic and inflammatory disorders. In its active tetrameric form, the enzyme features a central pore, in which four active sites are located. A cluster of negatively charged amino acids is found at the entrance to this channel. With the help of the library it could be shown that tetrameric ligands with complementary charge are able to block the entrance to this channel and thus prevent the substrate from accessing the active sites. As a consequence, the enzyme is inhibited non-competitively. The best inhibitors combined aromatic and cationic amino acids and the overall best sequence, with an excellent nanomolar activity ($K_i = 170$ nM), featured the amino acid sequence Arg-Trp-Lys (from AA^1 to AA^3). Furthermore, a clear multivalency effect was observed: the analogous one-armed ligand was less active by a factor of 1,800 ($K_i = 306$ μM). Finally, the thus identified inhibitors were selective for β-tryptase and did not influence similar enzymes, such as trypsin or chymotrypsin.

In conclusion, multivalent protein–ligands have proven to be highly efficient in recognizing protein surfaces. Up till now these recognition events were mainly based on charge–charge interactions, which allowed the design of ligands featuring

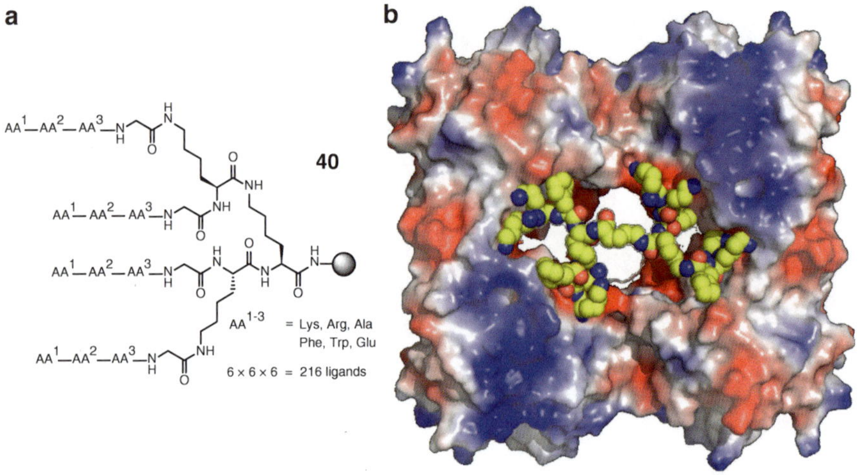

Fig. 14 (**a**) Tetravalent β-tryptase inhibitor library **40**. The most efficient inhibitor features the amino acid sequence Arg-Trp-Lys (from AA^1 to AA^3). (**b**) The tetrameric form of β-tryptase features a central pore with a cluster of negatively charged amino acids at its entrance, which gets blocked by the cationic, tetrameric ligands

dissociation constants in the lower millimolar or even upper nanomolar range. Furthermore, initial studies regarding the influence of these systems on the biological functions of their target enzymes have shown that it is in principal possible to interfere with enzymatic activity. Hence, protein surface recognition has been demonstrated to be a promising and active field of research for the development of novel therapeutic approaches to be used for the treatment of diseases for which drug therapies do not yet exist.

References

1. Lehn J-M (1995) Supramolecular chemistry: concepts and perspectives. VCH, Weinheim
2. Sewald N, Jakubke H-D (2002) Peptides: chemistry and biology. Wiley-VCH, Weinheim
3. Schmuck C (2001) Von der molekularen Erkennung zum Design neuer Wirkstoffe. Chem uns Zeit 35:356–366
4. Mirsky VM, Yatsimirsky A (2010) Artificial receptors for chemical sensors. Wiley-VCH, Weinheim
5. Williams DH, Bardsley B (1999) Die Vancomycin-Antibiotica und der Kampf gegen resistente Bakterien. Angew Chem 111:1264–1286
6. Xu R, Greiveldinger G, Marenus LE, Cooper A, Ellman JA (1999) Combinatorial library approach for the identification of synthetic receptors targeting Vancomycin-resistant bacteria. J Am Chem Soc 121:4898–4899
7. Schmuck C, Heil M (2006) One-armed artificial receptors for the binding of polar tetrapeptides in water: probing the substrate selectivity of a combinatorial receptor library. Chem Eur J 12:1339–1348

8. Westerlund B, Korhonen TK (1993) Bacterial proteins binding to the mammalian extracellular matrix. Mol Biol 9:687–694
9. Schmuck C, Heil M, Scheiber J, Baumann K (2005) Charge interactions do the job: a combined statistical and combinatorial approach to finding artificial receptors for binding tetrapeptides in water. Angew Chem Int Ed 44:7208–7212
10. Perrin DD (1972) Dissociation constants of organic bases in aqueous solution, Supplement. Butterworths, London
11. Schmuck C, Wich P (2006) Sequence-dependent stereoselectivity in the binding of tetrapeptides in water by a flexible artificial receptor. Angew Chem Int Ed 45:4277–4281
12. Schmuck C, Wich P (2006) Combinatorial receptor finding: large and random vs. small and focused libraries. New J Chem 30:1377–1385
13. Chen CW, Whitlock HW Jr (1978) Molecular tweezers: a simple model of bifunctional intercalation. J Am Chem Soc 100:4921–4922
14. Dower SK, DeLisi C, Titus JA, Segal DM (1981) Mechanism of binding of multivalent immune complexes to Fc receptors. 1. Equilibrium binding. Biochemistry 20:6326–6334
15. Chen H, Privalsky ML (1995) Cooperative formation of high-order oligomers by retinoid X receptors: an unexpected mode of DNA recognition. Proc Natl Acad Sci USA 92:422–426
16. Boyce R, Li G, Nestler HP, Suenaga T, Still WC (1994) Peptidosteroidal receptors for opioid peptides. Sequence-selective binding using a synthetic receptor library. J Am Chem Soc 116:7955–7956
17. Wennemers H, Conza M, Nold M, Krattiger P (2001) Diketopiperazine receptors: a novel class of highly selective receptors for binding small peptides. Chem Eur J 7:3342–3347
18. Wennemers H, Nold MC, Conza MM, Kulicke KJ, Neuburger M (2003) Flexible but with a defined turn – influence of the template on the binding properties of two-armed receptors. Chem Eur J 9:442–448
19. Löwik DWPM, Weingarten MD, Broekema M, Brouwer AJ, Still WC, Liskamp RMJ (1998) Tweezers with different bite: increasing the affinity of synthetic receptors by varying the hinge part. Angew Chem Int Ed 37:1846–1850
20. Monnee MC, Brouwer AJ, Liskamp RMJ (2004) Synthesis, screening and evaluation of a combined library of tweezer and tripodal synthetic receptors. QSAR Comb Sci 23:546–559
21. Kuchelmeister HY, Schmuck C (2009) An efficient synthesis of an orthogonally protected aromatic diamine as scaffold for tweezer receptors with two different arms. Eur J Org Chem 2009:4480–4485
22. Kuchelmeister HY (2011) Molecular recognition of biologically relevant targets – from anion binding motifs to application in cell biology. Dissertation, University of Duisburg-Essen
23. Shepherd J, Gale T, Jensen KB, Kilburn JD (2006) Synthesis of unsymmetrical tweezer receptor libraries and identification of receptors for Lys-D-Ala-D-Ala in aqueous solution. Chem Eur J 12:713–720
24. Davies M, Bonnat M, Guillier F, Kilburn JD, Bradley M (1998) Screening an inverted peptide library in water with a guanidinium-based tweezer receptor. J Org Chem 63:8696–8703
25. Voet D, Voet JG, Pratt CW (2002) Fundamentals of biochemistry. Wiley-VCH, Weinheim
26. Peczuh MW, Hamilton AD (2000) Peptide and protein recognition by designed molecules. Chem Rev 100:2479–2494
27. Verdine GL, Walensky LD (2007) The challenge of drugging undruggable targets in cancer: Lessons learned from targeting the BCL-2 family members. Clin Cancer Res 13:7264–7270
28. Kodadek T, Reddy MM, Olivos HJ, Bachhawat-Sikder K, Alluri PG (2004) Synthetic molecules as antibody replacements. Acc Chem Res 37:711–718
29. Yin H, Hamilton AD (2005) Strategies for targeting protein-protein interactions with synthetic agents. Angew Chem Int Ed 44:4130–4163
30. Gordo S, Giralt E (2009) Knitting and untying the protein network: modulation of protein ensembles as therapeutic strategy. Protein Sci 18:481–493
31. Tsou LK, Dutschman GE, Elizabeth AG, Telpoukhovskaia M, Cheng Y-C, Hamilton AD (2010) Discovery of a synthetic dual inhibitor of HIV and HCV infection based on a tetrabutoxy-calix[4]arene scaffold. Bioorg Med Chem Lett 20:2137–2139

32. Margulies D, Opatowsky Y, Fletcher S, Saraogi I, Tsou LK, Saha S, Lax I, Schlessinger J, Hamilton AD (2009) Surface binding inhibitors of the SCF-KIT protein–protein interaction. Chembiochem 10:1955–1958
33. Rodriguez JM, Nevola L, Ross NT, Lee G, Hamilton AD (2009) Synthetic inhibitors of extended helix-protein interactions based on a biphenyl 4,4′-dicarboxamide scaffold. Chembiochem 10:829–833
34. Hayashida O, Ogawa N, Uchiyama M (2007) Surface recognition and fluorescence sensing of histone by dansyl-appended cyclophane-based resorcinarene trimer. J Am Chem Soc 129:13698–13705
35. Hayashida O, Uchiyama M (2007) Multivalent macrocyclic hosts: histone surface recognition, guest binding, and delivery by cyclophane-based resorcinarene oligomers. J Org Chem 72:610–616
36. Martinell M, Salvatella X, Fernández-Carneado J, Gordo S, Feliz M, Menéndez M, Giralt E (2006) Synthetic ligands able to interact with the P53 tetramerization domain. Towards understanding a protein surface recognition event. Chembiochem 7:1105–1113
37. Wich PR, Schmuck C (2010) Reversible and noncompetitive inhibition of β-tryptase by protein surface binding of tetravalent peptide ligands identified from a combinatorial split-mix library. Angew Chem 122:4207–4210

Antimicrobial Peptides for Detection and Diagnostic Assays

Lisa C. Shriver-Lake, Stella H. North, Scott N. Dean, and Chris R. Taitt

Abstract Considered gold standards for biodetection, immunoassays and nucleic acid-based assays are sensitive, highly selective, and well characterized. However, they are capable of detecting only those targets for which specific reagents (such as antibodies or nucleic acid primers or probes) have been developed. Furthermore, new, emerging, and unexpected pathogens may not be detected. To address the challenge of detecting both known and unknown microbes, assays utilizing antimicrobial peptides (AMPs) are being developed for integration into both biosensors and high-throughput platforms. AMP-based detection represents a new paradigm in sensing—namely, the ability to screen a sample for the presence of many different microbes without target-specific reagents, and to provide broad classification information on the species detected.

Keywords Antimicrobial peptide, Biosensor, Immobilization, Semi-selective binding

Contents

L.C. Shriver-Lake, S.H. North, and C.R. Taitt (✉)
Center for Bio/Molecular Science & Engineering, US Naval Research Laboratory, Washington, DC 20375, USA
e-mail: chris.taitt@nrl.navy.mil

S.N. Dean
NRL/NREIP Intern Resident at the US Naval Research Laboratory, Washington, DC, USA

S.A. Piletsky and M.J. Whitcombe (eds.), *Designing Receptors for the Next Generation of Biosensors*, Springer Series on Chemical Sensors and Biosensors (2013) 12: 85–104
DOI 10.1007/5346_2012_19,
Published online: 27 June 2012

Abbreviations

AMP	Antimicrobial peptide
Cfu	Colony-forming units
HTS	High-throughput screening
LPS	Lipopolysaccharide
LTA	Lipoteichoic acid
OAKs	Oligoacyllysines
PCR	Polymerase chain reaction
PMB	Polymyxin B
QCM-D	Quartz crystal microbalance with dissipation monitoring
SPR	Surface plasmon resonance

1 Conventional Approach for Pathogen Detection

The rapid and reliable detection of microbial pathogens, etiologic agents of chronic or infectious disease is an important part of research in many fields, including medical diagnosis, public health applications, environmental health control, and biodefense. In recent years, many types of detection and diagnostic systems have been developed for rapid, analytical detection and identification of pathogenic microbes based on specific immunological or genetic characteristics of the target organism. Antibody- and PCR-based assays are currently the most commonly employed molecular techniques [1]. Antibodies provide a powerful analytical tool for a wide range of targets. Immunoassays are highly specific, versatile, and the antibodies bind strongly and stably to their respective antigen. Immunoassays are perhaps the only approach that have been successfully employed for the detection of bacterial cell, spores, viruses, and toxins alike [2]. PCR-based detection methods are founded on the premise that each species of pathogen carries a unique DNA or RNA signature that differentiates it from other organisms. Real-time PCR is highly sensitive and allows for quantitation of microorganisms at any level of specificity (i.e., strain, species, genus), and because the approach detects the organism by amplifying the target DNA rather than the signal, it is less likely to produce false positives [3]. These current approaches for identifying pathogens are sensitive and selective but are subject to certain limitations. First, they rely on *known* characteristics that may be altered by growth conditions (antigenic determinants) or genetic engineering (DNA/RNA sequence). In addition, when multiplexing assays for multiagent detection, background and nonspecific signals increase, limiting the number of targets that can be detected.

1.1 New Paradigm for Pathogen Detection

A significant challenge at the forefront of research in the field of biodetection is multiplex detection of unknown or uncharacterized pathogens or biologically engineered microbes. Current systems are incapable of identifying these threat agents because the specific assays have not been developed. In recent years, an alternative approach to pathogen detection has emerged, in which the presence and classification of microorganisms are determined by targeting generic factors. This methodology is part of an emerging direction in the area of differential sensing, which focuses on pattern-based detection sensors [4, 5]. This trend is underscored by advances in material science, biotechnology, and engineering that are providing broader spectrum recognition capabilities (e.g., aptamers, peptides, engineered proteins, and biomimetic materials) in complement to conventional antibodies and DNA/RNA oligonucleotide sequences.

Natural and engineered antimicrobial peptides (AMPs) represent an enabling technology for pattern-based detection systems. AMPs are key players in the innate immune system that organisms use to kill infectious microorganisms. They recognize molecular patterns on the surface of pathogens, e.g., lipopolysaccharide (LPS), lipoteichoic acid (LTA), peptidoglycan, and phospholipids, in order to maximize their antimicrobial efficacy [6]. Most AMPs share fundamental structural motifs that are essential for their broad spectrum activity; however, there are considerable differences in peptide sequences, lengths, and charges that differentially modulate lipid membrane association. It has been proposed that cross-reactive AMP arrays, which interact with overlapping but nonidentical binding specificities to different microbial targets, can be exploited for broad spectrum discrimination in pathogen detection assays [7, 8]. Natural and engineered AMPs represent a new genre of recognition elements that address the critical technology gap comprising the ability to detect and categorize unknown, emerging, or engineered pathogenic bacteria, viruses, and other disease-causing agents. Such a capability is central to appropriate medical countermeasures and decontamination initiatives.

2 AMPs as Recognition Molecules

AMPs are host defense effector molecules found in organisms throughout the evolutionary spectrum. They are implicated in the killing of microbes by phagocytic, epithelial, and other immunologically relevant cells of mammals, while intricately regulating immune responses [9]. These molecules carry out their multitude of functions by utilizing differences that exist between microbial and mammalian cells. Significant distinctions between microbes and mammalian systems include membrane composition, higher transmembrane potential, and the presence of capsular polysaccharides, LPS, peptidoglycan, and LTA, some of which can be stimuli that induce AMP expression [9]. The primary determinant of AMPs' selective toxicity has been suggested to be the phospholipid composition relevant to the anionic character and net positive charge of the peptides [10].

While AMPs' antimicrobial action is apparent under *in vitro* conditions, microbial killing is generally inhibited under physiological conditions among high concentrations of monovalent ions, as well as in the presence of host and bacterial proteases. However, the critical abilities of AMPs as effector molecules of the immune system (i.e., the induction of wound repair, chemotaxis of for phagocytic cells, and initiation of inflammatory responses) remain [11]. Under physiological conditions, some AMPs have been shown to retain the ability to degrade microbial biofilm, as well as inhibit other factors of virulence [12]. The ability of certain peptides to bind LPS and LTA has also been implicated as their mechanism to protect against sepsis [13]. A heptamer-peptide has been demonstrated as an inhibitor of device-associated staphylococcal infection in rats [14], while numerous studies have determined that structurally related peptides can have targets existing beneath the cytoplasmic membrane as necessary antimicrobial mechanisms [15, 16]. AMPs have been shown to affect nucleic acid and protein synthesis, inhibit enzymatic activity, and modulate genes related to virulence, such as biofilm formation, motility, and secrete exotoxin without cell death [12, 15].

AMPs are a unique and diverse group of molecules, which are categorized with consideration of their synthesis, composition, and secondary structure. A prominent group contains cationic peptides that are relatively short, lack cysteine residues, and may have a proline hinge mid-sequence; this group is perhaps the best characterized with respect to sequence, structure, and mechanisms of binding and killing. In aqueous solutions, these peptides are generally disordered, seen in circular dichroism as "random coil," but in the presence of sodium dodecyl sulfate, phospholipid vesicles, or LPS, the molecule takes on an α-helical conformation. The percentage of α-helicity seen in membrane models correlates with antimicrobial activity against bacteria *in vitro* [17, 18]. Included in this group are the magainins, the cecropins, melittin, parasin, and the cathelicidins. The cathelicidins are a diverse family of peptides and are identified based on a conserved N-terminal domain of the peptide precursor [19, 20]. Found in the granules of natural killer lymphocytes, neutrophils, and in the epithelia of the skin, gut, and lungs, this class of peptide is activated upon secretion by proteases [9].

A second group is comprised of cationic peptides that are rich in certain hydrophobic, charged, or aromatic residues. This group includes proline- and arginine-rich peptides, such as the bactenecins and PR-39. Other notable members of this group are indolicidin, which is rich in tryptophan residues and has a low net positive charge (+3), and prophenin, which is rich in proline and phenylalanine residues. These peptides lack cysteine residues and are linear and unstructured, or form extended coils. While capable of binding to LPS [21], these peptides have a mode of antimicrobial action that is distinct from the classical model of cytoplasmic membrane disruption; they are hypothesized to disrupt bacterial septum formation [15].

A third, large group comprises peptides that contain cysteine residues and form stable β-sheets. This group includes protegrin, which forms toroidal pores as its antimicrobial action [15] and the defensins. The α-defensins are produced by human neutrophils and typically possess six cysteines that form three disulfide bonds [19], in a 1–6, 2–4, 3–5 pattern. The α-defensins can inhibit nucleic acid and protein synthesis of bacteria. The β-defensins include six cysteines that form three disulfide bonds in a 1–5, 2–4, 3–6 pattern.

An additional group are those that are produced non-ribosomally by bacteria and fungi. Synthesized by multienzyme complexes, peptides in this group often contain unusual or modified amino acids. Cyclization, unusual linkages, and mixtures of D- and L-amino acids are also frequently encountered. This group includes the polymyxins, bacitracin, gramicidin, and alamethicin.

AMPs carry out their various antimicrobial actions via many different target molecules; however, the mode of action is largely dependent on interaction with the bacterial membrane. Peptides are attracted through electrostatic interactions between their cationic side chains and anionic components of the outer bacterial membrane, such as phospholipid head groups, LPS, and LTA. At low peptide concentrations, parallel-oriented adsorption of AMPs to the membrane occurs, resulting in thinning of the membrane and disruption of the natural distribution of membrane phospholipids in a concentration-dependent manner. Peptides assume an increasingly perpendicular orientation as their concentrations increase, allowing for insertion into the lipid bilayer, forming a transmembrane pore. This intermediate event — pore formation — has led to the development of several models for explanation of membrane permeabilization. These models are not necessarily mutually exclusive, nor competing; they are suggested as being related methods of action [15] on a spectrum of membrane disruption where any one model may be unable to explain all of the events that occur [16].

After a peptide-specific threshold concentration is reached, membrane permeabilization may commence through the mechanism termed the "barrel stave" model. Peptides self-aggregate and insert deeper into the hydrophobic membrane core, with the hydrophobic interface of the amphipathic pointed outward [22]. The noncharged alamethicin peptide has been shown to form a consistent barrel stave pore with an aggregate of eight monomers [11] where translocation of the peptides to the inner leaflet of the lipid bilayer causes severe membrane disruption and leakage of cytoplasmic contents.

In the "toroidal pore" model, a similar sequence of events occurs whereby aggregates of peptides associate with the membrane in a perpendicular orientation; upon reaching a necessary threshold concentration, they then insert into the cytoplasmic membrane. Inducing local positive curvature strain, the peptides form a pore lined with peptides and phospholipid head groups in an alternating fashion. With polar side chains interfacing with the polar lipid head groups, a rounding of the monolayer joins the leaflets of the membrane resulting in pore formation without lipid tails exposed to peptide [15, 22]. This mechanism allows for cationic peptides such as LL-37, melittin, protegrin-1, and magainin-2 to reach intracellular targets through a relatively large pore.

In a third prominent model of action, the "carpet mechanism" proposes peptide accumulation on the lipid bilayer in a parallel orientation. After reaching a threshold concentration, cationic peptides disintegrate the membrane and form micelles,

While redistribution of lipids and membrane thinning can result at low peptide concentrations in the carpet model, capable of causing significant global physiological disturbance [23], many cationic AMPs will form pores through the carpet mechanism if a high peptide concentration is applied. The human cathelicidin, LL-37, in addition to the toroidal pore, and cecropin peptides have been suggested as using the carpet mechanism of antimicrobial activity through surface adsorbance and electrostatic interactions with membrane phospholipids and cationic regions of the peptide [18].

3 Broad-Based Sensing Using AMPs

AMP-based detection represents a new paradigm in sensing — namely, the ability to screen a sample for the presence of many different microbes without target-specific reagents, and to provide broad classification information on the species detected. The concept for such a detection strategy is simple: a series of AMPs with overlapping, but not identical, binding specificities binds to and detects a broad spectrum of microbial targets. Based on the pattern of which AMPs bind the target, the user can glean information regarding what type of microbe is present, although an unequivocal identification is generally not possible (Fig. 1). By using this strategy, target-specific reagents (e.g., antibodies, PCR primers) do not need to be developed or be available on-site, and no knowledge of what might (or might not) be present is needed. In this manner, unexpected or unknown microbial species can be detected and sufficient information provided by the binding patterns to allow more targeted follow-on testing for species identification.

Affinity-based biosensors often utilize a sandwich-type format, using a surface-immobilized recognition molecule to capture the target from solution, and a "tracer" component used to quantify the target bound to the surface. We will describe integration of AMPs as both solution-phase and surface-immobilized components in detection assays. Although the successful combination of AMP-tracer and AMP-capture reagents in detection assays has not been reported to date, we believe that optimization of both components separately will lead to a system utilizing no affinity reagents other than AMPs for biological recognition.

3.1 Solution-Phase Detection Using AMPs

Use of AMPs as solution-phase "tracers" should be considered as distinct from their application as antimicrobial or decontaminating agents. While the latter applications rely on their microbiocidal properties, the former use requires cell recognition and binding *without* killing. Furthermore, AMP tracers require modification with a label that can be measured by an appropriate detector. Most importantly, this tag must not interfere with the AMP's recognition and binding characteristics.

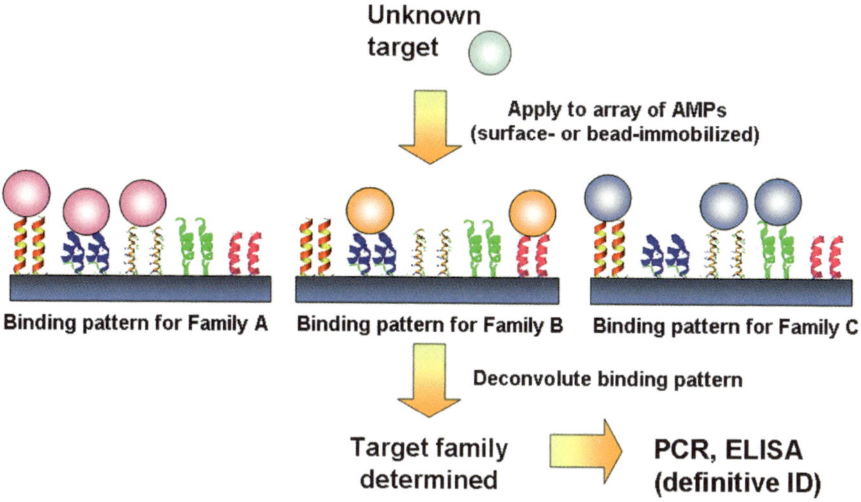

Fig. 1 Concept of AMP-based screening for detection and categorization of microbial species. The semi-selective binding characteristics of AMPs are used to detect many different species of microbes. The information provided by the overall pattern of binding is used to classify the detected species into different families of targets. Once the appropriate category of target is known, follow-on testing (such as immunoassay, PCR) can be used to identify which species is present

Using a 3-pronged approach encompassing both *in vitro* and multiple *in vivo* screening models, Welling's group at Leiden University has led the field in discovery, development, and use of radiolabeled AMPs for *in vivo* detection and imaging of microbial infections [24, 25]. They have identified a number of promising technetium-99m (Tc)-labeled peptides with different spectra of binding and pharmokinetic properties when tested in different *in vivo* models [26]. The most promising Tc-labeled compound could detect Gram-negative, Gram-positive, and fungal infections *in vivo* and discriminate them from sterile inflammations induced by LPS or heat-killed bacteria [24–26]. This material has been used in several clinical studies with promising results as a negative predictor of infection [27, 28]. Derivatives with improved pharmokinetic properties are currently under development [29].

While radiolabeled AMPs are intended as imaging agents for point-of-care use, AMPs tagged with *other* labels (e.g., $Ru(bpy)_3$- or fluorophore-labeled) offer tremendous potential for use as "tracer" molecules in biosensors. Olstein and Albert developed copolymers with repeating polymyxin B (PMB) "recognition" units and biotin "label" units and demonstrated binding to Gram-negative, but not Gram-positive cells. Improved sensitivity was demonstrated by increasing the "label" moieties in the copolymer [30]. Rocco [31] conjugated PMB directly to a fluorophore and demonstrated broadly specific binding to different bacteria with minimal binding/staining of eukaryotic cells. Lim's group used an analogous PMB-dye conjugate in biosensor assays where PMB was also used for microbial capture [32]. While *Escherichia coli* binding to immobilized PMB was detected at 10^7 colony-forming units (cfu) per mL using an antibody tracer, Cy5-PMB failed to detect their surface-bound *E. coli* even at tenfold higher concentrations. Our own studies have shown that PMB and other AMPs are often modified on *multiple* side chains even when using a 1:1 molar ratio of amine-specific labels [33]. Presumably, the loss of *multiple* positive charges impacts both the affinity and specificity of the modified AMPs [22, 34, 35].

To circumvent the pitfalls associated with amine-targeted modification, Mello's group has utilized AMPs synthesized with a C-terminal cysteine [36, 37]; this unique thiol can be targeted by thiol-specific fluorophores without affecting overall peptide charge or presumably structure. Indeed, in assays utilizing an antibody for capture and fluorescent cysteinyl-AMPs as tracers, an improvement in assay sensitivity was observed with Cy5-cysteine-cecropin P1, albeit not with all cysteinyl-labeled AMPs tested. In contrast to Lim's observations [32], no decrement in assay sensitivity was apparent for any of these labeled AMPs.

3.2 Biosensor-Based Detection Using Immobilized AMPs

Employing AMPs as the capture molecule in biosensors for the detection of bacteria is another application. They can be used to bind broad categories of bacteria. Both competitive assays and sandwich immunoassays have been demonstrated as sensors for the presence of bacteria or their components (i.e., LPS). The main issue in this area is to immobilize the AMP to a solid support such as fibers or planar waveguides without losing their ability to bind to bacteria. As with the solution-phase AMPs used as tracers, the chief requirement is the AMP's ability to bind to the bacteria, not its biocidal ability.

The first example of an immobilized AMP for detection as a capture molecule in a biosensor was described by James et al. at the Naval Research Laboratory; the system used was an evanescent wave-based biosensor, where fluorescent signals are read only from fluorescent molecules bound to the surface [38]. In this study, PMB was covalently immobilized on the surface of a tapered optical fiber using mercaptopropyl silane and an amine-reactive cross-linker. A known quantity of fluorophore-labeled LPS competed with unlabeled LPS for binding to PMB. As the concentration of unlabeled LPS increased, the fluorescence signal decreased. LPS was used as the target as this is a major component of Gram-negative bacteria and has been proposed as being part of the AMP–bacteria interaction. In this competitive fluorescence-based assay, unlabeled LPS was detected at 12.5 ng/mL in buffer and 25 ng/mL in plasma. Though the assay was run for 2 min, binding was completed within 30 s. Optimization of the AMP concentration used for immobilization was examined; the authors found that 10 mg/mL worked best, with higher concentrations as well as the lower concentrations reducing sensitivity.

This work has since been expanded at the Naval Research Laboratory by increasing the number of AMPs immobilized on a surface and developing multiplexed assays using these AMP arrays. The system used in these later studies is also an evanescent wave-based biosensor where fluorescent signals obtained are limited to those molecules/cells captured on the sensor substrate. As mentioned earlier, different AMPs have different affinities for different bacteria. Based on their previous work with antibodies, Kulagina and coworkers took advantage of the ability to immobilize different capture molecules in specific spots on an optical waveguide and immobilized several different AMPs (magainin I, cecropin, polymyxin, parasin) via their primary amines in stripes across the slide [8, 39, 40]. Various bacterial solutions were run through channels perpendicular to the patterned AMPs, followed by fluorescently labeled antibodies. When laser light was launched into the waveguide forming an evanescent wave region, only the areas in the evanescent wave with the full complex of AMP–bacteria-fluorescent antibody fluoresced. Dose–response curves were generated and compared to those obtained with immobilized antibodies on the same slides. The limits of detection were similar for *E. coli*, *Salmonella*, and *Bacillus* spp., but a few showed improved detection limits (killed *Francisella tularensis*, killed *Brucella*, and killed *Yersinia pestis*). Importantly, while distinguishable binding patterns were observed for each AMP and each species tested, similar species (e.g., alpha Proteobacteria, gamma Proteobacteria, Firmicutes) produced similar, but not identical, patterns. These observations support the hypothesis that AMP arrays can be used to classify detected microbes into their appropriate phylum, class, order, and possibly family and genus, based on their patterns of binding. In addition to both Gram-positive and -negative bacteria, viruses were also detected; interestingly, the patterns of AMP binding for the two viruses tested were virtually identical, again supporting the potential for detection and classification based on AMP arrays.

Recently, Manoor and coworkers developed a microcapacitive electrode biosensor for the detection of *E. coli* and *Salmonella* for water monitoring and pharmaceutical use [41]. This system incorporated a single peptide — magainin I — which was modified such that it contained a cysteine on either the C- or N-terminus for immobilization to gold electrodes. They demonstrated detection down to 10^3 bacteria/mL with selectivity for Gram-negative and pathogenic organisms. They were able to demonstrate bacterial selectivity between Gram-positive and Gram-negative bacteria, as well as between *E. coli* and *Salmonella*.

A major requirement for AMP integration on these and other biosensors is the retention of activity after immobilization. Though not pursuing microbial detection per se, Mello and coworkers have immobilized AMPs onto surfaces by various means and assessed for activity, structure, and presentation/orientation; surface plasmon resonance (SPR), sum frequency generation vibrational spectroscopy, and quartz crystal microbalance with dissipation monitoring (QCM-D) were used [42–45]. Their work published to date has been on the AMPs that have been cysteine-modified at the C-terminus for immobilization. This work can be used to further develop biosensors that employ gold surfaces for transduction. They demonstrated that the immobilized cecropin was bound to the surface and maintained its biocidal abilities. To demonstrate binding of LPS to surfaces, Ansorena's group immobilized PMB via mercaptoundecanoic or mercaptopropanoic acid on QCM-D and SPR surfaces [46]. They found that the PMB was essential to achieve high binding of LPS. Two excellent reviews have recently been published on covalent immobilization or tether protocols of AMPs onto solid surfaces, with emphasis on the development of antimicrobial materials [47, 48].

Mor's group has also investigated the activity of a synthetic "AMPs"—oligoacyllysines (OAKs)—immobilized on an SPR chip [49]. OAKs are synthetic copolymers with repeats of acyllysines as a mimic for naturally occurring AMPs. A resin-linked OAK bead was able to capture about 3,000 bacterial cells for use as a concentration step. Captured *E. coli* was confirmed as still viable with real-time PCR. Mor and coworkers were able to regenerate the beads multiple times with an ethanol wash.

3.3 *High-Throughput Screening Platforms*

High-throughput screening (HTS) capabilities are important and necessary for the rapid detection of multiple agents in a single reaction with minimal sample processing. Here we discuss two current efforts to transition AMP-based detection assays, previously demonstrated in low-density slide arrays, to commercial microtiter plate- and bead-based HTS platforms.

3.3.1 Microtiter Plates

Microtiter plates are effective solid-phase platforms for multiplexed, HTS and analysis. The multiwell format, available in 96, 384, 1,536, and even 3,456 or 9,600-well plates, provides ease of automation, high capacity for paralleled data collection, and versatile application of technologies. The most commonly used approach for bioimmobilization to microtiter plates is noncovalent adsorption. However, given the aforementioned importance of structure and presentation of AMPs for cell capture, it is imperative that surface attachment strategies take into consideration native conformation and molecular orientation to ensure functionally active peptide display. Mello and coworkers first used amine-directed peptide immobilization on maleic anhydride-activated microtiter plates to demonstrate the ability of cecropin P1 to capture and immobilize bacterial cells on solid surfaces [50]. Subsequently, AMPs were synthesized with C-terminal cysteine residues for site-directed immobilization to maleimide-activated microtiter plates. The surface-bound AMPs demonstrated preferential binding behavior for Gram-negative *E. coli* O157:H7 [51]. Mello's group has continued to focus on thiol-targeted chemistry for AMP immobilization, which has been highly productive for large-scale AMP screening for biodetection of pathogens.

The usefulness of AMPs for detection and diagnostic purposes relies heavily on the ability to immobilize them on the surface of a detection platform in a reliable and predictable manner that supports target capture. Therefore, we investigated the

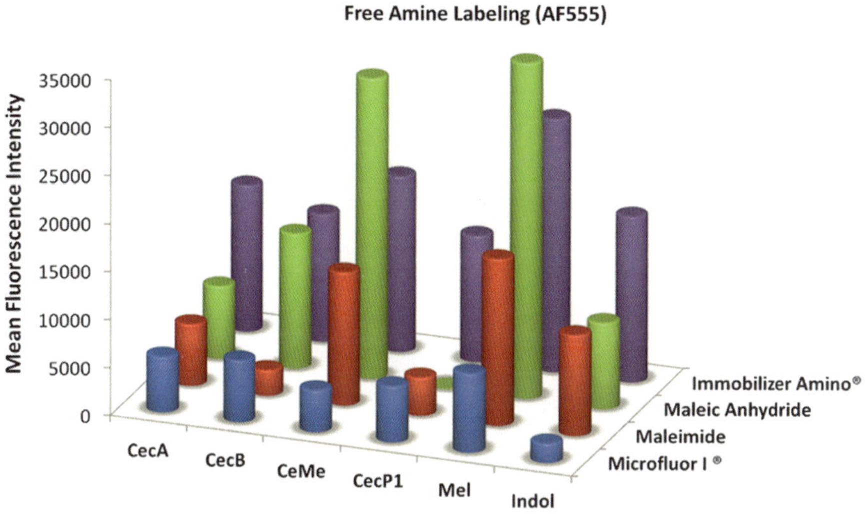

Fig. 2 *CecA* (cecropin A); *CecB* (cecropin B); *CeMe* (cecropin A (1–8)-melittin (1–18) hybrid); *CecP1* (cecropin P1); *Mel* (melittin); *Indol* (indolicidin)

effects of commercial microtiter plate chemistry on AMP immobilization and their subsequent interaction with bacterial biomarker, LPS. AMPs were immobilized to a series of commercial microtiter plates—two plates with amine-reactive functionality (Immobilizer Amino®, maleic anhydride), a maleimide-activated plates for thiol-directed coupling, and a hydrophobic polystyrene plate (Microfluor I®). We evaluated the effects of the different immobilization chemistries on peptide presentation and orientation by determining the relative amount of free primary amines on each peptide. As the primary amines are the potential points of attachment, the fluorescence labeling efficiency of the peptides after surface immobilization can provide information on the degree of multipoint attachment as well as the availability of these amine residues for binding to target analytes. We observed dramatic differences in the labeling efficiency of surface-bound AMPs that were both peptide- and immobilization chemistry-dependent (Fig. 2). These results are consistent with previous studies that have shown that the method used for immobilization greatly affects AMPs' functionality in rapid detection assays [39, 52, 53].

Interestingly, we did not observe a correlation between peptide display and LPS binding activity amongst the different AMPs tested. In addition, we could not identify optimal immobilization chemistry for target capture that could be applied to all AMPs. For example, the cecropin A (1–8)-melittin (1–18) hybrid appears to exhibit the most flexible configuration when coupled to the maleic anhydride microtiter plate; however, it demonstrates the highest LPS binding affinity when immobilized to the maleimide-activated microtiter plate (Fig. 3). In contrast, indolicidin exhibits the greatest number of free primary amines when coupled to the Immobilizer Amino plate yet shows the best binding affinity to LPS when coupled to the maleic anhydride microtiter plate. Even analogous methods of

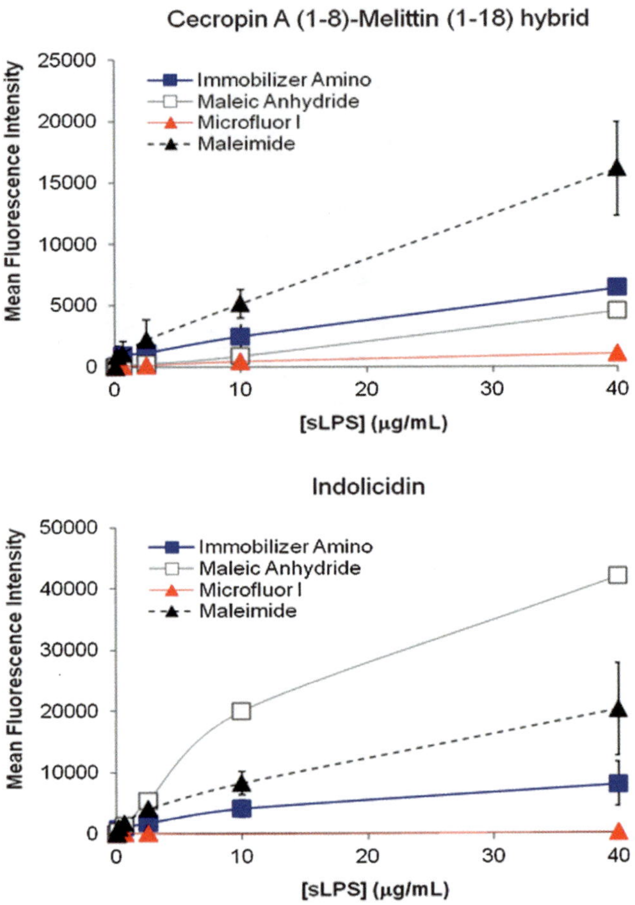

Fig. 3 Binding of Cy5-labeled LPS to cecropin A (1–8)-melittin (1–18) hybrid (*top panel*) and indolicidin (*bottom panel*) peptides immobilized to four commercial microtiter plates

immobilization (e.g., amine-directed linkage to Immobilizer Amino® and maleic anhydride-activated microtiter plates) result in significant differences in the labeling efficiency of free primary amines and target capture. These results suggest that the presentation and orientation of surface-bound AMPs are highly sensitive to the choice of immobilization chemistry used to attach the peptides to microtiter plates. Moreover, for surface-bound AMPs, peptide display is an insufficient predictor of the effectiveness of target capture.

To meet the need for a high density, HTS AMP-based detection platform, we developed a silane-based method for facile and controllable covalent immobilization of AMPs to microtiter plates [52]. This approach allows for a wide variety of peptide immobilization schemes to be applied to a single microtiter plate. Briefly, inert polystyrene microtiter plates were modified with plasma treatment to generate reactive hydroxyl moieties on the plate surface. The microtiter plates were then

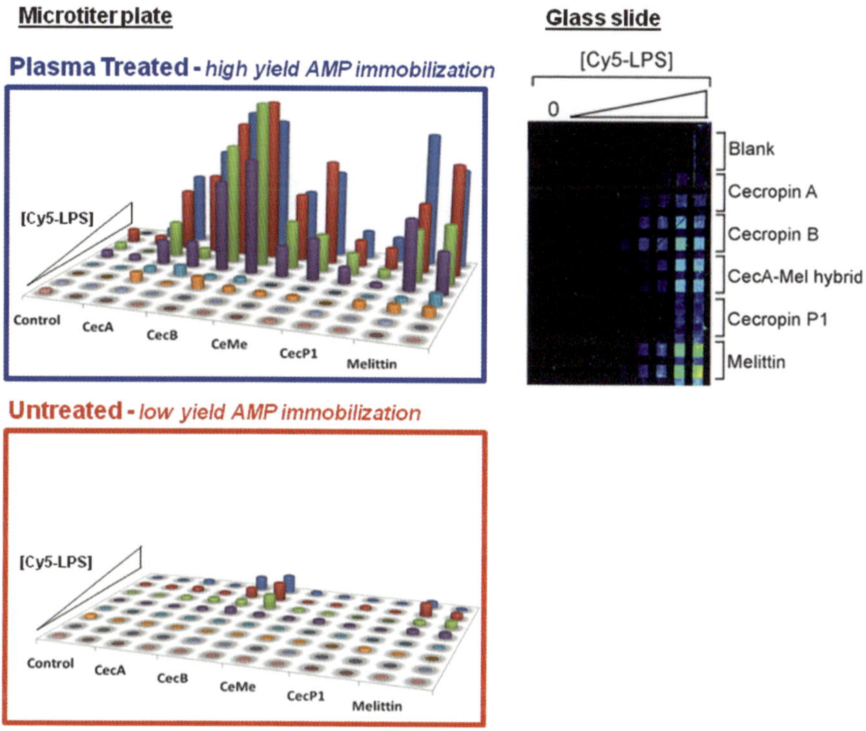

Fig. 4 Cy5-LPS fluorescence from plasma-treated (*top left panel*) and -untreated (*bottom left panel*) microtiter plates indicating the binding of Cy5-conjugated bacterial biomarker, LPS, to a variety of AMPs. Cy-5 LPS binding to AMPs immobilized onto glass slide (*right*)

functionalized with silane and cross-linkers that could provide tailored covalent immobilization for different AMPs. We observed that plasma treatment promoted AMP immobilization efficiency that was comparable to commercial microtiter plates [52] and increased the target capture yield by up to 40 times compared to untreated plates (Fig. 4, microtiter plate). In addition, when plasma-treated plates were processed using analogous chemistry as used on glass slides, the binding pattern observed were similar to that of the gold standard glass slides (Fig. 4, compared microtiter plate to glass slides). This novel method for microtiter plate processing allows for optimal immobilization chemistry for each AMP to ensure AMP functionality, maximize signature extraction, and improve sensing capability.

3.3.2 Bead-Based Formats

A second HTS technology to which AMPs have been applied is bead-based arrays. Exemplified by Luminex100 and MAGPIX systems, bead-based arrays capitalize on the vast multiplexing capabilities of individually interrogable microspheres to examine large numbers of samples for up to 500 targets simultaneously. Each bead set can be separately coated with a different recognition molecule (e.g., AMPs) and then mixed together to create user-tailored, multiplexed assays. Bead arrays have been widely used for clinical diagnostics, transcriptional and expression profiling, autoantibody screening, and food, water, and environmental monitoring [54–61].

We have recently demonstrated that AMPs could be immobilized onto carboxylated polystyrene beads and used to detect *E. coli*, *Salmonella*, *Bacillus*, and *Listeria* in Luminex assays [61]. Detection limits generally ranged from 10^4 to 10^5 cfu/mL when attaching AMPs using amine-targeted linkages. Sensitivities were generally tenfold poorer when AMPs were immobilized via C-terminal cysteines using thiol-targeted chemistry. Initial studies looking at amine-directed linking showed similar effects to those observed on other substrates and in other systems. Effectiveness of target capture was highly dependent on the nature of the surface to which the AMPs were attached (e.g., hydrophobic/hydrophilic, passivated/unpassivated with protein scaffolds, basic/acidic surfaces). Surprisingly, linkers with long flexible "tether" regions were not the most effective for retention of binding activity, but AMP density was not measured or accounted for in these studies.

More recently, we have adopted a more systematic approach for discriminating between the relative effects of initial surface characteristics, linker "tether" regions, and AMP density. Using commercial heterobifunctional linkers terminated with maleimide (thiol-reactive) and *N*-hydroxysuccinimidyl ester (amine-reactive) groups, density of both functional groups and immobilized AMPs could be accurately determined; moreover, the individual effects of various "tether" characteristics (e.g., hydrophobicity, flexibility, length) on AMP affinity and selectivity could be separately quantified. Interestingly, when comparing linkers with similar tether properties, we observed that increased freedom of movement (increased length, flexibility) often resulted in lower binding activities. These differences were striking when comparing binding activities of two helix-hinge-helix peptides, cecropin A and melittin, for the closely related species, *E. coli* and *Salmonella*. Both melittin and cecropin A exhibited similar patterns of binding when immobilized by linkers incorporating aliphatic tethers. However, by simply integrating a phenyl or cyclohexane moiety in the linker, cecropin A became highly specific for *Salmonella* (versus *E. coli*); this effect could be reversed by increasing the chain length of the hydrophobic linker, again yielding binding at similar levels as *E. coli* (Fig. 5). These intriguing results point to the potential of custom-tailoring surfaces to suit the user's desired specificity for detection by simply altering the tether used to link each AMP to the surface.

4 Issues and Challenges

The ability to detect a broad variety of microbial targets is desired for many applications: diagnostics, environmental monitoring, industrial process monitoring, food quality and safety assessments, and biodefense. As such, the more robust,

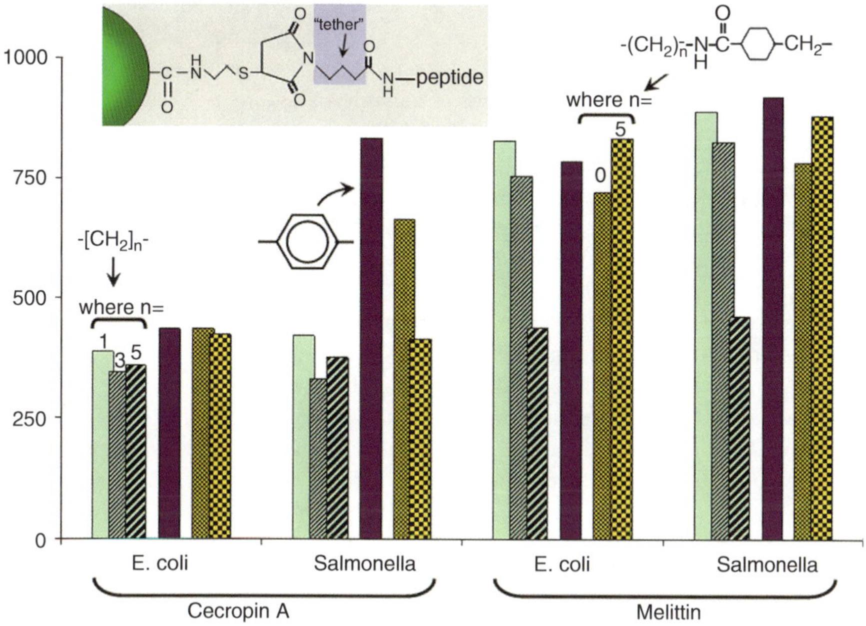

Fig. 5 Effect of "tether" region on binding activity of AMPs immobilized on Luminex microspheres. Cecropin and melittin were immobilized onto thiol-decorated microspheres using heterobifunctional (maleimide/*N*-hydroxysuccinimidyl ester) cross-linkers with different "tether" regions (shown)

rapid, and broad spectrum the sensor, the better suited it will be for on-site use. Integration of AMPs into biosensors has the potential to expand the number of bacterial, viral, and fungal targets detectable in a single test, while also providing a more heat-, solvent-, pH-, and salt-stable biorecognition species than most antibodies currently used in biosensors.

QCM and SPR-based systems utilizing AMPs have the advantage of "reagentless" detection; binding of microbial targets can be directly detected, without the need for a labeled species. Systems such as those described by Mannoor [41] and Mello [43, 51] therefore offer tremendous potential for such broad spectrum, rapid microbial detection. Unfortunately, instrumentation for these and other reagentless platforms has generally not yet evolved with sufficient robustness for routine use outside of a highly controlled, pristine environment, and their multiplexing capacities are still limited.

Biosensors intended for outside-the-lab use, on the other hand, often utilize electrochemical, optical, or other tags that can be used to discriminate detected target from external sources of signal. These biosensors therefore most often required use of labeled "tracer" species (e.g., antibody, AMP, alternative ligand) for target detection. To take full advantage of the broad specificity of AMP-based biosensors, this tracer should be capable of tagging all bound targets equally, such that the only components imparting assay specificity are the immobilized AMP-capture reagents. One potential solution is through use of a nonspecific, whole cell stain [62]; in our laboratory, however, we have observed significant staining of the AMP-modified surfaces in the absence of any microbial targets. Potentially, AMP-based tracers can also be used to detect bound microbes, as shown by Arcidiano [36, 37]. However, detection of bound targets is dependent on the specificity of the AMP moiety of the tracer *as well as* the AMP-capture species, potentially confounding efforts to deconvolute the binding patterns for classification purposes. Smith's group at Notre Dame has developed a novel compound that could likewise be used as a tracer in AMP-based assays. They have shown that a Zn-dipicolinic acid-squaraine rotaxane conjugate stains a wide variety of bacterial species and apoptotic mammalian cells, but does not appear to bind to healthy mammalian cells [63]; further, when used as an imaging agent, this compound was able to discriminate between bacterial infections and sterile inflammations in mice [64].

Integration of a series of peptide "beacons" as capture reagents would completely eliminate the need for a labeled tracer species, much as molecular beacons have become more commonplace for reagentless detection of nucleic acids [65]. While simple in concept, there have been only a handful of studies demonstrating specific detection using peptide beacons [66, 67]. The dearth of well-characterized beacons based on native peptides is likely an indication of the challenges in creating beacon-type structures while accounting for the inherent complexities in peptide structure/function relationships.

A significant challenge remains in the ability to transfer AMP-based assays from platform to platform. It has been well recognized within the antimicrobial coatings community that the method of AMP immobilization affects their biocidal activities (reviewed recently by Costa et al. [47, 48]). It stands to reason that similar effects will be encountered in AMP surfaces used for detection. Indeed, we and others have observed significant differences in affinities and specificities by AMPs immobilized onto different platforms through presumably analogous chemical linkages [51, 52, 68, 69]. This poses a significant challenge if each detection platform has its own specific chemistry required for attachment. In our own laboratory, we have developed a process for attaching AMPs to organic polymers such that patterns of binding to different AMPs are similar to those observed on silanized silicate surfaces [52, 68], but this process may not be applicable to all sensor substrates. Moreover, although binding patterns were similar, they were not identical. Until a universal methodology to allow identical AMP density and presentation on all surfaces is developed, we anticipate that deconvolution of binding patterns—used to classify detected microbes—will require re-optimization on each new platform.

Clearly, AMP-based detection is still a work in progress. The groundwork of putting AMPs into biosensors and detecting microbes and microbial markers has been shown; in a limited number of cases, the potential benefits of using AMPs (such as stability in a broad range of pHs and salt concentrations) have been demonstrated. Although solution-phase systems have aptly demonstrated the promise for AMP-based detection in clinical samples, significant effort is still required to provide evidence that biosensor-based tests are useful with complex samples such as foods, clinical fluids, and environmental specimens. AMPs and AMP-based materials can then be downselected to those with minimal matrix effects.

We anticipate that information technology will play a leading role in the development of future AMP-based assays. With increasing numbers of AMPs being used in multiplexed assays, the ability to store libraries of binding patterns and the ability to retrieve information necessary to deconvolute data from unknown samples will be imperative. Furthermore, we foresee the day when biosensor assays can be custom tailored to virtually any desired specificity (e.g., order-, family-, genus-, or species-level discrimination) by selecting different sets of AMPs (and linker chemistry) with appropriate recognition characteristics. Such AMP-based biosensors have the potential to expand detection capabilities significantly beyond what is currently possible by rapid screening of unknown samples to detect not only well-characterized microbial targets, but those that are unknown or newly emergent.

Acknowledgments This work has been funded by Joint Science & Technology Office for Chemical & Biological Defense/Defense Threat Reduction Agency and the Office of Naval Research (ONR). SND was supported by the Naval Research Enterprise Intern Program through ASEE and ONR. The views expressed here are those of the authors and do not represent the opinions of the U.S. Navy, the U.S. Department of Defense, or the U.S. Government.

References

1. Velusamy V, Arshak K, Korostynska O, Oliwa K, Adley C (2010) An overview of foodborne pathogen detection: in the perspective of biosensors. Biotechnol Adv 28(2):232–254. doi:10.1016/j.biotechadv.2009.12.004
2. Iqbal SS, Mayo MW, Bruno JG, Bronk BV, Batt CA, Chambers JP (2000) A review of molecular recognition technologies for detection of biological threat agents. Biosens Bioelectron 15(11–12):549–578. doi:10.1016/s0956-5663(00)00108-1
3. Simon T (1999) PCR and the detection of microbial pathogens in water and wastewater. Water Res 33(17):3545–3556. doi:10.1016/s0043-1354(99)00071-8
4. Margulies D, Hamilton AD (2010) Combinatorial protein recognition as an alternative approach to antibody-mimetics. Curr Opin Chem Biol 14(6):705–712. doi:10.1016/j.cbpa.2010.07.017
5. Umali AP, Anslyn EV (2010) A general approach to differential sensing using synthetic molecular receptors. Curr Opin Chem Biol 14(6):685–692. doi:10.1016/j.cbpa.2010.07.022
6. Harris M, Mora-Montes HM, Gow NAR, Coote PJ (2009) Loss of mannosylphosphate from *Candida albicans* cell wall proteins results in enhanced resistance to the inhibitory effect of a cationic antimicrobial peptide via reduced peptide binding to the cell surface. Microbiology 155(4):1058–1070. doi:10.1099/mic.0.026120-0
7. Ngundi MM, Kulagina NV, Anderson GP, Taitt CR (2006) Nonantibody-based recognition: alternative molecules for detection of pathogens. Expert Rev Proteomics 3(5):511–524. doi:10.1586/14789450.3.5.511
8. Kulagina NV, Shaffer KM, Ligler FS, Taitt CR (2007) Antimicrobial peptides as new recognition molecules for screening challenging species. Sens Actuators B Chem 121 (1):150–157

9. Lai Y, Gallo RL (2009) AMPed up immunity: how antimicrobial peptides have multiple roles in immune defense. Trends Immunol 30(3):131–141
10. Matsuzaki K (2009) Control of cell selectivity of antimicrobial peptides. Biochim Biophys Acta 1788(8):1687–1692
11. Jenssen H, Hamill P, Hancock RE (2006) Peptide antimicrobial agents. Clin Microbiol Rev 19 (3):491–511
12. Overhage J, Campisano A, Bains M, Torfs EC, Rehm BH, Hancock RE (2008) Human host defense peptide LL-37 prevents bacterial biofilm formation. Infect Immun 76(9):4176–4182
13. Scott MG, Gold MR, Hancock RE (1999) Interaction of cationic peptides with lipoteichoic acid and gram-positive bacteria. Infect Immun 67(12):6445–6453
14. Kiran MD, Adikesavan NV, Cirioni O, Giacometti A, Silvestri C, Scalise G, Ghiselli R, Saba V, Orlando F, Shoham M, Balaban N (2008) Discovery of a quorum-sensing inhibitor of drug-resistant staphylococcal infections by structure-based virtual screening. Mol Pharmacol 73 (5):1578–1586
15. Brogden KA (2005) Antimicrobial peptides: pore formers or metabolic inhibitors in bacteria? Nat Rev 3(3):238–250
16. Melo MN, Ferre R, Castanho MA (2009) Antimicrobial peptides: linking partition, activity and high membrane-bound concentrations. Nat Rev 7(3):245–250
17. Park CB, Yi KS, Matsuzaki K, Kim MS, Kim SC (2000) Structure-activity analysis of buforin II, a histone H2A-derived antimicrobial peptide: the proline hinge is responsible for the cell-penetrating ability of buforin II. Proc Natl Acad Sci U S A 97(15):8245–8250
18. Wang G (2008) Structures of human host defense cathelicidin LL-37 and its smallest anti-microbial peptide KR-12 in lipid micelles. J Biol Chem 283(47):32637–32643
19. De Smet K, Contreras R (2005) Human antimicrobial peptides: defensins, cathelicidins and histatins. Biotechnol Lett 27(18):1337–1347
20. Gennaro R, Zanetti M (2000) Structural features and biological activities of the cathelicidin-derived antimicrobial peptides. Biopolymers 55(1):31–49
21. Cirioni O, Giacometti A, Silvestri C, Della Vittoria A, Licci A, Riva A, Scalise G (2006) In vitro activities of tritrpticin alone and in combination with other antimicrobial agents against *Pseudomonas aeruginosa*. Antimicrob Agents Chemother 50(11):3923–3925
22. Yeaman MR, Yount NY (2003) Mechanisms of antimicrobial peptide action and resistance. Pharmacol Rev 55(1):27–55. doi:10.1124/pr.55.1.2
23. Qian S, Heller WT (2011) Peptide-induced asymmetric distribution of charged lipids in a vesicle bilayer revealed by small-angle neutron scattering. J Phys Chem B 115(32):9831–9837. doi:10.1021/jp204045t
24. Lupetti A, Welling M, Pauwels E, Nibbering P (2003) Radiolabelled antimicrobial peptides for infection detection. Lancet Infect Dis 3:223–229
25. Welling M, Paulusma-Annema A, Balter H, Pauwels E, Nibbering P (2000) Technetium-99m labelled antimicrobial peptides discriminate between bacterial infections and sterile inflammations. Eur J Nucl Med 27:292–301
26. Sarda-Mantel L, Saleh-Mghir A, Welling MM, Meulemans A, Vrigneaud JM, Raguin O, Hervatin F, Martet G, Chau F, Lebtahi R, Le Guludec D (2007) Evaluation of ^{99m}Tc-UBI 29–41 scintigraphy for specific detection of experimental *Staphylococcus aureus* prosthetic joint infections. Eur J Nucl Med Mol Imaging 34(8):1302–1309. doi:10.1007/s00259-007-0368-7
27. Meléndez-Alafort L, Rodríguez-Cortés J, Ferro-Flores G, Arteaga De Murphy C, Herrera-Rodríguez R, Mitsoura E, Martínez-Duncker C (2004) Biokinetics of ^{99m}Tc-UBI 29–41 in humans. Nucl Med Biol 31(3):373–379. doi:10.1016/j.nucmedbio.2003.10.005
28. Akhtar M, Qaisar A, Irfanullah J, Iqbal J, Khan B, Jehangir M, Nadeem M, Khan M, Afzal M, ul-Haq I, Imran M (2005) Antimicrobial peptide ^{99m}Tc-ubiquicidin 29–41 as human infection-imaging agent: clinical trial. J Nucl Med 46:567–573
29. Meléndez-Alafort L, Nadali A, Pasut G, Zangoni E, De Caro R, Cariolato L, Giron MC, Castagliuolo I, Veronese FM, Mazzi U (2009) Detection of sites of infection in mice using

^{99m}Tc-labeled PN_2S-PEG conjugated to UBI and ^{99m}Tc-UBI: a comparative biodistribution study. Nucl Med Biol 36(1):57–64. doi:10.1016/j.nucmedbio.2008.10.011

30. Olstein AD, Albert R (1998) Method of rapid analyte detection. US Patent 5,750,357
31. Rocco R (2000) Method and compound for detecting low levels of organisms. US Patent 6,051,395, 18 Apr 2000
32. Mercado Q, Kramer M, Lim D (2006) Use of polymyxin B for capture of Gram negative bacteria in rapid detection of gram-negative bacteria using a fiber-optic biosesensor (abstract). Paper presented at the American Society for Microbiology, Orlando, FL
33. Lassman M, Kulagina N, Taitt C (2004) Fragmentation of biotinylated cyclic peptides. Rapid Commun Mass Spectrom 18:1277–1285
34. Bessalle R, Haas H, Goria A, Shalit I, Fridkin M (1992) Augmentation of the antibacterial activity of magainin by positive-charge chain extension. Antimicrob Agents Chemother 36:313–319
35. Wiese A, Munstermann M, Gutsmann T, Linder B, Kawahara K, Zähringer U, Seydel U (1998) Molecular mechanisms of polymyxin B-membrane interactions: direct correlation between surface charge density and self-promoted transport. J Membr Biol 162(2):127–138
36. Arcidiacono S, Pivarnik P, Meehan A, Mello C, Senecal A (2008) Labeled antimicrobial peptides for detection of microorganisms. In: Proc Army Sciences Conference, Orlando, FL, US Army
37. Arcidiacono S, Pivarnik P, Mello C, Senecal A (2008) Cy5 labeled antimicrobial peptides for enhanced detection of *Escherichia coli* O157:H7. Biosens Bioelectron 23:1721–1727
38. James E, Shriverlake L, Ogert R, Wiesmann W, Cross A, Ligler F (1993) Versatility of a fiber optic biosensor. Biophys J 64(2):A338
39. Kulagina NV, Lassman ME, Ligler FS, Taitt CR (2005) Antimicrobial peptides for detection of bacteria in biosensor assays. Anal Chem 77(19):6504–6508. doi:10.1021/ac050639r
40. Kulagina NV, Shaffer KM, Anderson GP, Ligler FS, Taitt CR (2006) Antimicrobial peptide-based array for *Escherichia coli* and *Salmonella* screening. Anal Chim Acta 575(1):9–15. doi:10.1016/j.aca.2006.05.082
41. Mannoor MS, Zhang SY, Link AJ, McAlpine MC (2010) Electrical detection of pathogenic bacteria via immobilized antimicrobial peptides. Proc Natl Acad Sci U S A 107(45):19207–19212. doi:10.1073/pnas.1008768107
42. Han XF, Soblosky L, Slutsky M, Mello CM, Chen Z (2011) Solvent effect and time-dependent behavior of C-terminus-cysteine-modified cecropin P1 chemically immobilized on a polymer surface. Langmuir 27(11):7042–7051. doi:10.1021/la200388y
43. Strauss J, Kadilak A, Cronin C, Mello CM, Camesano TA (2010) Binding, inactivation, and adhesion forces between antimicrobial peptide cecropin P1 and pathogenic *E. coli*. Colloid Surface B 75(1):156–164. doi:10.1016/j.colsurfb.2009.08.026
44. Uzarski JR, Tannous A, Morris JR, Mello CM (2008) The effects of solution structure on the surface conformation and orientation of a cysteine-terminated antimicrobial peptide cecropin P1. Colloid Surface B 67(2):157–165. doi:10.1016/j.colsurfb.2008.07.011
45. Ye SJ, Nguyen KT, Boughton AP, Mello CM, Chen Z (2010) Orientation difference of chemically immobilized and physically adsorbed biological molecules on polymers detected at the solid/liquid interfaces in situ. Langmuir 26(9):6471–6477. doi:10.1021/la903932w
46. Ansorena P, Zuzuarregui A, Perez-Lorenzo E, Mujika M, Arana S (2011) Comparative analysis of QCM and SPR techniques for the optimization of immobilization sequences. Sens Actuators B Chem 155(2):667–672. doi:10.1016/j.snb.2011.01.027
47. Costa F, Carvalho IF, Montelaro RC, Gomes P, Martins MCL (2011) Covalent immobilization of antimicrobial peptides (AMPS) onto biomaterial surfaces. Acta Biomater 7(4):1431–1440. doi:10.1016/j.actbio.2010.11.005
48. Onaizi SA, Leong SSJ (2011) Tethering antimicrobial peptides: current status and potential challenges. Biotechnol Adv 29(1):67–74. doi:10.1016/j.biotechadv.2010.08.012
49. Rotem S, Raz N, Kashi Y, Mor A (2010) Bacterial capture by peptide-mimetic oligoacyllysine surfaces. Appl Environ Microbiol 76(10):3301–3307. doi:10.1128/aem.00532-10
50. Gregory K, Mello CM (2005) Immobilization of *Escherichia coli* cells by use of the antimicrobial peptide cecropin P1. Appl Environ Microbiol 71(3):1130–1134. doi:10.1128/aem.71.3.1130-1134.2005

51. Soares JW, Kirby R, Morin KM, Mello CM (2008) Antimicrobial peptide preferential binding of *E. coli* O157:H7. Protein Peptide Lett 15(10):1086–1093. doi:10.2174/092986608786071049
52. North SH, Lock EH, Cooper CJ, Franek JB, Taitt CR, Walton SG (2010) Plasma-based surface modification of polystyrene microtiter plates for covalent immobilization of biomolecules. ACS Appl Mater Interfaces 2(10):2884–2891. doi:10.1021/am100566e
53. Taitt CR, North SH, Kulagina NV (2009) Antimicrobial peptide arrays for detection of inactivated biothreat agents. Methods Mol Biol 570:233–255
54. Stinton LM, Swain M, Myers RP, Shaheen AA, Fritzler MJ (2011) Autoantibodies to GW bodies and other autoantigens in primary biliary cirrhosis. Clin Exp Immunol 163(2):147–156. doi:10.1111/j.1365-2249.2010.04288.x
55. Mera S, Tatulescu D, Cismaru C, Bondor C, Slavcovici A, Zanc V, Carstina D, Oltean M (2011) Multiplex cytokine profiling in patients with sepsis. APMIS 119(2):155–163. doi:10.1111/j.1600-0463.2010.02705.x
56. Wong KL, Tai JJY, Wong WC, Han H, Sem X, Yeap WH, Kourilsky P, Wong SC (2011) Gene expression profiling reveals the defining features of the classical, intermediate, and nonclassical human monocyte subsets. Blood 118(5):E15–E30. doi:10.1182/blood-2010-12-326355
57. Kim HO, Kim HS, Youn JC, Shin EC, Park S (2011) Serum cytokine profiles in healthy young and elderly population assessed using multiplexed bead-based immunoassays. J Transl Med 9. doi:10.1186/1479-5876-9-113
58. Zajac P, Pettersson E, Gry M, Lundeberg J, Ahmadian A (2008) Expression profiling of signature gene sets with trinucleotide threading. Genomics 91(2):209–217. doi:10.1016/j.ygeno.2007.10.012
59. Lee WM, Grindle K, Pappas T, Marshall DJ, Moser MJ, Beaty EL, Shult PA, Prudent JR, Gern JE (2007) High-throughput, sensitive, and accurate multiplex PCR-microsphere flow cytometry system for large-scale comprehensive detection of respiratory viruses. J Clin Microbiol 45(8):2626–2634. doi:10.1128/jcm.02501-06
60. Andersson H, Hartmanova B, KuoLee R, Ryden P, Conlan W, Chen WX, Sjostedt A (2006) Transcriptional profiling of host responses in mouse lungs following aerosol infection with type A *Francisella tularensis*. J Med Microbiol 55(3):263–271. doi:10.1099/jmm.0.46313-0
61. Taitt CR, Shriver-Lake LC, Anderson GP, Ligler FS (2011) Surface modification and biomolecule immobilization on polymer spheres for biosensing applications biomedical nanotechnology. Methods Mol Biol 726:77–94
62. Ligler F, Shriver-Lake L, Wijesuriya D (1996) Optical immunoassay for microbial analytes using non-specific dyes. US Patent 5,496,700, 5 Mar 1996
63. DiVittorio KM, Leevy WM, O'Neil EJ, Johnson JR, Vakulenko S, Morris JD, Rosek KD, Serazin N, Hilkert S, Hurley S, Marquez M, Smith BD (2008) Zinc(II) coordination complexes as membrane-active fluorescent probes and antibiotics. ChemBioChem 9(2):286–293. doi:10.1002/cbic.200700489
64. Leevy W, Gammon S, Johnson J, Lampkins A, Jiang H, Marquez M, Piwinica-Worms D, Smith B (2008) Non-invasive optical imaging of *Staphylococcus aureus* bacterial infection in living mice using a bis-dipicolyamine-zinc(II) affinity group conjugated to a near infrared fluorophore. Bioconjug Chem 19:686–692
65. Li Y, Zhou X, Yi D (2008) Molecular beacons: an optimal multifunctional biological probe. Biochem Biophys Res Commun 373:457–461
66. Oh K, Cash K, Hugenberg V, Plaxco K (2007) Peptide beacons: a new design for polypeptide-based optical biosensors. Bioconjug Chem 18:607–609
67. Thurley S, Roglin L, Seitz O (2007) Hairpin peptide beacon: dual labeled PNA-peptide-hybrids for protein detection. J Am Chem Soc 129(42):12693–12695
68. North S, Lock E, Walton S, Taitt C (2010) Processing microtitre plates for covalent immobilization chemistries. USA Patent Application no. 12/948,114, filed 15 Nov 2010. PCT Application no. PCT/US10/56960
69. North S, Wojciechowski J, Chu V, Taitt CR (2012) Surface immobilization chemistry influences peptide-based detection of lipopolysaccharide and lipoteichoic acid. J Pept Sci 18:366–372

Plastic Antibodies

Alessandro Poma, Michael Whitcombe, and Sergey Piletsky

Abstract Molecular recognition plays an important role in diagnostics, catalysis, separation, and drug development. For years these tasks have been entrusted to antibodies, but recently some viable alternatives are starting to emerge. This review provides a brief overview of natural and synthetic recognition systems, highlighting their advantages and drawbacks. Examples of synthetic receptors include aptamers, engineered proteins, and molecularly imprinted polymers (MIPs). The focus of the present work is on the development and application of a novel class of synthetic receptors—MIP nanoparticles ("plastic antibodies").

Keywords Enzyme mimics, Molecularly imprinted polymer, Nanoparticles, Sensing, Separation

Contents

A. Poma, M. Whitcombe, and S. Piletsky (✉)
Cranfield Health, Cranfield University, Bedfordshire MK43 0AL, UK
e-mail: s.piletsky@cranfield.ac.uk

S.A. Piletsky and M.J. Whitcombe (eds.), *Designing Receptors for the Next Generation of Biosensors*, Springer Series on Chemical Sensors and Biosensors (2013) 12: 105–130
DOI 10.1007/5346_2012_24,
Published online: 4 July 2012

1 Introduction: Antibodies—All That Glisters Is Not Gold

Molecular recognition plays an important role in diagnostics, catalysis, separation, and drug development. Enzymes, antibodies, nucleic acids, and receptors are widely applied in the fundamental study of molecular recognition phenomena, as well as in the development of practical therapeutic or diagnostic systems [1]. Antibodies largely dominate in most of the commercial applications. To provide an example, the global monoclonal antibodies (mAbs) market in 2009 was evaluated as $40 Billion, with $30 Billion related to therapeutic applications [2, 3]. The global in vitro diagnostics (IVD) market, another application area for antibodies and enzymes, has been estimated in 2010 as $44 Billion and is expected to reach $52 Billion by the end of 2013. Among the key constituents of the IVD market, the point-of-care (POC) segment holds the major part, followed by immunochemistry and molecular diagnostics [2, 4].

Antibodies are proteins which the immune system synthesizes to detect and neutralize "non-self" substances (e.g., bacteria, viruses, and toxins), also known as antigens [5]. The most commonly used immunoglobulins G (IgGs) possess a Y-shape resulting from the arrangement of two longer ("heavy") chains and two shorter ("light") chains, all stabilized by disulfide bonds, with an average molecular weight of 150–160 kDa. The lower part of the "Y" is referred to as the Fc region, and its role is to confer stability and drive the interactions with other components of the immune system (e.g., effector mechanisms). The upper part of the "Y" is known as the Fab region and contains the variable domains at which the antigen recognition and binding take place [6]. The antibody–antigen interaction is driven by a precise combination of electrostatic, hydrogen bonding, van der Waals, and/or hydrophobic forces, which results in extremely strong affinity [5]. Antibodies are undoubtedly highly specific and selective for several kinds of chemical and biological moieties and can be produced on a large scale [7]. Their industrial production relies on the cultivation of modified mammalian cell lines [e.g., those from Chinese-hamster ovary (CHO) and human embryonic kidney (HEK)-2930]. The manufacturing process is logistically difficult and expensive [6, 8, 9]. Furthermore, antibody production against small molecules requires chemical coupling between haptens and a carrier protein in order to generate an immune response in animals [7] and their purification involves several steps (especially for applications in therapy), which affects the manufacturing costs [10, 11]. Moreover, it is difficult to generate antibodies against molecules such as immunosuppressant drugs or toxins, because these chemicals act directly on the immune system and prevent its natural response [12, 13]. Additionally, being proteins, the characteristic problems related to their usage are low stability and poor performance in organic solvents, at low and high pH and at high temperature [14–16]. All these factors may alter their recognition properties, thus shortening their shelf-life [17, 18]. Finally, biomolecules may be difficult to immobilize on suitable supports for use in assays and sensors [15, 19], which is an extremely important feature for developing diagnostic devices. For all these reasons, other affinity tools, such as engineered binding proteins, aptamers, and molecularly imprinted polymers (MIPs), have gained interest as potential antibodies substitutes. Their attractive features include enhanced stability, efficient selection and screening procedures, and cost-effective production methods. In the following sections we provide a brief overview of engineered binding proteins, aptamers, and MIP nanoparticles as affinity tools for different applications such as diagnostics, therapeutics and drug delivery, as well as separation and catalysis.

1.1 Engineered Binding Proteins

As already stated above, the recognition ability of antibodies relies on a limited variable region structurally embedded in a more conserved framework. In the same way, proteins capable of binding to a certain target might be selected from a random library, characterized by a constant structural peptide framework and randomized variable binding regions. However, a large amount of peptide derivatives have to be generated and screened to successfully design and engineer a binding protein scaffold. This can be achieved by performing high-throughput screening based on molecular display technology, which establishes a physical link between phenotype and genotype. The most commonly used display technology is phage display in which genes encoding proteins of interest are fused to a gene that encodes a phage coat protein. In this way, phage particles can be made to display peptides of interest on their surface [20, 21]. *Escherichia coli* (*E. coli*) cells are infected with the members of the phage library to produce many copies of each of the library members displaying the variant proteins. This library is screened against the immobilized target molecule and the phages with appropriate specificity and affinity are separated and collected in a process known as biopanning. The collected high-affinity phages are used to re-infect *E. coli* cells and the process is repeated iteratively (usually three to five rounds) using more stringent washing steps. Eventually, monoclonal phages are selected, and the high-affinity protein scaffolds identified by sequencing the DNA of the corresponding phage [6]. Scaffolds that have good stability are required to have a sufficiently long shelf-life, which is important from a commercial point of view. Among the successful examples of engineered proteins there are fibronectin type III domain, which has a certain degree of similarity with the structure of an immunoglobulin G variable domain, and designed ankyrin repeat proteins (DARPins) [22–24]. However, developing scaffolds for a certain application is not easy; in particular due to the unpredictable, costly, and time-consuming nature of the screening procedure, hence the commercialization of these products is still at early stage [6].

1.2 Aptamers

Nucleic acids ligands (aptamers) can also be exploited as affinity tools. The term "aptamer" derives from the Latin word "aptus," which means "fitting," and the Greek word "meros," which means "particle." Aptamers are short (15–60 nucleotides) single-stranded nucleic acid (DNA or RNA) oligomers with a specific and complex three-dimensional shape, which allows them to recognize a variety of targets ranging from small organic molecules to large protein complexes [6, 25]. Aptamers can exhibit affinities down to the nanomolar range, but in contrast to mAbs they are produced entirely in vitro through the generation of combinatorial libraries (10^{14}–10^{15} synthetic nucleic acid sequences) and the subsequent stringent selection process with the immobilized target. The selected sequences are amplified by polymerase chain reaction (PCR) and used in several selection/amplification cycles (6–20) with increasingly stringent selection conditions in a process called SELEX (systemic evolution of ligands by exponential enrichment). Eventually, these aptamers are cloned, sequenced, and tested for the intended application [6, 25].

Aptamers with molecular weight 5–20 kDa are smaller than antibodies and can be used in high-density arrays. Furthermore, thanks to their robustness, aptamers can be chemically modified by, e.g., through biotinylation or by addition of fluorescent labels. They are exploited in ELISA assays or as detection elements in biosensors. Target binding may significantly alter the structure of an aptamer in a reversible way and such an event could be exploited to detect molecules of interest, either fluorescently or electrochemically. Moreover, unlike antibodies, aptamers are easy to regenerate, either at high temperature or high salt concentration which can be used in affinity purification of proteins. Another important advantage is that aptamers can be generated for virtually any target, even those for which antibodies cannot easily be raised (such as toxins or poorly immunogenic molecules). The SELEX process can also be performed under conditions similar to those used in the assay for which the aptamer has been developed. In this way it can be ensured that the oligonucleotide will retain its structure and recognition ability in the final process for which it was intended [26].

SELEX processes are, however, quite lengthy and labor intensive [27]. In addition attempts to automate SELEX procedure have so far proved to be unsuccessful. Moreover, despite their claimed robustness, aptamers are prone to degradation. In addition, commercialization of aptamer technology has been hindered by exclusive ownership of IP by a small number of companies [6, 28].

1.3 Molecularly Imprinted Polymers

The process of molecular imprinting is schematically represented in Fig. 1. The synthesis of MIPs involves monomers which possess functional groups capable of interacting with the target molecule (template), either covalently or through

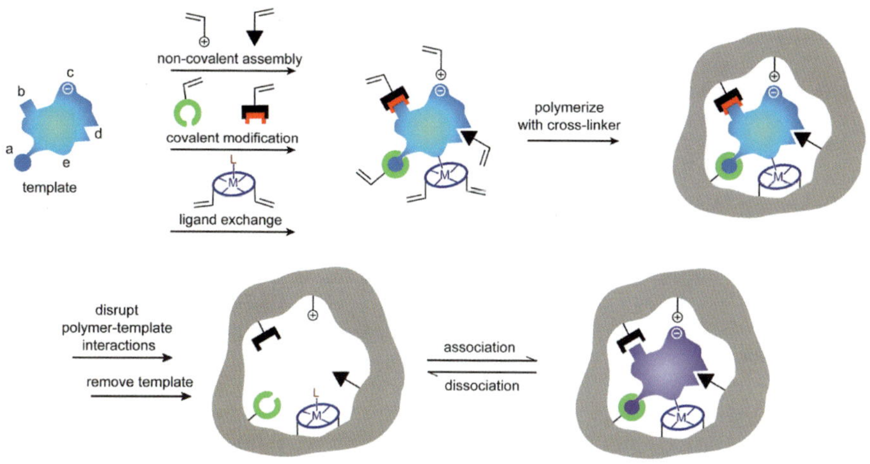

Fig. 1 Schematic representation of molecular imprinting. Template and the polymerizable functional monomer may interact through: (**a**) reversible covalent bonds, (**b**) covalently attached polymerizable binding groups that are activated for non-covalent interaction by template cleavage, (**c**) electrostatic interactions, (**d**) hydrophobic or van der Waals interactions, (**e**) metal–ion mediated interactions (adapted with permission from [30])

non-covalent interactions. The reaction mixture includes a cross-linker and a porogenic solvent. Polymerization is initiated either thermally or by UV light, leading to highly cross-linked polymers. The template is removed from the polymer by washing, leaving behind binding sites that are both spatially and chemically complementary to the template molecules, and capable of rebinding either the template or its structural analogues [29, 30].

In contrast to biomolecules, MIPs are usually stable at low and high pH, pressures and temperatures (<150°C) [15, 31–36]. Moreover, they are less expensive than biomolecules and easier to obtain, and they can be used in organic solvents. Finally, they can be synthesized for diverse classes of substances, such as ions [37], nucleic acids [38], proteins [39, 40], drugs [41–43], and even yeast cells and erythrocytes [44].

The number of published papers in the MIP area in the last 10 years has tripled, which reflects the growing interest in these materials [45]. However, MIPs are also burdened with some limitations, mainly connected with the methods of their production and the final format of the polymer. One such limitation is linked to MIPs prepared in the "bulk" format, which require grinding and sieving to obtain a fraction of particles with a narrow range of sizes. This is a lengthy process and is impractical or unsuitable for many applications [46–48]. Furthermore free radical addition polymerization is exothermic and the bulk format prevents efficient heat exchange with the exterior. This can lead to rapid increases in the temperature of the polymerization mixture and consequently increased pressure within the reaction mixture, which may adversely affect MIP properties and can lead to explosions for

Table 1 Comparative properties of MIPs produced in bulk and nanoscale formats

Bulk MIPs	MIP nanoparticles
Broad distribution of binding sites with varying affinity, high level of nonspecific binding	"Monoclonal" binding sites, one to two per particle, two to three orders of magnitude difference between specific and non-specific binding
Affinity in the range 10^{-9}–10^{-3} M depending on template	Affinity in the range 10^{-10}–10^{-6} M, possibility of using affinity chromatography for the fractionation of high-performance nanoparticles
Insoluble material, difficult to process	Soluble particles which can be treated as standard reagents
Substantial batch-to-batch variability	Better control of manufacturing process, using chemical reactors
Possibility of template leaching from the polymer	Traces of template can be relatively easily removed using dialysis or affinity separation
Limited prospects for in vivo applications	MIP nanoparticles with biological activity can be produced

anything other than small-scale reactions [29, 32, 36]. Complete removal of the template is often difficult with bulk MIPs, which can result in the potential for leaching of residual template from the matrix interfering with the intended application, particularly when used for trace analysis [49]. Integration of bulk MIPs with sensors and assay protocols is also complicated [15, 29, 30]. A more promising approach lies in the synthesis of MIPs in the form of nanoparticles.

2 MIP Nanoparticles: True "Plastic Antibodies"?

In contrast to bulk materials, MIP nanoparticles (MIP NPs) show improved characteristics (Table 1).

MIP NPs have higher surface-to-volume ratios and greater total active surface areas per unit weight of polymer. Imprinted cavities are more easily accessible to the template, which improves binding kinetics and facilitates the template removal process, thus enhancing their overall performance [50–52]. Nanoparticles can be dosed precisely and allow conjugation with probes and enzymes for ELISA-type assays [13, 53]. Several methods have been used to obtain MIP nanoparticles [54]. MIP nanoparticles have been used in catalysis [55, 56], in drug delivery [41, 42], in binding assays [7, 9, 53, 57], in capillary electrochromatography (CEC) [58–61], and in sensors [62–64]. Each of these applications will be discussed in detail in the following sections by analyzing some examples from the literature in an attempt to highlight the advantages and disadvantages arising from the use of MIP nanoparticles for these purposes.

3 Applications of MIP Nanoparticles

3.1 Drug Delivery

The cross-linked nature and affinity properties of MIPs make them suitable reservoirs for controlled drug release. This is particularly useful for drugs with a low therapeutic index (e.g., theophylline), which might cause adverse effects if their concentration is not kept below a certain threshold value. Chemically or physically triggered release also can be achieved using MIPs, e.g., when the polymer interacts with the specific imprinted target moiety, such as a cell surface receptor overexpressed in a tumor [45]. Due to their dimensions and high-surface area, MIP nanoparticles could represent a very interesting solution for these applications. In a pioneering work by Ciardelli and coauthors, MIP nanoparticles were used as a drug delivery system for the controlled release of theophylline (THO) [42]. Particles of 200 nm diameter were synthesized using a modification of the precipitation polymerization approach of Ye and coauthors [65]. However drug release properties of MIP NPs were not so easy to predict and understand. The release pattern depended on the fine balance of the strength of monomers–template interaction, concentration of monomers and polarity of the particles. Jantarat and coauthors fabricated composite cellulose membranes embedded with MIP nanoparticles for the transdermal enantioselective release of racemic propranolol [66]. The *S*-enantiomer of this β-blocker is 100–130 times more potent than its *R*-isomer, and it would make sense to develop delivery system that would allow specific delivery of enantiomerically pure compounds [67]. The authors applied suspension polymerization in liquid perfluorocarbons to obtain MIP NPs with diameters in the range 300–500 nm, directly attached to the surface of 3–10 μm diameter microspheres [68]. The release of the *S*-enantiomer from the composite membrane was 1.7 times more rapid than the *R*-enantiomer which proves the principal feasibility of this approach. However more optimization is required for this approach to be used as a practical method for clinical applications. The same group also prepared MIP NPs for *S*-omeprazole to be used in the fabrication of an orally administered drug delivery system [69]. Authors exploited polymerizable cinchona alkaloids methacryloyl quinine and methacryloyl quinidine as functional monomers to provide strong anchoring groups. Composite membranes containing MIP NPs were additionally covered with polyhydroxyethylmethacrylate and polycaprolactone triol to provide gastroresistant properties. The corresponding delivery devices exhibited gastroresistant properties and a selective release of *S*-omeprazole with an *S*/*R* enantiomeric ratio of 2. MIP hydrogel nanospheres with diameter 270 nm were prepared and used as drug delivery systems for 5-fluorouracil (5-FU) (Fig. 2, left) [41]. The use of this drug is hampered by its short half-life and relatively high toxicity; hence carefully controlled daily injections are needed to maintain the therapeutic activity [70]. MIP nanoparticles

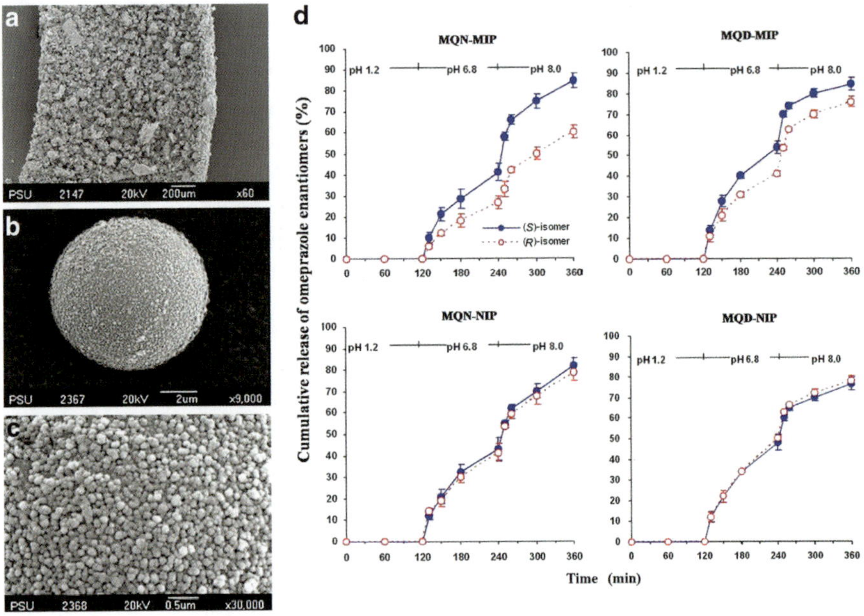

Fig. 2 (**a**) Scanning electron microscopy (SEM) image of a cross-section of the controlled delivery device containing the MIP nanoparticles-on-microspheres. (**b**, **c**) SEM images of the prepared MIPs microspheres composed of nanoparticles on the surface at 9,000-fold magnification (**b**) and 30,000-fold magnification (**c**). (**d**) In vitro dissolution profile of omeprazole enantiomers from MIP- and NIP-loaded delivery systems in dissolution medium changed every 2 h with pH 1.2, 6.8 and 8.0, respectively (mean ± SD, $n = 6$) (adapted with permission from [69])

exhibited very low cross-reactivity, even for the analogue uracil, which differs from the template only by a hydrogen atom replacing fluorine. The release of 5-FU from imprinted MIP nanospheres in simulated plasma fluid showed a sustained release over 50 h (65% of the total amount of drug loaded), while non-imprinted polymers completed the release after 5 h (Fig. 2, right).

Magnetic MIP nanoparticles for drug delivery have been developed by Kan and coauthors, who grafted an aspirin-imprinted MIP shell onto 12 nm diameter silane-modified magnetic cores, obtaining 500 nm diameter MIP nanoparticles [71]. MIP NPs exhibited good selectivity for the template in comparison with its structural analogues such as salicylic acid or *o*-aminobenzoic acid. When tested in vitro, during the first 2 h magnetic MIPs released about 50% of the loaded drug, while non-imprinted nanoparticles released about 85%. Due to their magnetic properties, MIP nanoparticles could be easily separated and manipulated. Theoretically, they could be used to target the drug release towards particular sites in the body by exploiting an external magnetic field [72]. It might be interesting to fabricate magnetic MIP nanosystems below 100 nm in size suitable for passing through altered capillary fenestrations in tissues such as sites of inflammation or tumors.

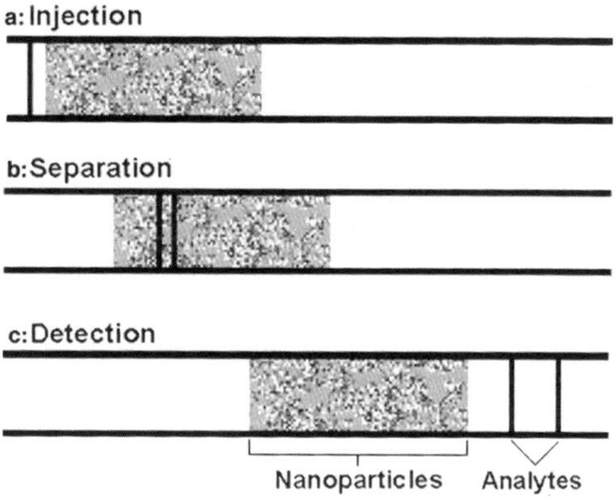

Fig. 3 Schematic of the partial filling technique (reproduced with permission from [79])

3.2 Capillary Electrochromatography

CEC is a hybrid separation technique that combines the high efficiency typical of capillary electrophoresis with the phase selectivity of a high-performance liquid chromatography [73]. An interesting aspect of this technique is the possibility to use a so-called pseudostationary phase (PSP). Different from a common stationary phase, PSPs are interaction phases that move with (or against) the mobile phase and are continuously replaced, without needing to be packed [74–77]. To be suitable for this purpose, MIP NPs have to possess certain properties, such as: (a) able to form stable suspensions and exhibit enough selectivity in the electrolyte solutions used as mobile phases; (b) be charged, in order not to co-elute with the electroosmotic flow; (c) have a uniform velocity to avoid peak broadening; (d) exhibit high-surface areas and low mass-transfer resistance; and (e) not interfere with the detection mechanism [78]. A "partial filling" technique has been exploited for use of MIP NPs in CEC, whereby a fraction of nanoparticles suspension is injected before the sample (Fig. 3a). The analytical sample will be separated during the run by passing through the more slowly moving "plug" of nanoparticles when the voltage is applied, arriving at the detection window before the nanoparticles (Fig. 3b), thus avoiding scattering from the PSP during UV detection (Fig. 3c) [74, 76, 79].

Schweitz and coauthors successfully synthesized and used 200–500 nm MIP nanoparticles in CEC separation of propranolol enantiomers [58]. The authors used UV-initiated precipitation polymerization at −26°C to strengthen the interactions between the template and the functional monomer. Under optimized conditions, racemic resolution was achieved in slightly more than 1 min, even for multiple racemic mixtures of atenolol, pindolol, and propranolol [76]. The same group performed the simultaneous CEC resolution of two different racemic analytes, propranolol and ropivacaine, either by injecting two different types of MIP nanoparticles at the same time, or by using MIP nanoparticles which were simultaneously imprinted for the two templates [59]. In the first case it was possible to separate the racemic mixtures of the two different templates in a single run (Fig. 4).

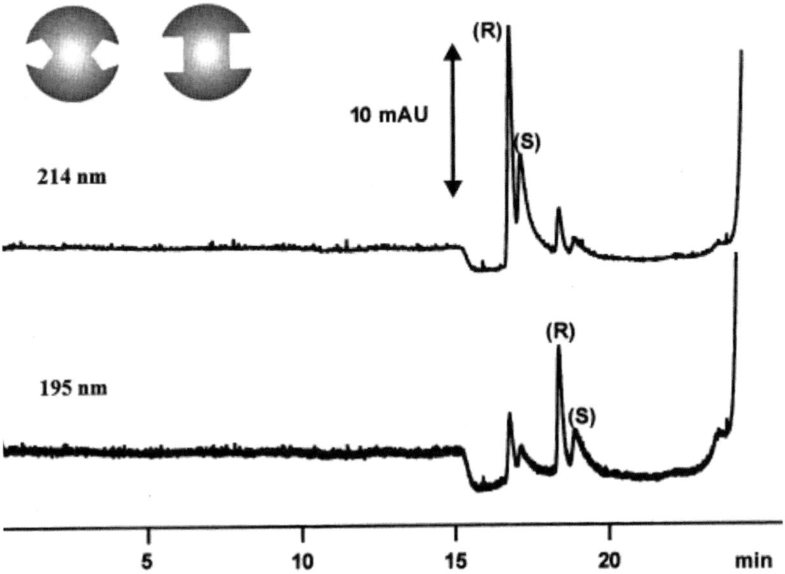

Fig. 4 Separation of ropivacaine and propranolol enantiomers in CEC by the partial filling technique using a plug composed of *S*-ropivacaine MIP and *S*-propranolol MIP. Detection was performed at 214 (*top*) and 195 nm (*bottom*). For experimental details refer to [59]. Reprinted with permission from [59]. Copyright (2003) American Chemical Society

However, in the case of multi-template imprinting, the amount of *S*-propranolol used in the nanoparticles synthesis strongly affected the recognition performance for *S*-ropivacaine. The 1:80 template-to-monomer ratio was found to be optimal to obtain MIP nanoparticles with twofold selectivity. Priego-Capote and coauthors have used modified mini-emulsion polymerization to prepare MIP NPs with binding sites located mainly onto the surface of the nanoparticles [61]. This in theory should improve mass transfer of the analytes and reduce peak tailing in CEC. The synthesized MIP nanoparticles were able to bind threefold more template than the non-imprinted ones. When tested in CEC analysis a racemic mixture of the template was separated without evident peak tailing. However the affinity for the template demonstrated by the novel MIPs was much lower than that obtained through other polymerization methods (e.g., precipitation polymerization). In addition the size of the particles obtained was not uniform, ranging from 30 to 150 nm. Qu and coauthors have used magnetic MIP NPs in a microfluidic CEC device for resolution of racemic ofloxacin [80, 81]. The MIP NPs were pumped into the capillary as a slurry and packed in the microfluidic device in a specific region next to magnets. In optimal conditions, a resolution value of 1.46 was achieved in slightly more than 3 min of analysis. This approach seems really promising for a fast and cheap qualitative analysis. The detection limits (5 μM for the template *S*-ofloxacin, and 1 μM for *R*-ofloxacin), however, were too high compared to conventional chromatographic methods, and hence system required further improvement.

3.3 Enzyme Mimics

Given their small size as well as their dispersibility/solubility characteristics, MIP nanoparticles are promising candidates for the development of enzyme mimics [82]. The first example of catalytic MIP nanoparticles has been reported by Markowitz and coauthors, which used micro-emulsion polymerization to imprint a surfactant-derivative of transition state analogue (TSA) of α-chymotrypsin [83]. To mimic the enzyme, the authors used silanes with a structure resembling the amino acids of the catalytic triad found in the active site of serine proteases. MIP NPs with diameters in the range 400–600 nm were obtained, which exhibited an enantioselective hydrolysis. Resmini and coauthors have synthesized soluble MIP NPs with hydrolytic activity, performing radical polymerization under high-dilution conditions [84–86]. The dimensions of these particles, together with their solubility, make them similar to a natural enzyme or antibody [82]. The authors used polymerizable derivatives of tyrosine and arginine as monomers in the imprinting of a phosphate-based TSA for the carbonate hydrolysis reaction. Catalytic low-cross-linked microgels produced in this way followed Michaelis–Menten kinetics, exhibited a turnover number (k_{cat}) 530 times higher than the uncatalyzed reaction and possessed a remarkable substrate selectivity. A lower degree of cross-linking provided sufficient flexibility to the active site residues necessary for the display of catalytic activity. Wulff and coauthors prepared highly efficient catalytic MIP NPs with 10–20 nm diameter by polymerization of an ionic complex between the diphenyl phosphate TSA and *N,N′*-diethyl-4-vinylbenzamidine [56]. The authors used a "post-dilution method" which involved carrying out the polymerization process at high concentrations of monomer, similar to those used in bulk polymerization, but with the addition of a large volume of solvent just before the point of macro-gelation. The increased degree of cross-linking that resulted from the method was beneficial for the catalytic activity of the synthesized material and for reduced polydispersity. The catalytic activity of the MIP NPs obeyed Michaelis–Menten kinetics with $k_{cat} \approx 3000$ times higher than the non-catalyzed reaction. Chen and coauthors prepared peroxidase-mimicking MIP NPs for the oxidation of homovanillic acid (HVA) using catalytic hemin groups [87, 88]. The polymerization process was performed under high-dilution conditions to avoid macro-gelation. The authors obtained 200 nm MIP NPs with a moderate degree of polydispersity and with ability to increase the oxidation rate of HVA ($k_{cat} = 4.56 \times 10^7\ M^{-1}\ s^{-1}$) threefold more rapidly than the non-imprinted control. In addition, MIP NPs exhibited a remarkable selectivity towards the oxidation of template analogues. The catalytic behavior was pH-sensitive and followed Michaelis–Menten kinetics. Despite all these advantages, problems such as a low yield (12.5%) and the long purification times limited the potential commercialization of these products. Huang and coauthors produced gluthatione peroxidase mimicking MIP NPs using micro-emulsion polymerization with allyl arginine and acryloyloxypropyl 3-hydroxypropyl telluride as monomers [89]. MIP NPs of 30–40 nm in diameter were able to increase the efficiency of the degradation of cumene hydroperoxide

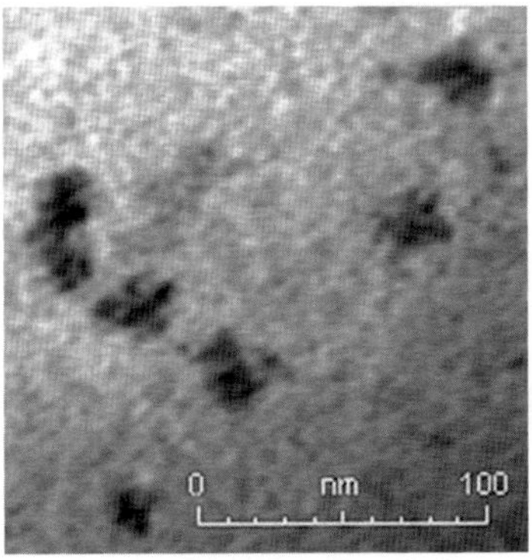

Fig. 5 Transmission electron microscopy (TEM) image of catalytic imprinted nanogels with aldolase type I activity (reproduced with permission from [55])

(CUOOH) by 600,000 times. However, the difference in catalytic behavior of imprinted and non-imprinted particles was not impressive. Recently Resmini and coauthors produced MIP NPs able to catalyze a Kemp elimination reaction, for which no natural enzymes are currently known to exist [90]. They imprinted an indolic structure as TSA using 4-vinylpyridine as a basic functional monomer. MIP NPs were completely soluble in water and exhibited an optimum activity at pH 9.4. The same group also synthesized MIP NPs capable of catalyzing a C–C bond formation reaction, and in particular a cross-aldol reaction between 4-nitrobenzaldeyde and acetone, thus mimicking the enamine-based mechanism of the natural aldolase type I enzymes [55]. In this case a covalent imprinting approach was exploited, bonding a diketone TSA to a polymerizable proline derivative as functional monomer. By using high-dilution radical polymerization, the authors obtained 20 nm MIP NPs with a very low polydispersity (Fig. 5).

The MIP NPs were not inhibited by product, exhibiting 20-fold higher catalytic activity compared to the non-imprinted NPs and a good enantioselectivity (62% enantiomeric excess). These results confirm that MIP NPs could be used when there are no suitable enzymes or whenever natural molecules have inadequate stability or high price.

3.4 Sensing Applications

Given their robustness and entirely synthetic nature, MIP nanoparticles are particularly suitable for application in sensors and assays. Reimhult and coauthors produced a quartz crystal microbalance (QCM) sensor by coating its surface with MIP NPs imprinted with *R*- or *S*-propranolol [63]. MIP NPs of 130 nm diameter were synthesized by precipitation polymerization, then dispersed in a solution of poly (ethylene terephthalate) and eventually spin-coated onto the surface of the QCM crystal. However it seems this treatment altered the accessibility of MIP binding sites, adversely affecting the polymer recognition properties. Rather than using MIP

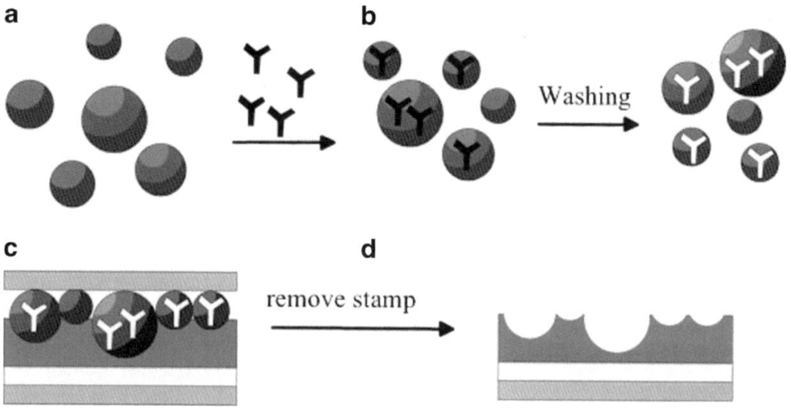

Fig. 6 Schematic representation of the QCM sensor based on artificial antibody replicas (reproduced with permission from [64])

nanoparticles directly as sensing elements, Schirhagl and coauthors exploited them as a "secondary template" to transfer their imprints onto a modified QCM wafer [64]. The authors first prepared MIP nanoparticles for IgG raised against human rhino virus 14 by precipitation polymerization (Fig. 6a), and following removal of the template (Fig. 6b) used them as stencils to imprint a secondary polymeric layer deposited on a QCM wafer (Fig. 6c, d).

The fabricated chemosensor responded to the virus generating signal 6 times higher than that of the sensor coated with natural antibodies. The mechanism of this phenomenon is not clear and typically antibodies-based sensor show higher response than MIP-based devices [91]. QCM sensor based on MIP NPs exhibited a faster and stronger response and higher selectivity than sensors made of bulk polymer [91]. Bompart and coauthors investigated the possibility of using micro-Raman spectroscopy to quantify the target molecule adsorbed by MIP NPs [92]. The authors obtained 200 nm NPs imprinted with *R*- or *S*-propranolol by precipitation polymerization. Micro-Raman spectroscopy of nanoparticles pre-equilibrated with the template allowed detection of propranolol with good selectivity at 1 μM concentration. However it was very difficult to achieve standardization of all measurement parameters and the detection level was not impressive. To increase the sensitivity of the system the authors used surface-enhanced Raman spectroscopy (SERS) performed on 400 nm composite gold core–shell MIP nanoparticles produced by seeded emulsion polymerization [93]. This allowed performing target molecule measurements on single MIP nanoparticles, reaching a detection limit of 0.1 μM. This detection capacity was retained even in the presence of a 100-fold excess of interfering compounds such as caffeine or acetylsalicylic acid. In addition, measurements could also be performed in spiked biological samples (diluted equine serum) down to a detection limit of 1 μM. The possibility of performing multiplexed measurements by using different MIP nanoparticles is indeed quite attractive and deserves further investigation. Sener and coauthors have developed a lysozyme sensor using protein-imprinted MIP NPs [94]. MIP NPs of 50 nm in diameter were synthesized by mini-emulsion polymerization and physically deposited on a QCM crystal by solvent evaporation. The imprinted sensor exhibited a detection limit of 1.2 ng mL^{-1} for lysozyme in the analysis of chicken egg-white samples. The system could be used in up to four consecutive measurements. The same lysozyme-imprinted MIP NPs were also used in a surface plasmon resonance (SPR) sensor [95]. The SPR sensor exhibited fast response times with a 1000-fold lower detection limit than the QCM sensor.

One of the first MIP NPs-based assays reported in the literature was described by Haupt and coauthors [96]. The MIP NPs were imprinted with 2,4-dichlorophenoxyacetic acid (2,4-D), and used with structurally related fluorescent, chemiluminescent or electrochemical probes to quantify the amount of bound analyte [97]. The detection limit achieved was 100 nM. However the assay showed significant cross-reactivity with 2,4-D structural analogues. Nevertheless, these first results paved the way for the use of MIP nanoparticles in pseudo-immunoassays. Ye and coauthors synthesized 200–300 nm MIP nanoparticles imprinted with THO and 17β-estradiol (E2) by precipitation polymerization and used them in a first MIP NPs-based radioimmunoassay [65, 98]. The authors were able to quantify templates down to concentrations of 0.1 and 0.01 μg mL^{-1} for THO and E2, respectively, even in the presence of competing structural analogues. The authors later improved this assay system by adding a "scintillation monomer" hydroxymethyl-2,5-diphenyloxazole acrylate and divinylbenzene into the preparation [99, 100]. The binding signal was generated by proximity energy transfer, which arose from the specific binding of a tritium-labeled template (*S*-propranol). In this approach there was no need to remove the unbound labeled ligand, because the latter was in solution and therefore too distant from the scintillation monomer to generate a signal. MIP nanoparticles exhibited discrete enantioselectivity. However, the MIP NPs did not perform well in aqueous environment and relied on "unpopular" radioactive isotopes. Surugiu and coauthors developed an ELISA-like assay based on MIP NPs imprinted with 2,4-D. The target analyte was labeled with tobacco peroxidase as an enzymatic probe and used to detect the template either colorimetrically or by chemiluminescence [101]. In both cases the competition between free template and analogue was monitored in solution, allowing quantification of 2,4-D to 1 μg mL^{-1}. Unfortunately the cross-reactivity of the assay to structural analogues was quite high. Later the same MIP NPs were immobilized on microtiter plate wells using polyvinyl alcohol (PVA) as glue. Detection of chemiluminescence from the competition reaction was performed using a CCD camera [102]. The detection limit of the assay was decreased to 34 nM. As in the previous cases, some cross-reactivity was exhibited, especially when compared to the same assay performed using antibodies. Pérez and coauthors designed an immunoprecipitation like assay for cholesterol using surface-imprinted MIP nanoparticles prepared by core–shell emulsion polymerization [103]. MIP nanoparticles exhibited uniform morphology and small diameters (60 nm). The addition of a multi-ligand template, polyethylene glycol (PEG)-bis-cholesterol, was able to flocculate the MIP particles, while mono-ligand template did not give rise to these effects (Fig. 7).

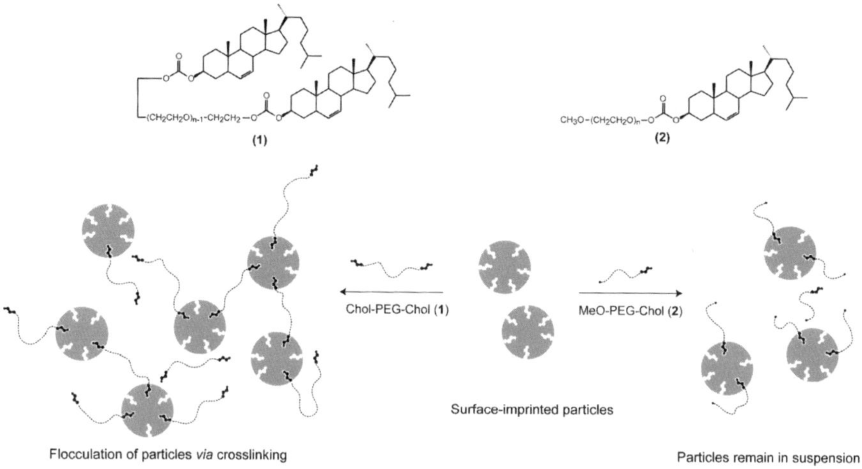

Fig. 7 "Immunoprecipitation-like" separation of surface-imprinted particles in the presence of (PEG)-bis-cholesterol. The addition of the multi-ligand template resulted in flocculation of MIP particles (adapted with permission from [103])

Another possible strategy to exploit MIP NPs in assays would be to make them fluorescent [104, 105]. Diltemiz and coauthors grafted CdS quantum dots with a MIP for guanosine [38]. MIP nanoparticles had an average diameter of 45 nm, and their intrinsic fluorescence was enhanced by binding of the template, proportionally to its concentration. MIP nanoparticles exhibited a high response to guanine and guanosine, while adenosine did not give rise to any change in fluorescence. Purely organic MIP nanoparticles with fluorescent sensing capability were recently prepared by Ivanova-Miteseva et al. who prepared a fluorescent core by partially modifying the peripheral amino groups of a poly(amido amine) (PAMAM) dendrimer with dansyl residues [106]. The remaining free amino groups were then modified with diethyldithiocarbamate (iniferter) groups capable of initiating photochemical polymerization of an imprinted polymer shell. The particles had an unusual cube-like shape and were 50 nm in size. The fluorescent MIP NPs (but not blank NPs) showed an enhancement of fluorescence in the presence of the template (acetoguanamine) with a detection limit of 30 nM, but did not respond to close structural analogues. Recently Li and coauthors have developed 350 nm MIP NPs with a double-layer core–shell structure made of a Fe_3O_4 nanoparticle core, an inner shell of fluorescein isothiocyanate and an outer MIP shell, for faster separation and recognition of E2 [107]. The MIP shell was produced using a controlled living RAFT polymerization. The fluorescent intensity of MIP NPs decreased with increasing concentrations of E2, showing a detection limit of 0.19 μM. They exhibited a discrete imprinting effect and very good selectivity. Such a system not only provided a source of fluorescence but also allowed magnetic separation to replace centrifugation and filtration steps during the experimental procedure.

3.5 Separation

The most popular MIP formats used in separation are membranes and microparticles. Nevertheless MIP NPs with high-surface area might offer advantages especially if they can be integrated with membranes or fibers to improve their performance. The first example of MIP NPs-containing composite membrane dates back to 2002 [108]. In this work Lehmann and coauthors prepared MIP NPs imprinted with Boc-L-phenylalanine anilide using mini-emulsion polymerization and used them to create composite membranes for enantiomeric separation. The authors, however, did not study the separation performance of the created membrane, but only its porosity and flow properties. Silvestri and coauthors prepared 174 nm MIP NPs imprinted with THO and caffeine by precipitation polymerization, and embedded them in poly(methyl methacrylate-*co*-acrylic acid) membranes using a solid-phase inversion method [109]. The binding characteristics of the membranes were quite good, with 40 times more THO bound to the MIP NP-based membrane compared to membranes without nanoparticles. Moreover, they exhibited a selectivity factor for THO versus caffeine of 10. The same authors later used this strategy to create composite membranes containing MIP NPs for cholesterol [110]. All the synthesized nanoparticles exhibited specific rebinding capacities for the template, both in ethanol and in phosphate buffer. This trend was mirrored by the membranes which possessed a specific binding capacity of 14 mg template g^{-1} of the composite system. Chronakis and coauthors incorporated MIP NPs imprinted with E2 and THO into composite nanofibers using the method of electrospinning [111]. MIP NPs were synthesized using precipitation polymerization, suspended in a solution of polyethylene terephthalate (PET) in dichloromethane and trifluoroacetic acid, and electrospun forming regular nanofibers of about 150–300 nm in diameter (Fig. 8).

The nanofibers produced in this way were able to accommodate up to 75% (w/w) of nanoparticles, exhibiting excellent binding properties. The same authors later incorporated NPs imprinted with propranolol into nanofibers prepared by electrospinning and used them to specifically extract and concentrate propranolol from spiked tap water samples [112]. The binding specificity was preserved even in the presence of other β-blockers. Piperno and coauthors have used cross-linking of PVA to create more stable and water-compatible electrospun nanofibers with incorporated MIP NPs imprinted with dansyl-L-phenylalanine [113]. The 400 nm MIP NPs were produced by precipitation polymerization. Electrospun cross-linked fibers retained their recognition activity even after multiple adsorption/desorption cycles, thus highlighting their stability and possibility to be used as solid-phase extraction (SPE) media. Zhu and coauthors synthesized 400 nm core–shell silica MIP nanoparticles by a sol–gel process that could be used directly in SPE applications for extraction of bisphenol A [114]. MIP NPs had an adsorption capacity 2.5-fold higher than non-imprinted particles, with very rapid rebinding kinetics, probably due to the binding sites being located near the surface of the particles. In addition, they recovered close to 100% of the template, even in the

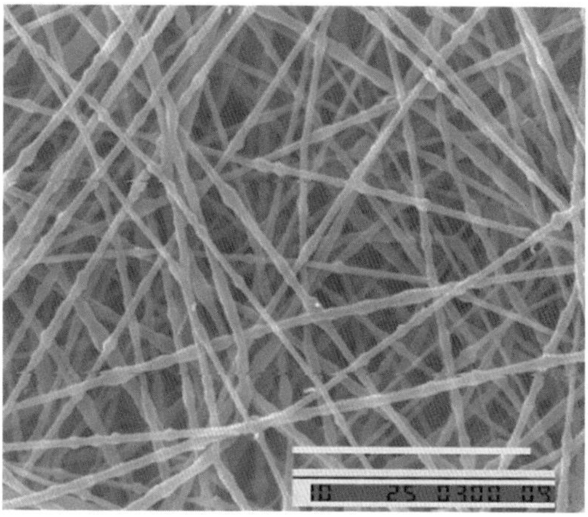

Fig. 8 SEM image of electrospun PET nanofibers containing 37.5% (w/w) of MIP-E2 nanoparticles. The scale bar is 10 μm (adapted with permission from [111]).

presence of structural analogues present in excess. In the SPE extraction of cosmetic samples spiked with bisphenol A, MIP nanoparticles performed better than commercial silica materials. The same group also synthesized 300–400 nm core–shell MIP NPs for SPE of the herbicide metsulfuron-methyl (MSM) [115]. A MIP layer was thermally grafted to silica cores modified with an acryloyl silane. The optimized MIP NPs had a maximum adsorption capacity 3.4 times higher than non-imprinted particles and 3.8 times higher than conventional C_{18} silica. In addition, rapid rebinding kinetics could be obtained. In the SPE extraction of samples of soil or from crops spiked with the template and its structural analogues, MIP nanoparticles allowed the pre-concentration of MSM to a much higher level than commercial silica or non-imprinted nanoparticles, thus allowing the subsequent HPLC quantification to be performed with minimal noise. In a similar way Gao and coauthors prepared core–shell nanoparticles imprinted with sulfamethoxazole (SMO) to be used in SPE applications [116]. The MIP NPs specifically rebound three times more SMO than bulk MIP or non-imprinted NPs, reaching the adsorption equilibrium in about 45 min. In the SPE pre-concentration of samples of eggs and milk spiked with sulfonamides, the recoveries ranged from 73 to 89% with relative standard deviations below 7.5%, allowing detection of SMO and of another structural analogue, sulfadiazine, by HPLC analysis of the concentrates. Shamsipur and coauthors prepared MIP NPs for copper ion using an anthraquinone derivative as the specific complexing monomer [37]. MIP nanoparticles of 60–100 nm diameter were prepared by precipitation polymerization. Under optimized adsorption conditions MIP NPs exhibited a binding capacity of 73.8 μmol g^{-1} and very good selectivity in the presence of other ions. MIP NPs could be reused for at least 20 consecutive times, without any loss of affinity. Furthermore, when tested with well and tap water samples spiked with Cu^{2+}, MIP nanoparticles recovered 95–105% of the ions present. Prasada Rao and coauthors prepared core–shell MIP NPs imprinted with uranyl ions [117]. The authors first prepared functionalized silica cores bearing an amino silane for the immobilization of quinoline-8-ol to improve the specific interaction with the uranyl ions. A polymeric shell was formed by precipitation polymerization, giving rise to MIP NPs of 50–80 nm diameter.

Under optimized rebinding conditions MIP NPs were able to remove selectively 500–1,000 ppb of uranyl ions from water samples, about 25% more than non-imprinted materials. When tested with spiked environmental waters, the recovery performance was 94% for ground water. However, for salt water samples, the recovery was lower (70%). Several groups also investigated the possibility of using MIP NPs with magnetic cores [118]. Wang and coauthors prepared Fe_3O_4 magnetic NPs coated with a silica layer imprinted for estrone [119]. MIP NPs rebound estrone about 3.6-times more effectively and specifically than the corresponding non-imprinted particles. However, the product was not tested in "real" samples to assess its performance as a rapid recovery affinity material. Li and coauthors prepared surface-imprinted magnetic polystyrene nanoparticles for bovine hemoglobin through a multi-stage core–shell polymerization process [120]. It involved the use of 3-aminophenylboronic acid (APBA) as functional and cross-linking monomer. In fact, thanks to its aqueous solubility and the variety of reversible interactions which it establishes with amino acids, APBA is particularly suitable for protein imprinting [121]. Magnetite core nanoparticles were coated with MIP shells created by the polymerization of APBA in the presence of the template. The final size of the coated particles reached a diameter of 480 nm with a 15–20 nm thick MIP film. The core–shell MIP NPs exhibited superparamagnetic properties suitable for facile separation in a magnetic field. Additionally, they showed rapid rebinding kinetics (30–120 min), good specificity and selectivity for the template, as well as a very high adsorption capacity of around 45.5 mg g^{-1}. Such a high capacity is unusual for this kind of material and together with their magnetic properties makes these imprinted nanoparticles very attractive for enrichment of low concentration proteins in proteomics. Another very interesting core–shell approach for preparing magnetic MIP NPs for proteins has been developed by Zhou and coauthors, who imprinted human hemoglobin in a polydopamine (PDA) layer synthesized on the surface of magnetic Fe_3O_4 nanoparticles [122]. The synthesized nanoparticles exhibited strong recognition affinity towards hemoglobin, with a dissociation constant of 18.13 μg mL^{-1}. The binding capacity of MIP NPs was 22.3 μg mg^{-1}. In addition, they had a very good selectivity evaluated against proteins such as myoglobin, horse radish peroxidase, and cytochrome *c*. The use of PDA seems to be particularly suitable for protein imprinting, because it is hydrophilic, biocompatible, and has amino and catechol groups which can help in establishing interactions with the macromolecular template. Moreover, the thickness of the PDA layer can be very easily tuned by changing the polymerization time [123].

3.6 The Future: Biologically Active MIP Nanoparticles

Undoubtedly the most interesting, albeit distant future application for MIP NPs, is the creation of biologically active systems that can be used as drugs, antibody, or enzyme substitutes in vivo. The first example of water-soluble MIP NPs demonstrating biological activity dates back to 2006 when Piletsky and coauthors investigated the possibility of enhancing the photosynthetic reaction using MIPs imprinted with thylakoid D1 protein [124]. For this purpose MIP and non-imprinted (NIP) bulk monoliths were prepared in water and extensively ground to produce MIP NPs. Affinity chromatography on immobilized D1 protein showed that 5–10 kDa MIP NPs had increased binding affinity to this protein as compared to non-imprinted polymers. In addition, MIP nanoparticles were able to increase the photosynthetic activity of chloroplasts. However, the yield of the product synthesized with this method was very low (about 0.2 mg per fraction). Nevertheless, this first work represents a milestone in the production of soluble MIP NPs with biological activity. Haupt and coauthors synthesized water-soluble MIP NPs able to inhibit the enzymatic activity of trypsin using a precipitation polymerization approach [125]. Their method relied on tailor-made "anchoring monomer," methacryloylaminobenzamidine (a polymerizable derivative of the trypsin inhibitor benzamidine), to complex the template with high affinity and locate the synthesis of the MIP NPs at the surface of the enzyme. The calculated inhibition constant (K_i) for the MIP NPs was 79 nM which is much lower than the K_i value for free benzamidine (18.9 μM), which proved the effectiveness of this imprinting strategy. Moreover, MIP NPs demonstrated negligible inhibition activity for enzymes such as chymotrypsin or kallikrein which demonstrated the selectivity of the MIP NPs. Shea and coauthors recently prepared MIP NPs by precipitation polymerization imprinted against the bee venom peptide melittin [39], and applied them as antidotes in living animals [40]. The authors optimized the composition of the polymerization mixture by the creation of a small combinatorial library of acrylamide functional monomers including *N*-isopropylacrylamide and *N,N'*-methylenebisacrylamide (cross-linker). Nanoparticles synthesized under the optimum conditions had a diameter of 50 nm, which is comparable in size to IgM. The dissociation constant (25 pM) was similar to that of natural antibodies for melittin (17 pM). Moreover, only slight cross-reactivity with other proteins was observed. MIP NPs were tested in vitro on fibrosarcoma cells and did not show any toxic effect. The authors injected mice with a lethal dose of melittin, followed 20 s later by an intravenous injection of MIP or non-imprinted NPs. MIP NPs halved the mortality and reduced the toxic effects of melittin. MIP NPs with adsorbed melittin were concentrated in the liver while melittin alone was distributed extensively throughout the body and bloodstream. Although these results are very promising, the toxicological implications of injecting MIP NPs need to be carefully investigated to assess the risks deriving from the use of these nanomaterials prior to their application in humans. In addition, suitable manufacturing protocols should be developed for large-scale manufacturing of MIP NPs, which would most likely rely on affinity separation, as in the case of natural antibodies [126, 127].

4 Conclusions and Outlook

In recent years we have witnessed a growth of activity in the development of alternative affinity materials to antibodies such as aptamers, engineered scaffold proteins, and MIP NPs. Recent developments in the synthesis and applications of

MIP NPs are particularly encouraging. A number of interesting practical applications for such materials were described and discussed in the present review. The process of replacing natural antibodies with their synthetic analogues, however, is hindered by the lack of suitable low costs protocols for large-scale manufacturing of such materials. Following further advances in polymer and synthetic chemistry, as well as in screening tools, this situation is set to change; motivating companies to put more investment into the development of novel synthetic receptors, thus leading to a new generation of superior affinity materials, readily available for routine diagnostic and industrial applications.

References

1. Yan M (2002) Molecularly imprinted polymers as antibody mimics: applications in immunoassays and recent developments. J Clin Ligand Assay 25:234–236
2. RNCOS (2011) Global in vitro diagnostic market analysis. 1–130. Rncos.com, Noida, India
3. Biocompare Surveys and Reports (2009) 2009 Antibody report – Market overview and industry survey – Executive summary. 1–32. Market Research.com, Rockville, USA
4. Espicom (2011) Point of care diagnostics: players, products & future market prospects. 1–396. Espicom, Chichester, UK
5. Hoshino Y, Shea KJ (2011) The evolution of plastic antibodies. J Mater Chem 21:3517–3521
6. Ruigrok VJB, Levisson M, Eppink MHM et al (2011) Alternative affinity tools: more attractive than antibodies? Biochem J 436:1–13
7. Lavignac N, Allender CJ, Brain KR (2004) Current status of molecularly imprinted polymers as alternatives to antibodies in sorbent assays. Anal Chim Acta 510:139–145
8. Shukla AA, Thömmes J (2010) Recent advances in large-scale production of monoclonal antibodies and related proteins. Trends Biotechnol 28:253–261
9. Ansell RJ (2004) Molecularly imprinted polymers in pseudoimmunoassay. J Chromatogr B Anal Technol Biomed Life Sci 804:151–165
10. Roque ACA, Lowe CR, Taipa MÂ (2004) Antibodies and genetically engineered related molecules: production and purification. Biotechnol Prog 20:639–654
11. Steinmeyer DE, McCormick EL (2008) The art of antibody process development. Drug Discov Today 13:613–618
12. Urraca JL, Moreno-Bondi MC, Orellana G et al (2007) Molecularly imprinted polymers as antibody mimics in automated on-line fluorescent competitive assays. Anal Chem 79:4915–4923
13. Haupt K, Dzgoev A, Mosbach K (1998) Assay system for the herbicide 2,4-dichlorophenoxyacetic acid using a molecularly imprinted polymer as an artificial recognition element. Anal Chem 70:628–631
14. Szenczi A, Kardos J, Medgyesi GA et al (2006) The effect of solvent environment on the conformation and stability of human polyclonal IgG in solution. Biologicals 34:5–14
15. Piletsky SA, Turner A (2006) A new generation of chemical sensors based on MIPs. In: Piletsky SA, Turner A (eds) Molecular imprinting of polymers. Eurekah.com/Landes Bioscience, Georgetown
16. Ahrer K, Buchacher A, Iberer G et al (2006) Thermodynamic stability and formation of aggregates of human immunoglobulin G characterised by differential scanning calorimetry and dynamic light scattering. J Biochem Biophys Methods 66:73–86
17. Omersel J, Žager U, Kveder T et al (2010) Alteration of antibody specificity during isolation and storage. J Immunoassay Immunochem 31:45–59

18. Hock B, Rahman M, Rauchalles S et al (1999) Stabilisation of immunoassays and receptor assays. J Mol Catal B Enzym 7:115–124
19. Butler JE (2000) Solid supports in enzyme-linked immunosorbent assay and other solid-phase immunoassays. Methods 22:4–23
20. Paschke M (2006) Phage display systems and their applications. Appl Microbiol Biotechnol 70:2–11
21. Sidhu SS, Koide S (2007) Phage display for engineering and analyzing protein interaction interfaces. Curr Opin Struct Biol 17:481–487
22. Grönwall C, Ståhl S (2009) Engineered affinity proteins – generation and applications. J Biotechnol 140:254–269
23. Hey T, Fiedler E, Rudolph R et al (2005) Artificial, non-antibody binding proteins for pharmaceutical and industrial applications. Trends Biotechnol 23:514–522
24. Binz HK, Amstutz P, Plückthun A (2005) Engineering novel binding proteins from nonimmunoglobulin domains. Nat Biotechnol 23:1257–1268
25. Stoltenburg R, Reinemann C, Strehlitz B et al (2007) SELEX-A (r)evolutionary method to generate high-affinity nucleic acid ligands. Biomol Eng 24:381–403
26. Danielsson B (2007) Artificial receptors. Adv Biochem Eng Biotechnol 109:97–122
27. Mairal T, Cengiz Özalp V, Lozano Sánchez P et al (2008) Aptamers: molecular tools for analytical applications. Anal Bioanal Chem 390:989–1007
28. Missailidis S, Hardy A (2009) Aptamers as inhibitors of target proteins. Expert Opin Ther Pat 19:1073–1082
29. Mayes AG, Whitcombe MJ (2005) Synthetic strategies for the generation of molecularly imprinted organic polymers. Adv Drug Deliv Rev 57:1742–1778
30. Alexander C, Andersson HS, Andersson LI et al (2006) Molecular imprinting science and technology: a survey of the literature for the years up to and including 2003. J Mol Recognit 19:106–180
31. Svenson J, Nicholls IA (2001) On the thermal and chemical stability of molecularly imprinted polymers. Anal Chim Acta 435:19–24
32. Piletsky SA, Piletska EV, Karim K et al (2002) Polymer cookery: influence of polymerization conditions on the performance of molecularly imprinted polymers. Macromolecules 35:7499–7504
33. Piletsky SA, Guerreiro A, Piletska EV et al (2004) Polymer cookery. 2. Influence of polymerization pressure and polymer swelling on the performance of molecularly imprinted polymers. Macromolecules 37:5018–5022
34. Piletsky SA, Mijangos I, Guerreiro A et al (2005) Polymer cookery: influence of polymerization time and different initiation conditions on performance of molecularly imprinted polymers. Macromolecules 38:1410–1414
35. Mijangos I, Navarro-Villoslada F, Guerreiro A et al (2006) Influence of initiator and different polymerisation conditions on performance of molecularly imprinted polymers. Biosens Bioelectron 22:381–387
36. Piletska EV, Guerreiro AR, Whitcombe MJ et al (2009) Influence of the polymerization conditions on the performance of molecularly imprinted polymers. Macromolecules 42:4921–4928
37. Shamsipur M, Besharati-Seidani A, Fasihi J et al (2010) Synthesis and characterization of novel ion-imprinted polymeric nanoparticles for very fast and highly selective recognition of copper(II) ions. Talanta 83:674–681
38. Diltemiz SE, Say R, Büyüktiryaki S et al (2008) Quantum dot nanocrystals having guanosine imprinted nanoshell for DNA recognition. Talanta 75:890–896
39. Hoshino Y, Kodama T, Okahata Y et al (2008) Peptide imprinted polymer nanoparticles: a plastic antibody. J Am Chem Soc 130:15242–15243
40. Hoshino Y, Koide H, Urakami T et al (2010) Recognition, neutralization, and clearance of target peptides in the bloodstream of living mice by molecularly imprinted polymer nanoparticles: a plastic antibody. J Am Chem Soc 132:6644–6645
41. Cirillo G, Iemma F, Puoci F et al (2009) Imprinted hydrophilic nanospheres as drug delivery systems for 5-fluorouracil sustained release. J Drug Target 17:72–77

42. Ciardelli G, Cioni B, Cristallini C et al (2004) Acrylic polymeric nanospheres for the release and recognition of molecules of clinical interest. Biosens Bioelectron 20:1083–1090
43. Cunliffe D, Kirby A, Alexander C (2005) Molecularly imprinted drug delivery systems. Adv Drug Deliv Rev 57:1836–1853
44. Jenik M, Seifner A, Krassnig S et al (2009) Sensors for bioanalytes by imprinting-polymers mimicking both biological receptors and the corresponding bioparticles. Biosens Bioelectron 25:9–14
45. mipdatabase.com http://www.mipdatabase.com/all_items.php. Accessed 23 May 2012
46. Brüggemann O, Haupt K, Ye L et al (2000) New configurations and applications of molecularly imprinted polymers. J Chromatogr A 889:15–24
47. Pérez-Moral N, Mayes AG (2006) MIP formats for analytical applications. In: Piletsky SA, Turner A (eds) Molecular imprinting of polymers, 1st edn. Eurekah.com/Landes Bioscience, Georgetown
48. Kandimalla VB, Ju H (2004) Molecular imprinting: a dynamic technique for diverse applications in analytical chemistry. Anal Bioanal Chem 380:587–605
49. Lorenzo RA, Carro AM, Alvarez-Lorenzo C et al (2011) To remove or not to remove? The challenge of extracting the template to make the cavities available in molecularly imprinted polymers (MIPs). Int J Mol Sci 12:4327–4347
50. Tokonami S, Shiigi H, Nagaoka T (2009) Review: micro- and nanosized molecularly imprinted polymers for high-throughput analytical applications. Anal Chim Acta 641:7–13
51. Gao D, Zhang Z, Wu M et al (2007) A surface functional monomer-directing strategy for highly dense imprinting of TNT at surface of silica nanoparticles. J Am Chem Soc 129:7859–7866
52. Tan CJ, Tong YW (2007) Molecularly imprinted beads by surface imprinting. Anal Bioanal Chem 389:369–376
53. Ge Y, Turner APF (2009) Molecularly imprinted sorbent assays: recent developments and applications. Chem Eur J 15:8100–8107
54. Poma A, Turner APF, Piletsky SA (2010) Advances in the manufacture of MIP nanoparticles. Trends Biotechnol 28:629–637
55. Carboni D, Flavin K, Servant A et al (2008) The first example of molecularly imprinted nanogels with aldolase type I activity. Chem Eur J 14:7059–7065
56. Wulff G, Chong B-O, Kolb U (2006) Soluble single-molecule nanogels of controlled structure as a matrix for efficient artificial enzymes. Angew Chem Int Ed 45:2955–2958
57. Yoshimatsu K, Reimhult K, Krozer A et al (2007) Uniform molecularly imprinted microspheres and nanoparticles prepared by precipitation polymerization: the control of particle size suitable for different analytical applications. Anal Chim Acta 584:112–121
58. Schweitz L, Spegel P, Nilsson S (2000) Molecularly imprinted microparticles for capillary electrochromatographic enantiomer separation of propranolol. Analyst 125:1899–1901
59. Spégel P, Schweitz L, Nilsson S (2003) Selectivity toward multiple predetermined targets in nanoparticle capillary electrochromatography. Anal Chem 75:6608–6613
60. Nilsson C, Birnbaum S, Nilsson S (2007) Use of nanoparticles in capillary and microchip electrochromatography. J Chromatogr A 1168:212–224
61. Priego-Capote F, Ye L, Shakil S et al (2008) Monoclonal behavior of molecularly imprinted polymer nanoparticles in capillary electrochromatography. Anal Chem 80:2881–2887
62. Piletsky SA, Turner APF (2008) Imprinted polymers and their application in optical sensors. In: Ligler FS, Rowe Taitt CA (eds) Optical biosensors: today and tomorrow, 2nd edn. Elsevier Science, Hungary
63. Reimhult K, Yoshimatsu K, Risveden K et al (2008) Characterization of QCM sensor surfaces coated with molecularly imprinted nanoparticles. Biosens Bioelectron 23:1908–1914
64. Schirhagl R, Lieberzeit PA, Dickert FL (2010) Chemosensors for viruses based on artificial immunoglobulin copies. Adv Mater 22:2078–2081
65. Ye L, Cormack PAG, Mosbach K (1999) Molecularly imprinted monodisperse microspheres for competitive radioassay. Anal Commun 36:35–38

66. Jantarat C, Tangthong N, Songkro S et al (2008) *S*-Propranolol imprinted polymer nanoparticle-on-microsphere composite porous cellulose membrane for the enantioselectively controlled delivery of racemic propranolol. Int J Pharm 349:212–225
67. Barrett AM, Cullum VA (1968) Lack of interaction between propranolol and mebanazine. J Pharm Pharmacol 20:911–915
68. Mayes AG, Mosbach K (1996) Molecularly imprinted polymer beads: suspension polymerization using a liquid perfluorocarbon as the dispersing phase. Anal Chem 68:3769–3774
69. Suedee R, Jantarat C, Lindner W et al (2010) Development of a pH-responsive drug delivery system for enantioselective-controlled delivery of racemic drugs. J Control Release 142:122–131
70. Johnson KR, Young KK, Weimin F (1999) Antagonistic interplay between antimitotic and G1-S arresting agents observed in experimental combination therapy. Clin Cancer Res 5:2559–2565
71. Kan X, Liu T, Zhou H et al (2010) Molecular imprinting polymer electrosensor based on gold nanoparticles for theophylline recognition and determination. Microchim Acta 171:423–429
72. Giri S, Trewyn BG, Stellmaker MP et al (2005) Stimuli-responsive controlled-release delivery system based on mesoporous silica nanorods capped with magnetic nanoparticles. Angew Chem Int Ed 44:5038–5044
73. Myers P, Bartle KD (2001) Capillary electrochromatography. The Royal Society of Chemistry, London
74. Nilsson C, Nilsson S (2006) Nanoparticle-based pseudostationary phases in capillary electrochromatography. Electrophoresis 27:76–83
75. Behnke B, Johansson J, Bayer E et al (2000) Fluorescence imaging of frit effects in capillary separations. Electrophoresis 21:3102–3108
76. Spégel P, Nilsson S (2002) A new approach to capillary electrochromatography: disposable molecularly imprinted nanoparticles. Am Lab 34:29–33
77. Nilsson C, Viberg P, Spégel P et al (2006) Nanoparticle-based continuous full filling capillary electrochromatography/electrospray ionization-mass spectrometry for separation of neutral compounds. Anal Chem 78:6088–6095
78. Göttlicher B, Bächmann K (1997) Application of particles as pseudo-stationary phases in electrokinetic chromatography. J Chromatogr A 780:63–73
79. Nilsson J, Spégel P, Nilsson S (2004) Molecularly imprinted polymer formats for capillary electrochromatography. J Chromatogr B Anal Technol Biomed Life Sci 804:3–12
80. Qu P, Lei J, Zhang L et al (2010) Molecularly imprinted magnetic nanoparticles as tunable stationary phase located in microfluidic channel for enantioseparation. J Chromatogr A 1217:6115–6121
81. Hayakawa I, Atarashi S, Yokohama S et al (1986) Synthesis and antibacterial activities of optically active ofloxacin. Antimicrob Agents Chemother 29:163–164
82. Biffis A, Graham NB, Siedlaczek G et al (2001) The synthesis, characterization and molecular recognition properties of imprinted microgels. Macromol Chem Phys 202:163–171
83. Markowitz MA, Kust PR, Deng G et al (2000) Catalytic silica particles via template-directed molecular imprinting. Langmuir 16:1759–1765
84. Maddock SC, Pasetto P, Resmini M (2004) Novel imprinted soluble microgels with hydrolytic catalytic activity. Chem Commun 10:536–537
85. Pasetto P, Maddock SC, Resmini M (2005) Synthesis and characterisation of molecularly imprinted catalytic microgels for carbonate hydrolysis. Anal Chim Acta 542:66–75
86. Pasetto P, Flavin K, Resmini M (2009) Simple spectroscopic method for titration of binding sites in molecularly imprinted nanogels with hydrolase activity. Biosens Bioelectron 25:572–578
87. Chen Z, Hua Z, Wang J et al (2007) Molecularly imprinted soluble nanogels as a peroxidase-like catalyst in the oxidation reaction of homovanillic acid under aqueous conditions. Appl Catal A Gen 328:252–258

88. Chen Z, Xu L, Liang Y et al (2010) pH-sensitive water-soluble nanospheric imprinted hydrogels prepared as horseradish peroxidase mimetic enzymes. Adv Mater 22:1488–1492
89. Huang X, Liu Y, Liang K et al (2008) Construction of the active site of glutathione peroxidase on polymer-based nanoparticles. Biomacromolecules 9:1467–1473
90. Servant A, Haupt K, Resmini M (2011) Tuning molecular recognition in water-soluble nanogels with enzyme-like activity for the Kemp elimination. Chem Eur J 17:11052–11059
91. Yaqub S, Latif U, Dickert FL (2011) Plastic antibodies as chemical sensor material for atrazine detection. Sens Actuators B Chem 160:227–233
92. Bompart M, Gheber LA, De Wilde Y et al (2009) Direct detection of analyte binding to single molecularly imprinted polymer particles by confocal Raman spectroscopy. Biosens Bioelectron 25:568–571
93. Bompart M, De Wilde Y, Haupt K (2010) Chemical nanosensors based on composite molecularly imprinted polymer particles and surface-enhanced Raman scattering. Adv Mater 22:2343–2348
94. Sener G, Ozgur E, Yilmaz E et al (2010) Quartz crystal microbalance based nanosensor for lysozyme detection with lysozyme imprinted nanoparticles. Biosens Bioelectron 26:815–821
95. Sener G, Uzun L, Say R et al (2011) Use of molecular imprinted nanoparticles as biorecognition element on surface plasmon resonance sensor. Sens Actuators B Chem 160:791–799
96. Haupt K, Mayes AG, Mosbach K (1998) Herbicide assay using an imprinted polymer-based system analogous to competitive fluoroimmunoassays. Anal Chem 70:3936–3939
97. Haupt K (1999) Molecularly imprinted sorbent assays and the use of non-related probes. React Funct Polym 41:125–131
98. Ye L, Cormack PAG, Mosbach K et al (2001) Molecular imprinting on microgel spheres. Anal Chim Acta 435:187–196
99. Ye L, Mosbach K (2001) Polymers recognizing biomolecules based on a combination of molecular imprinting and proximity scintillation: a new sensor concept. J Am Chem Soc 123:2901–2902
100. Ye L, Surugiu I, Haupt K (2002) Scintillation proximity assay using molecularly imprinted microspheres. Anal Chem 74:959–964
101. Surugiu I, Ye L, Yilmaz E et al (2000) An enzyme-linked molecularly imprinted sorbent assay. Analyst 125:13–16
102. Surugiu I, Danielsson B, Ye L et al (2001) Chemiluminescence imaging ELISA using an imprinted polymer as the recognition element instead of an antibody. Anal Chem 73:487–491
103. Pérez N, Whitcombe MJ, Vulfson EN (2001) Surface imprinting of cholesterol on submicrometer core-shell emulsion particles. Macromolecules 34:830–836
104. Pérez-Moral N, Mayes AG (2004) Noncovalent imprinting in the shell of core-shell nanoparticles. Langmuir 20:3775–3779
105. Pérez-Moral N, Mayes AG (2007) Molecularly imprinted multi-layer core-shell nanoparticles - a surface grafting approach. Macromol Rapid Commun 28:2170–2175
106. Ivanova-Mitseva PK, Guerreiro A, Piletska EV et al (2012) Cubic molecularly imprinted polymer nanoparticles with a fluorescent core. Angew Chem Int Ed 51:5196–5199
107. Li Y, Dong C, Chu J et al (2011) Surface molecular imprinting onto fluorescein-coated magnetic nanoparticles via reversible addition fragmentation chain transfer polymerization: a facile three-in-one system for recognition and separation of endocrine disrupting chemicals. Nanoscale 3:280–287
108. Lehmann M, Brunner H, Tovar GEM (2002) Selective separations and hydrodynamic studies: a new approaoch using molecularly imprinted nanosphere composite membranes. Desalination 149:315–321
109. Silvestri D, Borrelli C, Giusti P et al (2005) Polymeric devices containing imprinted nanospheres: a novel approach to improve recognition in water for clinical uses. Anal Chim Acta 542:3–13

110. Ciardelli G, Borrelli C, Silvestri D et al (2006) Supported imprinted nanospheres for the selective recognition of cholesterol. Biosens Bioelectron 21:2329–2338
111. Chronakis IS, Jakob A, Hagström B et al (2006) Encapsulation and selective recognition of molecularly imprinted theophylline and 17β-estradiol nanoparticles within electrospun polymer nanofibers. Langmuir 22:8960–8965
112. Yoshimatsu K, Ye L, Lindberg J et al (2008) Selective molecular adsorption using electrospun nanofiber affinity membranes. Biosens Bioelectron 23:1208–1215
113. Piperno S, Tse Sum Bui B, Haupt K et al (2011) Immobilization of molecularly imprinted polymer nanoparticles in electrospun poly(vinyl alcohol) nanofibers. Langmuir 27:1547–1550
114. Zhu R, Zhao W, Zhai M et al (2010) Molecularly imprinted layer-coated silica nanoparticles for selective solid-phase extraction of bisphenol A from chemical cleansing and cosmetics samples. Anal Chim Acta 658:209–216
115. Peng Y, Xie Y, Luo J et al (2010) Molecularly imprinted polymer layer-coated silica nanoparticles toward dispersive solid-phase extraction of trace sulfonylurea herbicides from soil and crop samples. Anal Chim Acta 674:190–200
116. Gao R, Zhang J, He X et al (2010) Selective extraction of sulfonamides from food by use of silica-coated molecularly imprinted polymer nanospheres. Anal Bioanal Chem 398:451–461
117. Milja TE, Prathish KP, Prasada Rao T (2011) Synthesis of surface imprinted nanospheres for selective removal of uranium from simulants of Sambhar salt lake and ground water. J Hazard Mater 188:384–390
118. Pérez N, Whitcombe MJ, Vulfson EN (2000) Molecularly imprinted nanoparticles prepared by core-shell emulsion polymerization. J Appl Polym Sci 77:1851–1859
119. Wang X, Wang L, He X et al (2009) A molecularly imprinted polymer-coated nanocomposite of magnetic nanoparticles for estrone recognition. Talanta 78:327–332
120. Li L, He X, Chen L et al (2009) Preparation of novel bovine hemoglobin surface-imprinted polystyrene nanoparticles with magnetic susceptibility. Sci China Ser B Chem 52:1402–1411
121. Bossi A, Piletsky SA, Piletska EV et al (2001) Surface-grafted molecularly imprinted polymers for protein recognition. Anal Chem 73:5281–5286
122. Zhou W-H, Lu C-H, Guo X-C et al (2010) Mussel-inspired molecularly imprinted polymer coating superparamagnetic nanoparticles for protein recognition. J Mater Chem 20:880–883
123. Lee H, Dellatore SM, Miller WM et al (2007) Mussel-inspired surface chemistry for multifunctional coatings. Science 318:426–430
124. Piletsky SA, Piletska EV, Sergeyeva TA et al (2006) Synthesis of biologically active molecules by imprinting polymerisation. Biopolym Cell 22:63–67
125. Cutivet A, Schembri C, Kovensky J et al (2009) Molecularly imprinted microgels as enzyme inhibitors. J Am Chem Soc 131:14699–14702
126. Guerreiro AR, Chianella I, Piletska E et al (2009) Selection of imprinted nanoparticles by affinity chromatography. Biosens Bioelectron 24:2740–2743
127. Hoshino Y, Haberaecker WW III, Kodama T et al (2010) Affinity purification of multifunctional polymer nanoparticles. J Am Chem Soc 132:13648–13650

Computational Approaches in the Design of Synthetic Receptors

Sreenath Subrahmanyam, Kal Karim, and Sergey A. Piletsky

Abstract Artificial receptors have been employed in molecular recognition for a variety of biological applications. They have been used as materials for sensors, affinity separation, solid-phase extraction, and for research into biomolecular interaction. There have been a number of publications relating to the application of molecular modeling in the characterization of their affinity and selectivity; there are very few publications that discuss the application of molecular modeling to the computational design of artificial receptors. This chapter discusses recent successes in the use of computational design for the development of artificial receptors, and touches upon possible future applications, further emphasizing an exciting group of synthetic receptors—molecularly imprinted polymers.

Keywords Artificial receptors, Molecular dynamics, Molecular imprinting, Molecular modeling, Sensors

Contents

S. Subrahmanyam (✉), K. Karim, and S.A. Piletsky
Cranfield Biotechnology Centre, Cranfield University, Bedfordshire MK43 0AL, UK
e-mail: sri@cranfield.ac.uk

S.A. Piletsky and M.J. Whitcombe (eds.), *Designing Receptors for the Next Generation of Biosensors*, Springer Series on Chemical Sensors and Biosensors (2013) 12: 131–166
DOI 10.1007/5346_2012_22,
Published online: 28 July 2012

Acronyms and Further Descriptions

"ab initio"	Latin term meaning "from the beginning"
ΔE	Binding energy
2-VP	2-Vinyl pyridine
4-VP	4-Vinyl pyridine
AA	Acrylic acid
Accelrys DS viewer	Modeling and simulation tools for drug discovery
Agile molecule	A three-dimensional molecular viewer which shows molecular models and provides geometry editing capabilities
AHLs	3-Oxo-C6-acyl-homoserine lactone
ALM	Allylamine
AMBER	Assisted model building with energy refinement refers to a MM force field for the simulation of biomolecules and a package of molecular simulation programs
AMPSA	2-Acrylamido-2-methyl-1-propanesulfonic acid
B3LYP	Becke 3-parameter, Lee, Yang and Parr, a density functional method
Bite-and-Switch	"Bite-and-Switch" is defined in terms of polymer's ability to bind the template (bite) and generate the signal (switch)
BLAs	β-Lactam antibiotics
B-Me	Biotin methyl ester
CAChe MOPAC	A general-purpose semiempirical molecular orbital package for the study of chemical structures and reactions
Cerius	A software to visualize structures, predict the properties and behavior of chemical systems refine structural models (Molecular Simulations Inc.)
Chem 3D	A software that provides visualization and display of molecular surfaces, orbitals, electrostatic potentials, charge densities, and spin densities (http://www.cambridgesoft.com/)
DFT	Density functional theory
Dielectric constant	A measure of the ability of a material to store a charge from an applied electromagnetic field and then transmit that energy
DMAEM	Dimethyl aminoethyl methacrylate
DOCK	Program that addresses the problem of "docking" molecules to each other. It explores ways in which two molecules, such as a drug and an enzyme or protein receptor, might fit together

DVB	Divinylbenzene
EGDMA	Ethylene glycol dimethacrylate
ELISA	Enzyme-linked immunosorbent assay
GAMESS	General Atomic and Molecular Electronic Structure System: a general ab initio quantum chemistry package that can compute wave functions ranging from RHF, ROHF, UHF, GVB, and MCSCF
Gibbs free energy	The chemical potential that is minimized when a system reaches equilibrium at constant pressure and temperature
GRID	A computational procedure for detecting energetically favorable binding sites on molecules of known structure. The energies are calculated as the electrostatic, hydrogen bond and Lennard Jones interactions of a specific probe group with the target structure (Peter Goodford, Molecular Discovery Ltd)
Gaussian	"Ab initio" electronic structure program that originated in the research group of People at Carnegie-Melon. Calculate structures, reaction transition states, and molecular properties (http://www.gaussian.com)
Gaussview	Graphical user interface (GUI) designed for use with Gaussian for easier computational analysis
HEMA	Hydroxyethyl methacrylate
His	Histidine
HOOK	Linker search for fragments placed by MCSS
HO-PCBs	Hydroxy polychlorinated biphenyls
HPLC	High performance liquid chromatography
HVA	Homovanillic acid
HyperChem	A molecular modeling package for windows
IA	Itaconic acid
k'	Retention factor
Leapfrog™	A component of the SYBYL™ software package (Tripos) and is a second-generation de novo drug discovery program that allows for the evaluation of potential ligand structures
LEGEND	Atom-based, stochastic search
Ligbuilder	General-purpose structure-based drug design program
LUDI	Fragment-based, combinatorial search
MAA	Methacrylic acid
Materials Studio	Software for modeling/simulation of crystal structure, polymer properties, structure–activity relationships (http://www.accelrys.com/products/mstudio)
MBAA	*N,N'*-Methylenebisacrylamide
MD	Molecular dynamics
MIC	Molecularly imprinted catalysis
MIP	Molecularly imprinted polymer

MM	Molecular mechanics
MMA	Methylmethacrylate
MMFF94	A tool for conformational searching of highly flexible molecules
MOE	Molecular Operating Environment is a software system designed for computational chemistry
Monte Carlo	An algorithm which computes based on repeated random sampling to arrive at results
MOPAC AM1	AM1 is used in the electronic part of the calculation to obtain molecular orbitals, the heat of formation and its derivative with respect to molecular geometry. MOPAC calculates the vibrational spectra, thermodynamic quantities, isotopic substitution effects and force constants for molecules, radicals, ions, and polymers
NAM	A scalable molecular dynamics code that can be run on the Beowulf parallel PC cluster for molecular dynamics simulations on selected molecular systems
NIP	Non-imprinted polymer
NVT-MD	Molecular dynamics performed under constant number of atom, volume, and temperature ensemble
OPA	*o*-Phthalic dialdehyde
OscailX	Molecular modeling software, National University of Ireland (http://www.ucg.ie/cryst/software.htm)
OTA	Ochratoxin A
PCFF	Polymer consistent force field
PCM	Polarizable continuum model
PCModel	Structure building, manipulation, and display program which uses molecular mechanics and semiempirical quantum mechanics to optimize geometry. Available on PC (DOS and Windows), Macintosh, SGI, Sun and IBM/RS computers (Kevin Gilbert, Serena Software)
PenG	Penicillin G
pK_a	Ionization constant
PRO-LIGAND	Fragment-based search
*Q*m	Mean absolute atomic charge
QM	Quantum mechanics
RECON	An algorithm for the rapid reconstruction of molecular charge densities and charge density-based electronic properties of molecules, using atomic charge density fragments precomputed from ab initio wave functions. The method is based on Bader's quantum theory of atoms in molecules
RESP	Atomic partial charge assignment protocol
SDIM	Sulfadimethoxine

SHAKE	A molecular dynamics algorithm
Simulated annealing	A method that simulates the physical process of annealing, where a material is heated and then cooled leading to optimization
SM_2	Sulfadimidine
SMZ	Sulfamethazine
SPROUT	Fragment-based, sequential growth, combinatorial search
SYBYL™	A molecular modeling and visualization package permitting construction, editing, and visualization tools for both large and small molecules (www.tripos.com)
T:M:X ratio	Template monomer cross-linker ratio
TAE	Transferable atom equivalent
TFMAA	2-(Trifluoromethyl) acrylic acid
THO	Theophylline
TQT1	ToxiQuant T1 System
UAHF	United atom Hartree–Fock
van der Waals	Weak intermolecular forces that act between stable molecules
VI	1-Vinylimidazole
VMD	Visual molecular dynamics

1 Introduction

Natural receptors are generally large protein molecules that form three-dimensional structures by highly specific intramolecular interactions. While the recognition sites offer a precise configuration and exhibit very efficient recognition processes, this specific recognition is achieved at the expense of having complex and fragile structures with high molecular weight. An alternative to this is the choice of artificial receptors that incorporate a combination of medium-sized organic building blocks to which functional groups for molecular recognition can be attached. Rational design of artificial receptors, which possess very high affinity and selectivity, is currently one of the most researched topics in molecular recognition.

Several artificial receptors such as crown ethers, cyclodextrins, cyclophanes, and calixarenes find applications in molecular recognition processes. Further, there have been several research publications that focus on computational design and analysis of recognition properties of artificial receptors such as cyclodextrins [1–5], dendrimers [6–10], crown ethers [11, 12], and calixarenes [13–15]. Typically computational approaches were previously used for the design of small molecules [16–35]. The computational design of supramolecular synthetic receptors (analogues of natural protein receptors) still remains very challenging. The only class of supramolecular synthetic receptors which has been designed using computational approach is molecularly imprinted polymers (MIPs). This review discusses recent developments with specific reference to modern molecular modeling approaches that have been employed to design MIPs.

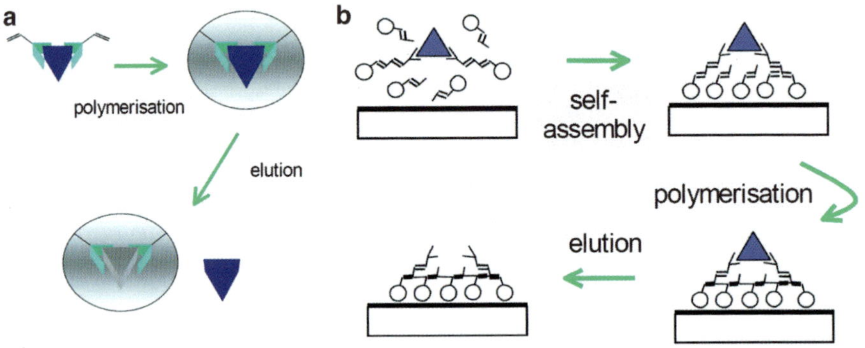

Fig. 1 Scheme of: (**a**) three-dimensional and (**b**) two-dimensional molecular imprinting (courtesy of VTT) [37]

Molecular imprinting can be defined as the process of template-induced formation of specific recognition sites (binding or catalytic) in a material where the template directs the positioning and orientation of the material's structural components by a self-assembling mechanism [36] (Fig. 1). The material itself could be oligomeric (a typical example is the DNA replication process), polymeric (organic MIPs and inorganic imprinted silica gels), or two-dimensional surface assembly (grafted monolayer).

MIPs have several advantages when compared to other synthetic receptors [38]:

1. High affinity and selectivity, which are similar to those of natural receptors
2. Very high stability, which is superior to that of natural biomolecules
3. Simplicity of their preparation and the ease of adaptation to different practical applications

A wide range of chemical compounds have been imprinted successfully, ranging from small molecules [39–41] to large proteins and cells [42]. MIPs have been developed for a variety of applications including chromatography [43, 44], solid-phase extraction (SPE) [45, 46], enzyme-like catalysis [47], sensor technology [43, 48, 49], biomimetic sensors [50–52], and immunoassays [53–55]. MIPs are robust, inexpensive, and, in many cases, possess affinity and specificity that are suitable for industrial applications. The high specificity and stability of MIPs render them as promising alternatives to enzymes, antibodies, and natural receptors for use in sensor technology [36, 47].

There have been several attempts aimed at the development of a generic procedure for MIP preparation as mentioned below; however, the method that has been in prime focus in recent years is computational design:

1. Rational approaches involving combinatorial methods: an array of MIPs was prepared that could be analyzed in situ by binding assays [56–60]
2. Use of a virtual library of functional monomers to assign and screen against the target template molecule [61–64]

3. Rational approaches that involve computation of total energies (*E*), energy differences (DE), and distances (*d*) of closest approach between the monomers and template using molecular dynamics [65]
4. The use of density functional theory (DFT) method to calculate the binding energy ΔE between a template and monomers as a measure of their interaction that facilitates the selection of the monomers [66, 67]
5. Rational design that involve conformation of template–functional monomer complexes employing semiempirical methods [68–73]
6. Chemometric approaches to optimizing monomer, template, and cross-linker ratios [74]
7. Predicting template:monomer complexes using neural network methods [75, 76]

The following sections discuss each of the above-mentioned computational methods for the rational design of MIPs. Table 1 summarizes the different computational procedures adopted for the rational design of MIPs for a variety of templates, analysis of polymer properties, and performance of the MIPs.

2 Computational Methods for Rational Design of MIPs

2.1 Rational Approaches That Involve Molecular Mechanics

One of the most established rational approaches in the design of imprinted polymers is combinatorial synthesis/screening [56, 59]. However the combinatorial approach has its limitations, such that a simple two-component system utilizing 100 monomers would require the preparation of several thousand polymers and even then would not take into account the possible different ratios of monomer mixtures. One potential solution to the problem of rational design of polymer lies in molecular modeling and performing thermodynamic computations using a patented protocol developed at Cranfield University [61]. Variations of this protocol are in use nowadays in many laboratories around the world.

When there is the requirement to perform structural analysis in large molecular systems, comprising hundreds of molecules, there is an inherently heavy demand on computational time and resources. In these cases, molecular mechanics (MM) is used. MM refers to a system that can be used for qualitative descriptions that include only potential energy which is essentially devoid of any quantum mechanical calculations. To facilitate calculations, MM considers atoms as balls of certain radius and the bonds between them as string. The exact values of atom sizes and bond geometry and strength originate from empirical data collected from X-ray crystallography and NMR experiments. Several MM software packages exist for a variety of general and specific applications. Some of the most widely used are AMBER, MOE, RasMol, QMol, Raster 3D, and AGM Build.

However, a major problem associated with the computational design of imprinted polymers is the difficulty in performing detailed thermodynamic

Table. 1 Computational procedures for the rational design of MIPs

Process	Application	Targets and references
Molecular mechanics (MM)		
LEAPFROG algorithm used to screen virtual library of functional monomers; simulation of complex formation done using simulated annealing (SYBYL)	Optimization of polymer composition; selection of best monomers leading to MIPs with high binding capacity for the template leading to synthesis of MIP	Creatinine [63]; microcystin-LR [64]; ochratoxin-A [64, 77]; abacavir [78]; biotin [79]; carbaryl [80]; cocaine, deoxyephedrine, methadone, morphine [81]; triazines [82, 83]; tylosin [84]; acyl-homoserine lactone [85]; aflatoxin-B1 [86]; nonylphenol [87]; cholic acid [88]; deoxynivalenol [89]; amiodarone [90]
MM calculations with AMBER 7 Docking software to map the energetic interactions	Prediction of binding affinity and selectivity	Theophylline and its derivative [91]
MM and MD calculations performed for the template–monomer complex using HyperChem	Analysis of the complex formation between template and monomer and possible structure of imprinting sites	L- or D-tryptophan methyl ester [92]; *N*-α-*t*-boc-L-histidine [93]
Interactions between template and monomer in MIPs analyzed using Amber MM method	Prediction of the ratios of template, functional monomer and solvent	Caffeine and theophylline [68]; ibuprofen [94]
3D chemical structures of the labeled BLAs mechanics using MOPAC, AM1 force field, and Chem3D Ultra 7.0 software	Analysis of recognition of the fluorescent analogues of template by the MIPs	Penicillin G [72]
Energy minimization by MM/QM to estimate enthalpies of formation, bond orders, intermolecular distances and ionization potentials using PCModel for windows	Analysis of enthalpies of complex formation between functional monomer and template	Dibenzothiophene sulfone [185]
Calculations of interaction energies using PCMODEL 8.0, MMFF94 and force field	Selection of monomers for synthesis of MIPs based on the interaction energies	Paracetamol [95]
pK_a calculations of template by Gaussian03W in vacuum	Correlation between molecular volumes, pK_a of templates and the retention factors	Hydroxy polychlorinated biphenyls [96]
Molecular dynamics (MD)		
Molecular models of template and monomers optimized by HyperChem 501	Selection of best monomers for MIPs with high binding capacity for the template	*N*,*O*-dibenzylcarbamate [97]

(continued)

Table. 1 (continued)

Process	Application	Targets and references
Intermolecular interactions for MIPs using *Cerius* version 410 and Materials Studio	Prediction of monomers specific for the template	Theophylline (derivatives) [98]; chemical warfare agents [65]
Intermolecular Monte Carlo analysis	Analysis of monomer:template complex	Biotin [57]
Virtual library was first created and then MM and QM were performed	Three best monomers selected using MM and QM to select the optimum monomer and solvent	Acetochlor [99]
Interactions with sulfadimidine were calculated using GROMACS 3.3 and Gaussian 03	Prediction of monomers specific for the template	Sulfadimidine [100]
Polymer topology studied using PRODRG and MD simulations using GROMACS 3.2	Predict interaction energy differences and identify active binding sites	Dimethoate [101]
GROMACS 3.1 was used for MD simulations	Screening of ligands that bind to template	Morphine [102]
Computations for the TrpOMe-n-DDP-complex using MD software—HyperChem 5.1	Chiral discrimination via a surface imprinting; catalysis of a hydrolysis reaction	Tryptophan methyl ester [92]; catalysis of hydrolysis [103]
NAMD molecular dynamics program	Prediction of behavior of MIP	Prediction of behavior MIP [104]
MD simulations using the AMBER 8	Imprinting and bulk effects from pre-polymerization mixtures	Dichlorophenoxyacetic acid [105]; naproxen [106]
Quantum mechanics (QM)		
HyperChemPro 60 to calculate low energy confirmations and electronic distributions	Analysis of the effects of the electric charge distribution and of the size of the molecules on the retention mechanism in SPE	Terbutylazine and ametryn [70, 107]
Energy minimization using MOPAC and WebLab ViewerLite	Analysis of interaction of fluorescent monomer with carboxamidrazone substrate	*N*1-benzylidene pyridine-2-carboxamidrazones [73]
Virtual library of intermediates using Chem3D Pro/ MOPAC	Optimization of monomer for MIP	Transesterification [71]
Electronic energies calculated by DFT using Gaussian 98	Choice of the best functional monomer and solvent	Homovanillic acid [108]
Binding energy of monomer/ template Gaussian 98	Screening of functional monomers for MIPs	Theophylline and its derivatives [67]

(continued)

Table. 1 (continued)

Process	Application	Targets and references
Gaussian 03/B3LYP used in monomer/template interaction	Influence of porogens on MIP performance	Nicotinamide [66]
DFT method with the hybrid B3LYP exchange-correlation	Interaction between template and monomers	Harmane [109]
Ab initio DFT calculations, methods	Interaction between template and functional monomers	TNT[110]
Quantum chemical methods		
Semiempirical AM1 method—MO calculations using CAChe and MOPAC.	Calculation of a complex between (*S*)-nilvadipine and 4-vinyl pyridine	(*S*)-nilvadipine [111]
Semiempirical AM1 method	Composition, binding energies calculated between template and monomers for MIP design	Cocaine[112]; *N,O* dibenzylcarbamate [97]; theophylline [67]; Quinine, arginine, ornithine, lysine citrulline [113]
Semiempirical AM1 method	Molecular geometries of buffers optimized for template	Metformin [114]
Optimization by Gaussian 94 quantum software and MP2/6-31G//HF/6-31G, PM3 methods, Gaussian 03, and B3LYP	Relationship for binding energies of complexes of templates and MMA correlated with retention times and imprinting factors	Picolinamide, nicotinamide, *iso*-nicotinamide [115]
B3LYP/6-311G method adopted based on geometries of the structures and Mulliken charges	Evaluation of the ligand recognition mechanisms using and Mulliken charges	*N*-(4-isopropylphenyl)-*N′*-butyleneurea [116]
Emperical PM3 method	Optimization of a monomer: template	Trichlorophenol [117]; naproxen [118]; picolinamid [119]
Emperical PM3 method	Computation of adsorption coefficient of bile acids	Sodium taurocholate [88]
Emperical PM3 method (HF/6-31G based thery)	Strengths of hydrogen bonds formed between polymer and ligand	Phenol and thiophenol [120] transesterification catalysis [71]
Emperical PM3 method	Modeling pre-polymerization complex and modeling binding site	Nicotinamide or *iso*-nicotinamide [115]
PWC/DNP calculations	Computation of binding sites on MIP	Rh–amine complex imprinted silicate system [121]
Chemometrics and neural network methods		
Chemometric Design Expert software used for factorial data	Optimization of the pre-polymerization mixture	Sulfonamide [74]

(continued)

Table. 1 (continued)

Process	Application	Targets and references
Neural network using the back propagation algorithm (WEKA)	Prediction of imprinting factor of MIPs and study of template monomer complexes	Atropine and Boc-L-Trp d-Brompheniramine [75, 76]
General chemometric design	Optimization for MIP preparation	Bisphenol-A [122]
Chemometric multivariate analysis study on screening and rebinding using radio-ligand counting methods	Extensive screening and more sophisticated response parameters for analysis	Propranolol [123]

calculations on multicomponent systems. While molecular modeling of complex systems and possible interactions of polymers with template, solvent, and other molecules are difficult because of the requirements of large computational workloads, the method developed at Cranfield University can achieve this by simplifying the model. Since the structure of the monomers–template complexes formed in the monomer mixture is preserved in the synthesized polymer, instead of modeling the polymer, modeling the monomer mixture and the interactions taking place in solutions between monomers, cross-linker, template, and solvent would be possible, which substantially reduces computational load [61–63].

The protocol developed at Cranfield starts with the design of a virtual library of molecular models of functional monomers and template (Fig. 2). The next step is to screen the virtual library against a template to determine the monomers that strongly bind to the template. Calculations are performed to estimate how the monomers bind to template using simulated annealing to determine optimum ratios of template to monomers. In effect, the strength and type of interactions, existing between monomers and template in monomer mixtures, which in theory, determine the recognition properties of the MIP, will be analyzed and used for optimization of polymer composition. Each of the steps has been described below.

2.1.1 Modeling of the Template Molecule

A molecular model of the template molecule is made and charges for each atom are calculated, and the structure of the template is refined using MM.

2.1.2 Construction of the Monomer Database

While there are about 4,000 polymerizable compounds that have been reported that could potentially be used as functional monomers, in reality many of them have similar properties and functions and hence it is assumed that it is sufficient to test possible interactions between a minimal library of functional monomers and a target template [124]. Several different software packages can be used for creating

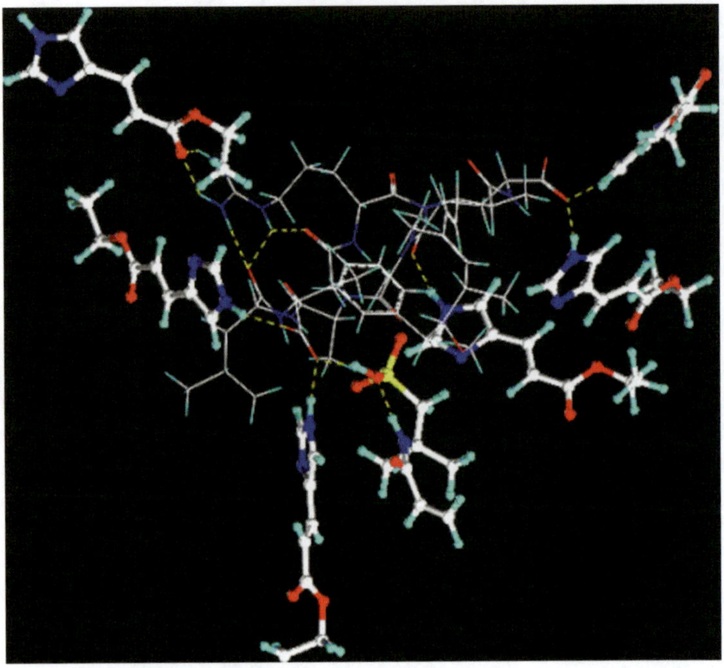

Fig. 2 Interactions between microcystin-LR and monomers. Microcystin-LR, in lines in the center of the picture, interacts with six molecules of urocanic acid ethyl ester (UAEE) and 1 molecule of AMPSA [64] (*right*) shown as *ball* and *stick*. Reproduced with permission

virtual library of monomers and molecular models of template. Examples include Agile Molecule, Sirius, SYBYL, Oscail X, and MOE.

2.1.3 Screening of the Virtual Library

The quantity and quality of MIP recognition sites that result due to a binding event are a direct function of the nature and extent of the monomer–template interactions present in the pre-polymerization mixture. The previous research directed toward understanding the physical basis of molecular recognition has shown that the extent of template complexation at equilibrium is governed by the change in Gibbs free energy of template–functional monomer interaction [125–127]. Andrews et al. [128] detailed an approach to calculate the average binding energies of ten common functional groups based on an analysis of structural factorization of the energetic contributions to binding. These approaches detailed the importance of each of the physical entities that govern a molecular recognition event. The general thermodynamic explanation which summarizes contribution of individual physical parameters in a binding event has been described by Williams [see Eq. (1)]:

$$\Delta G_{\mathrm{bind}} = \Delta G_{t+r} + \Delta G_r + \Delta G_h + \Delta G_{\mathrm{vib}} + \sum \Delta G_p + \Delta G_{\mathrm{conf}} + \Delta G_{\mathrm{vdW}}. \quad (1)$$

This equation can be used to describe template–monomer interactions as well as template–MIP binding events [125, 126, 129]. The Gibbs free energy changes are: ΔG_{bind}, complex formation; ΔG_{t+r}, translational (energy associated with the motion) and rotational (energy associated with rotation); ΔG_{r}, restriction of rotors upon complexation; ΔG_{h}, hydrophobic interactions; ΔG_{vib} residual soft vibrational modes; $\sum \Delta G_{\text{p}}$, the sum of interacting polar group contributions; ΔG_{conf}, adverse conformational changes; and ΔG_{vdW}, unfavorable van der Waals interactions.

This or similar equations lie at the cornerstone of practically all screening/ modeling packages used in the design of MIPs. In practice the screening of virtual libraries is performed using the Leapfrog algorithm (Tripos Inc.). LeapFrog is used in drug development for screening of new, potentially active, ligand molecules against the structure of known receptor-binding sites. LeapFrog can also generate new compounds by repeatedly making small structural changes, evaluating the binding energy of the new compound, and keeping or discarding the changes based on the results [130, 131].

The first step in MIP design using LeapFrog is the identification of the binding site points on the surface of the template molecule. LeapFrog samples the environment immediately surrounding the template and determines its average electrostatic, steric, and lipophilic characteristics. Then each of the monomers is placed in close proximity of the template-binding site. The second step is the calculation of binding energy and once the binding site is well defined, the "fit" is assessed. Since many possible hits arise, each has to be scored to decide which one of those hits is most promising. There are a variety of scoring techniques employed by different programs that exist such as LEGEND [132], LUDI [133], SPROUT [134], HOOK [135], and PRO-LIGAND [136]. The scoring functions these programs employ however vary from (a) H-bond placement, (b) constraints that are due to steric effects, (c) explicit force fields, and (d) empirical or knowledge-based scoring methods. Programs such as GRID and LigBuilder set up a grid in the binding site and then assess interaction energies by placing probe atoms or fragments at each grid point [137].

Scoring functions guide the growth and optimization of structures by assigning fitness values to the sampled space. Scoring functions attempt to approximate the binding free energy by substituting the exact physical model with simplified statistical methods. Force fields such as that used by Leapfrog involve more computation than some other types of scoring functions. Leapfrog calculates major components of the binding energy such as steric, electrostatic, and hydrogen-bonding enthalpies. Other methods, such as the one used in the GRID4 program (Tripos Inc.) [138, 139] which enhance the rate of calculation, are also in use.

2.1.4 Computation of Monomer Template Ratio

The next step in the protocol is the computation of monomer template ratio performed by simulated annealing using a molecular dynamics approach (for detailed description see Sect. 2.2).

2.2 Examples of MM Methods Used in MIP Design

In world's first successful demonstration of rational design of MIP using the above-mentioned protocol [63], researchers at Cranfield University established a "proof of concept" by demonstrating that the screening of a virtual library of monomers led to the development of an optimized MIP composition specific for creatinine [63]. When this polymer was synthesized in the laboratory, it demonstrated superior selectivity in comparison to an MIP that was prepared using a traditional functional monomer, MAA. In this work Cranfield researchers combined the above-described computational procedure for rational design of MIPs with a "Bite-and-Switch" approach for the detection of polymer–template interaction [140]. In what could be considered as one of the best examples of the rational design using the Cranfield protocol, a highly selective MIP for the cyanobacterial toxin microcystin-LR was designed and demonstrated [64].

Two MIPs for microcystin-LR were then synthesized, one using a functional monomer with the best binding score (which shows the capability of forming strongest complexes with the template), 2-acrylamido-2-methyl-1-propanesulfonic acid (AMPSA) (Fig. 2), and the other using a "traditional" functional monomer MAA. The optimal MIP formulation synthesized had affinity and sensitivity (studied using ELISA), comparable with those of polyclonal antibodies and superior chemical and thermal stabilities compared with those of antibodies. The computationally designed MIP also showed higher affinity in comparison with the MAA-MIP. It was also found that MIPs had much lower cross-reactivity for microcystin-LR analogues than both polyclonal and monoclonal antibodies.

While it is proven that MIPs perform well in organic solvents, the practical applications of MIPs are hindered due to their poor performance in polar media. Although it is desirable to achieve affinity separation and sensing in water, MIPs usually do not work well in aqueous media. This is because of the disruption of hydrogen bonds and competition between solvent and template molecules for the binding sites. A significant contribution to the loss of polymer affinity originates from the potential difference in the structure of the polymer-binding sites in organic solvent (traditionally used for polymer preparation) and in water due to differences in polymer swelling. In an effort to develop MIPs compatible with water, the group of Piletsky imprinted biotin [79], having identified those monomers that provide strong binding to the template in water using their computational screening method. In order to mimic aqueous conditions and to obtain stable confirmation, the energy minimization of monomers and template was performed using the dielectric constant of water is ($\varepsilon = 80$). The results of the modeling confirmed that monomers MAA, TFMAA, and AMPSA formed a strong complex with the template molecule in water through ionic and hydrogen bonds. This was the first demonstration of the use of molecular modeling for the rational selection of monomers capable of template recognition in water. The designed MIP was successfully grafted to the polystyrene surface in an aqueous environment and demonstrated high affinity for biotin in water.

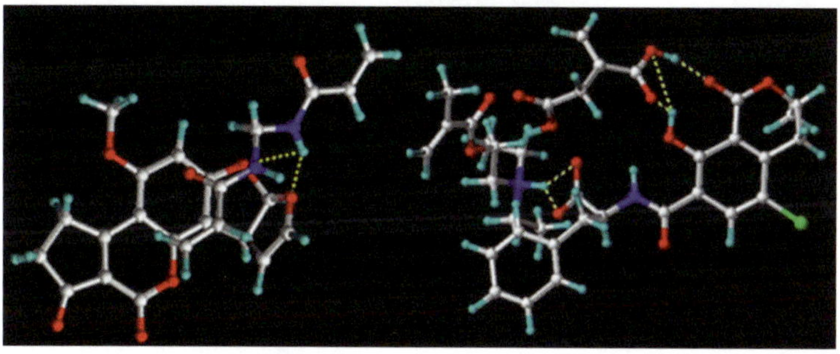

Fig. 3 3D molecular complex between negatively charged OTA and functional monomers DEAEM (*left*); 3D molecular complex of AFB1 with MBAA functional monomer (*right*) (oxygen atoms are shown in *red*, carbon atoms are *white* and the *light blue* atoms are hydrogen). Reproduced with permission [86]

The same team also synthesized MIPs capable of the controlled release of simazine in water [82]. Leapfrog was used to identify a list of monomers used in the production of polymers with different affinities and correspondingly different release profiles of the herbicide. The speed of release correlated with the calculated binding characteristics. The high affinity, MAA-based polymer released ~2% and the low-affinity HEMA-based polymer released ~27% of the template over 25 days.

In an interesting study, computationally designed methacrylate-based synthetic polymers that inhibit QS by sequestering the bacterial signal molecules were developed by Piletska et al. Biofilm production and expression of virulence factors were correlated with Quorum Sensing (QS), a density-dependent regulation of gene expression controlled by specific signal molecules produced by bacteria. The polymers were able to absorb 0.1–0.3 mg (per gram of polymer) of *N*-acyl-homoserine lactones (AHLs). This work has implications in aquaculture, where these polymers can sequester a (a test bacterium) signal molecule of *Vibrio fischeri* and prevent QS-controlled phenotypes, thus representing a new solution for controlling disease outbreaks [85].

Polymers with affinity to two of the most abundant mycotoxins AFB1 and OTA were designed by the computational approach (Fig. 3) for application in the ToxiQuant T1 System (TQT1). The principle of quantification of AFB1 and OTA using the TQT1 instrument consisted of fluorimetric analysis of mycotoxins adsorbed on the polymer upon exposure to UV light. High affinity of the resins allowed the adsorption of both toxins as discrete bands on the top of the cartridge with detection limit as low as 1 ng of mycotoxin. Based on the computational modeling, MBAA was selected for the preparation of the polymer specific for AFB1, and a mixture of DEAEM and IA was selected for preparation of a polymer specific for OTA. DMF was used as the porogen in both cases [86].

A similar method was previously employed to prepare an MIP with high affinity for nonylphenol, a xenobiotic used in the manufacture of antioxidants, lubricating oil additives, and surfactants (Table 2). Nonylphenol is degraded in wastewater plants leading to the formation of lipophilic nonylphenol derivatives [87]. Chromatographic

Table 2 Monomers with highest binding for nonylphenol [87]

Highest binding monomers for nonylphenol	Binding energy (kcal/mol)
Itaconic acid	−38.06
TFMAA	−27.17
Urocanic acid	−27.16
Methacrylic acid	−21.63
Vinyl imidazole	−16.00
DEAEM	−12.66
2-Vinyl pyridine	−6.60

Reproduced with permission

tests of the computationally designed MIP demonstrated higher affinity toward nonylphenol, and both NIP and MIP were observed to be suitable for removal and pre-concentration from contaminated water samples with 99% efficiency of the recovery (231 mg/g for NIP and 228 mg/g for MIP). A comparative test under the same conditions but using the commercial resins PH(EC) (Biotage) and C18 (Varian) showed recovery rates <84%. The synthesized materials can be used for sample pre-concentration and environmental analysis of this class of compounds [141].

The MM method is the fastest and least expensive method in computational terms, and hence is an ideal choice for studies on analysis of structural parameters and stable conformation for a variety of molecules. Optimization steps are often carried out to confirm that the molecules are in their lowest energy state so that the calculated results can be compared to those obtained experimentally. However, since MM does not deal directly with electrons and orbitals, it cannot be used to study, e.g., chemical reactivity of functional monomers.

2.3 *Rational Approaches to MIP Synthesis Involving Molecular Dynamics*

MD simulation provides better descriptions of interactions (generally electrostatic and van der Waals) because they reflect the effect which the surrounding environment has on the properties of molecules. MD simulations in general are powerful tools to investigate complex systems made of thousands of atoms. A good understanding of intermolecular interactions, mechanisms of imprinting, and properties in molecular imprinting processes requires advanced state-of-the-art computational tools which will help in investigations on the molecular clusters and prediction of interaction energies (Table 1).

2.4 *Examples of MD Methods Used in MIP Design*

One of the successful approaches that are used for the computation of monomer template ratio is simulated annealing. Simulated annealing is a Monte Carlo approach for minimizing multivariate functions. The term simulated annealing

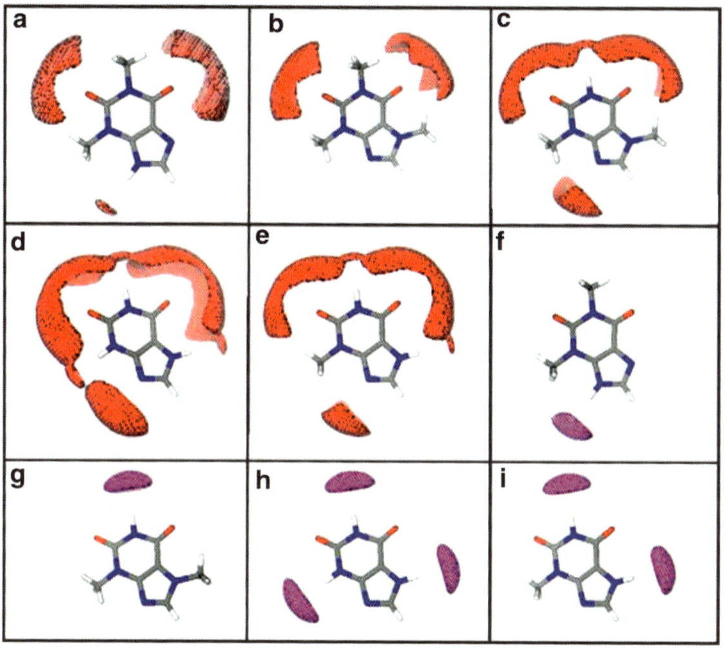

Fig. 4 Contour maps of the molecular interaction fields produced: by OH probe at −4.5 kcal/mol for THO (**a**), CAF (**b**), theobromine (**c**), xanthine (**d**), 3-methylxanthine (**e**), by O probe at −3.0 kcal/mol for THO (**f**), theobromine (**g**), xanthine (**h**), 3-methylxanthine (**i**) [91]. Reproduced with permission

derives from a physical process of heating and then slowly cooling a substance to obtain a crystalline structure. The simulated annealing process lowers the temperature by slow stages until the system is "frozen or at a standstill" and no further changes occur. At each temperature the simulation must proceed long enough for the system to reach a steady state or equilibrium. This is known as thermalization. The sequence of temperature changes and the number of iterations applied to thermalize the system at each temperature comprise an annealing schedule [142]. To apply simulated annealing, the system is initialized with a particular configuration. A new configuration is constructed by imposing a random displacement. If the energy of this new state is lower than that of the previous one, the change is accepted unconditionally and the system is updated. If the energy is greater, the new configuration is accepted probabilistically. This procedure permits an environment to proceed toward lower energy states, at the same time keeping open the option of escaping out of local minima due to the probabilistic acceptance of some upward moves. Logarithmic decrease of temperature in simulated annealing generally assures an optimal solution.

Monti et al. [91] detailed a protocol that combined MD, MM, docking, and site mapping to simulate the formation of possible imprints for THO using MAA and MMA as monomers in acetonitrile solution (Fig. 4). MM calculations and MD simulations were carried out with AMBER 7, as it previously showed satisfactory performance of this force field in the evaluation of the stability of hydrogen bonded as well as van der Waals adducts [143–146]. The structures of THO, MAA, and MMA molecules were firstly optimized using DFT [147–149], and their atomic charges were determined with RESP [150, 151]. All the simulations were performed in the NPT ensemble using Berendsen thermostat and barostat [152] with temperature set to 310 K and pressure to 1 atm. THO molecule was surrounded by functional monomer shells and solvated, creating around it a rectangular parallelepiped acetonitrile box [153, 154].

Docking procedures, used to find favorable orientations of the ligands inside the polymer cavity, were performed using the DOCK5.0 program [155–157], and the GRID program (GRID, 2004) was used to map the energetic interactions. The created model was able to predict binding affinity and selectivity when considering THO analogues, such as caffeine, theobromine, xanthine, and 3-methylxanthine [91].

The entire modeling study was performed in four different phases: first, a *non-covalent phase* where the template and the functional monomers form non-covalent complexes in solution prior to polymerization; second, a *locking phase* where the non-covalent monomers–template complexes are cross-linked and the binding site is generated with appropriately oriented functional monomers and model polymer structures are created and selected; third, a *validation phase* where the polymer specificity and recognition capabilities are tested; and finally the *mapping phase* where the characteristics of the binding cavities are analyzed. This work showed that the MD simulations were able to predict the selectivity and binding affinity of the MIP, and when complemented with experimental data gave a clearer picture of the system and the type of interactions in the complex (Fig. 4).

Similarly, selective adsorption properties of dimethoate imprinted polymers were studied through a MD simulation [101]. The MD modeling was carried out to investigate the recognition mechanism by predicting the interaction energy differences and indicating the active site groups, which confirmed that the MIP, based on butyl methacrylate (BMA) functional monomer, had the most selective recognition for dimethoate compared to other functional monomers, including methyl methacrylate (MMA) and ethyl methacrylate (EMA) (Fig. 5).

Two parameters: the imprinting factor indicator and the competitive factor indicator were calculated, based on the interaction energy differences for the template in relation to structurally related organophosphorus pesticides (OPs). The methodology was interesting, in that PRODRG software was used to analyze polymer topology and MD simulations were performed with the GROMACS-3.2 in the NPT ensemble with GROMOS 96 force field. A van der Waals cutoff of 1.4 nm and PME were used to describe nonbonded interactions [158–161]. However, the influence of the cross-linking agent was neglected, apparently to simplify the modeling process. A similar experiment was performed by other researchers [95]. In an effort to obtain the optimized conformation of MIP, the authors saturated the system of polymer chain and dimethoate with the solvent THF and performing simulation for 20 ns. Computational prediction was verified through chromatography.

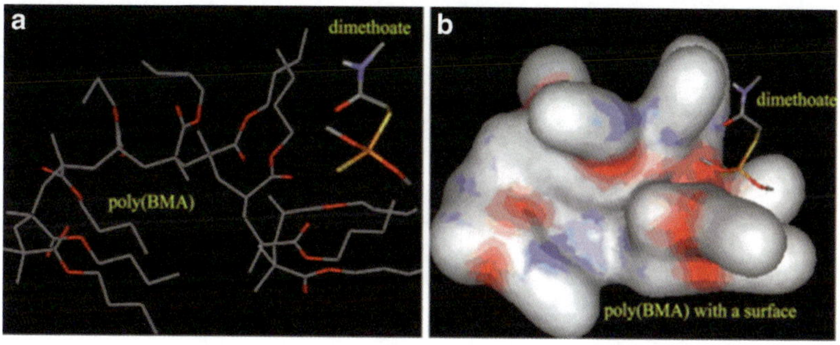

Fig. 5 Structures of a typically simulated equilibrated conformation of dimethoate interacting with poly(BMA) (NPT at 300 K, 1 atm) [101]. Reproduced with permission

In another example of MD simulation, Benito-Peña et al. [72] analyzed binding of seven novel fluorescent labeled β-lactam antibiotics (BLAs) with a library of six polymers imprinted with penicillin G (PenG). The 3D chemical structures of the labeled BLAs have been modeled followed by energy minimization by molecular dynamics (MOPAC, AM1 force field) using Chem3D Ultra 7.0 software (Cambridge-Soft, MA). The results of molecular modeling showed that recognition of the fluorescent analogues of PenG by the molecularly imprinted material is due to a combination of size and shape selectivity.

Yoshida et al. [92] employed HyperChem and performed MD calculations to verify the recognition mechanism of the MIP they synthesized for the separation of optically active tryptophan methyl ester. The computational modeling proved that the enantiomeric selectivity is conferred by electrostatic and hydrogen-bonding interactions between the functional molecule and the target tryptophan methyl ester, along with the chiral space formed on the polymer surface.

Li et al. [100] demonstrated a detection method for veterinary drugs, wherein they employed MD simulations and screening to identify functional monomers capable of interacting with sulfadimidine (SM_2). The authors prepared a library comprising monomers and calculated their interactions with SM_2 in acetonitrile using GROMACS 3.3 (Fig. 6). The GROMOS-96 force fields of the functional monomers, SM_2 and acetonitrile molecule were presented and computationally evaluated the PRODRG SEVER 2.5. The NVT-MD simulations were performed at 300 K, and the functional monomer and template with ratio 1:1 were immersed in acetonitrile. Molecular dynamics simulations in explicit solvent were performed for 1 ns with time steps of 0.002 ps. The authors performed energy calculations of the simulated system using GROMACS software package and visual MD (VMD). The surface molecularly imprinted silica with SM_2 as template was prepared by a surface-imprinting technique prepared using MAA and DVB in acetonitrile. The molecular recognition of SM_2 was analyzed in detail using Gaussian 03. The experimental validation reported in this work is in conjunction with the theoretical calculations.

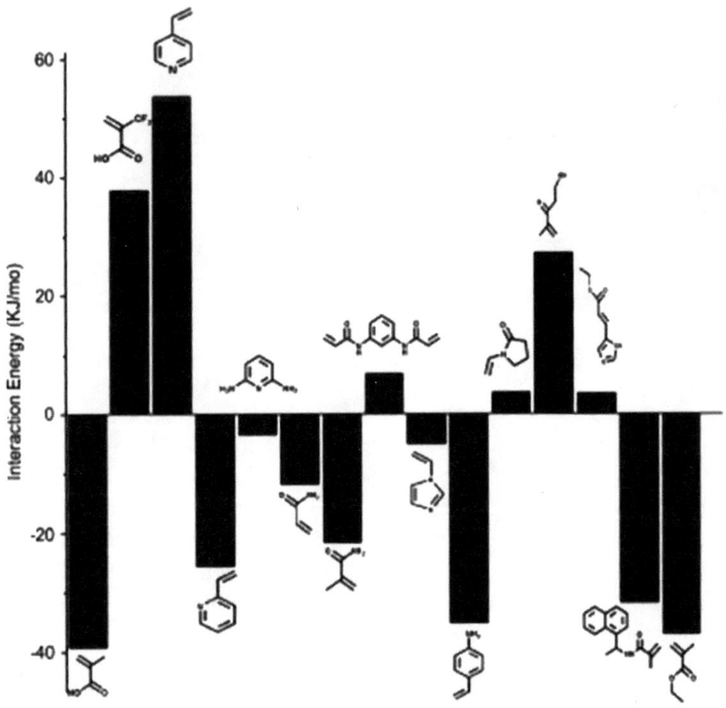

Fig. 6 (continued)

Pavel and Lagowski [65, 98, 162] studied the intermolecular interactions in molecular imprinting of theophylline (THO). The minimized structures of five ligands, THO and its derivatives (theobromine, theophylline-8-butanoic acid, caffeine and theophylline-7-acetic acid) were employed in MD simulation using Cerius2 version 4.10 software designed by Accelrys, Inc. (San Diego, CA, USA). The polymer consistent force field (PCFF) was employed, as it was found to be very suitable and reliable for the molecular simulation of organic molecular clusters of monomers and polymers [163–165]. The forces acting on each atom of a model polymer were calculated. The initial molecular clusters of the simulated monomers and polymers were optimized, giving information about total energies (E), energy differences (DE), and distances (d) between the monomers and different ligands in a given cluster. Using the same MD simulations, Pavel et al. [65] designed monomers for MIPs specific for chemical warfare agents. They showed successful predictions of interaction energies, the closest approach, distances, and the active site groups.

Several research groups have used the PM3 method for the analysis of template–functional monomer complexes [166]. From a computational perspective, PM3 provides improved modeling of non-covalent interactions such as hydrogen-bonding and van der Waals interactions.

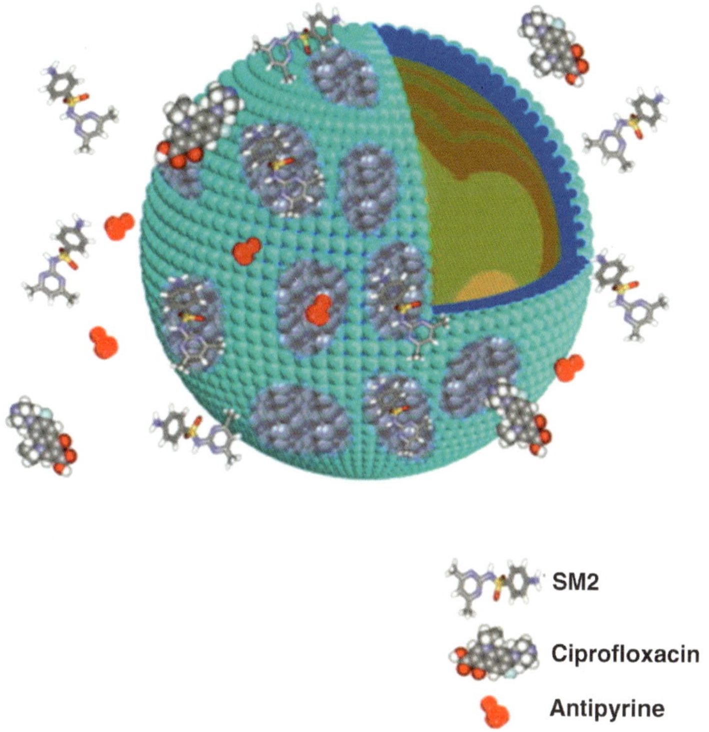

Fig. 6 Snapshot of the interaction energy between SM_2 and monomers (*left*) and a schematic diagram of surface imprinting and recognition mechanism (*right*) [100]. Reproduced with permission

Since MD simulations are numerical and generally include a large number of particles, the simulation time required for modeling is very substantial. However, contrary to MM, MD molecules and their complexes have the ability to adapt to their environment, which provides a closer representation of reality.

2.5 *Rational Approaches to MIP Design Involving Quantum Mechanics*

Quantum Mechanics (QM) is a field of quantum chemistry that uses a mathematical basis to represent chemical phenomena at the molecular level. It uses a complex mathematical expression called a wave function with which the energies and properties of atoms and molecules can be computed. While for simple model systems wave functions can be analytically determined, for complex systems such as those that involve molecular modeling, approximations have to be made. One of the most commonly employed approximation methods is that of *Born and Oppenheimer* [167]. This approximation is based on the idea that it is not necessary to develop a wave

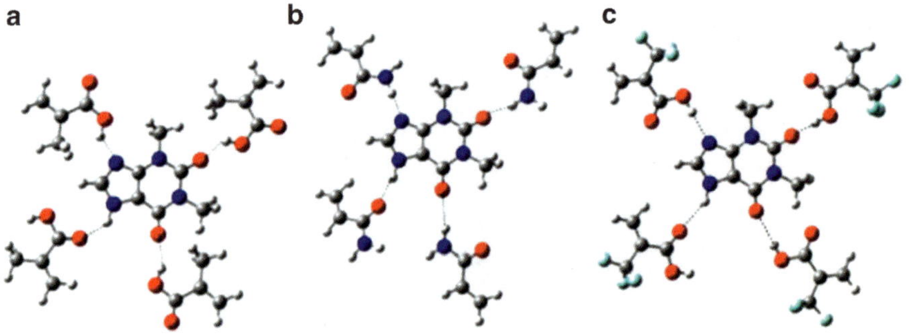

Fig. 7 The complex formed between THO and MAA (**a**), AA (**b**) and TFMAA (**c**), respectively [67]. Reproduced with permission

function description for both the electrons and the nuclei at the same time. The nuclei are heavier and move much more slowly than the electrons, and therefore can be regarded as stationary while the electronic wave function is computed. By computing the QM of electronic motion, the energy changes for different chemical processes, vibrations, and chemical reactions can be understood [168].

2.6 Examples of QM Methods Used in MIP Design

Dong et al. [67] employed this method to screen monomers using the binding energy, ΔE, of a template molecule and a monomer as a measure of their interaction. In this study THO was chosen as the template molecule, and MAA, AA, and TFMAA were the functional monomers (Fig. 7).

The calculation of ΔE was performed using DFT with the Gaussian 98 software [169]. First, the conformations of THO, MAA, AA, and TFMAA were optimized, and the energy of the molecules with the optimized conformation was calculated. Then the energy calculation was applied to the complex formed between THO and MAA, or AA or TFMAA, respectively. The MIP synthesized using TFMAA as monomer showed the highest selectivity to THO while the MIP from AA gave the lowest, as predicted from the calculation of binding energy (ΔE), [Eq. (2)].

$$\Delta E = E\,(\text{complex}) - E\,(\text{THO}) - E\,(\text{monomer}). \tag{2}$$

Wu et al. [66] showed that a similar technique could be employed to determine the influence of the solvents used as porogen on the affinity and selectivity of MIPs. The interaction energy values between NAM and MAA were modeled with methanol, acetonitrile, chloroform, and toluene as the porogens. Gaussian 03 [170] was adopted as the software to carry out the simulation, and B3LYP [147, 171] was selected as the calculation method. B3LYP is a DFT method, which takes electronic correlation energy into consideration. This provides results of weak interaction system compared with Hartree–Fock method. The retention factors and selectivity factors of NAM and its analogues were evaluated, and good correlations were found between the interaction energies and the retention factors. When the porogens had poor hydrogen-bonding capacity, the interaction energy was mainly influenced by the dielectric constant of the solvent, and when the porogen had strong capacity in forming hydrogen bond, both the dielectric constant of the solvent and the hydrogen-bonding interference affected the formation of the template–monomer complex and the corresponding interaction energy.

A computational optimization of the monomer formulation of molecularly imprinted catalysts (MIC) for lipase-catalyzed transesterification process was demonstrated [71]. The authors screened the intermediates of the lipase-catalyzed transesterification process commonly containing "catalytic triad" motif made up of compounds such as serine, histidine, and aspartic acid [172–174]. To construct the virtual intermediates, *p*-nitrophenyl acetate was used as substrate and monomers containing carboxylate moieties as molecular recognition elements. The energy of each intermediate was then minimized using the semiempirical MOPAC method with a minimum RMS gradient of 0.100, which specifies the convergence criteria for the gradient of the potential energy surface. AM1 theory was used with a closed shell function to calculate heat of formation (ΔHf) of the intermediates, which represents the gas-phase heat of formation at 298 K of 1 mol of the intermediate from its elements in their standard state. The result of this work has been utilized successfully for the design of artificial lipases.

QM are perhaps the most accurate approaches currently used in the field of molecular modeling, because the modeling method involves fewer "assumptions" and the results depend entirely on the accuracy of performed calculations. However, the sheer size of the required computations currently does not allow realistic modeling of supramolecular systems.

2.7 Rational Approaches to MIP Design Involving Quantum Chemical Methods

QC methods have been widely employed, mainly for the reason that the computational cost is kept to a minimum. Further, several studies have shown that they also lead to accurate estimations. This is because QC methods employ calculations of electronic structures of molecules leading to more reliable results of non-covalent interactions, more specifically of pre-polymerization mixtures (Table 1).

2.8 Examples of Quantum Chemical Methods Used in MIP Design

QC approaches have been successfully used to perform optimization studies through selection of appropriate functional monomers. This was shown by the design of MIPs that recognized theophylline [67], *N,O*-dibenzylcarbamate [175], and Metformin [114]. Ideally, elaborate use of B3LYP/6-31+G//B3LYP/3-21G level of theory for theophylline [67] has been shown to be more extensive as a study, because of elaborate calculations that provide better representations of binding between functional monomers and the template. Using PM3 calculations and MP2/6-31G//HF/6-31G studies, the importance of the influence of binding energies between template and functional monomers with relation to the resultant capacity factor of the MIP was explained in a previous study [176]. QC approaches can also compute binding energies for templates that differ from the target structure, as has been shown before [73]. Here the authors describe the design of a MIP that mimicked cytochrome CYP2D6, the computations of which were based on PM3, giving optimized template molecule geometries, which resulted in the formation of a MIP with good recognition properties (Table 1). An earlier example of the calculation of binding energy [112], using an AM1 method for several complexes of cocaine with MMA and 4VP leading to an optimized polymerization mixture, proved to be a success, as further validation with NMR and rebinding studies confirmed.

Chiral recognition was examined for a MIP synthesized for (*S*)-nilvadipine using MAA, TFMAA, 2-VP or 4-VP as a functional monomers and EGDMA as cross-linker [111]. Molecular computations were done with CAChe MOPAC version 94 implemented in CAChe programs23. Molecular geometries of (*S*)-nilvadipine and 4-VP were optimized by the AM1 method. The simulation was performed on the hydrogen-bonding complex model with the dihydropyridine and pyridine rings of (*S*)-nilvadipine and 4-VP molecules, respectively. Molecular modeling revealed a one-to-one hydrogen-bonding-based complex formation of (*S*)-nilvadipine with 4-VP in chloroform and that (*S*)-nilvadipine imprinted EGDMA polymers should recognize the template molecule by its molecular shape, and that hydrophobic and hydrogen-bonding interactions seem to play important roles in the retention and chiral recognition of nilvadipine on the 4-VP-*co*-EGDMA polymers in hydro-organic mobile phases. The (*S*)-nilvadipine-imprinted 4-VPY-*co*-EGDMA polymers indeed gave the highest resolution for nilvadipine among the MIPs prepared.

QC methods employing PM3 methods have also computed adsorption coefficients of bile acids on a film of overoxidized polypyrrole imprinted with sodium taurocholate [88]. PM3 methods are considered fairly simple and straightforward, and their ability to compute adsorption coefficients successfully confirms the computational power of such methods. Further electronic structure computations for the prediction of polymer properties were successfully demonstrated [176], where interaction energies were calculated for pre-polymerization complexes of nicotinamide or *iso*-nicotinamide with methacrylic acid as functional monomer.

2.9 Rational Approaches to MIP Design Involving Chemometrics and Neural Network Methods

Chemometrics uses mathematical and statistical methods to select the optimal experimental procedures and for the extraction of data for the analysis [177].

The design of experiments follows a mathematical framework for changing several selected factors simultaneously to predict the optimum conditions, thus reducing the number of experiments necessary [178]. In effect the goal is to plan and perform experiments in order to extract the maximum amount of information in the fewest number of trials. Chemometrics has various applications in science, such as optimization of experimental parameters, design of experiments, data retrieval and statistical analysis, analysis of structure–property relationship estimations, signal processing, pattern recognition, and modeling.

2.10 Examples of Chemometrics Methods Used in MIP Design

Steinke et al. [179] used the chemometrics approach to investigate the effect of variables such as type and quantity of monomers, cross-linker, porogens, initiator, type of initiation (UV or thermal), polymerization pressure, temperature, reaction time, and reaction vial dimensions have on the properties of synthesized polymers. Davies et al. [74] have used chemometrics design expert software to optimize the composition of a MIP for sulfonamide, in particular the template:monomer:cross-linker ratio. MAA was selected as functional monomer and EGDMA as cross-linker. They selected the amounts of template, monomer, and cross-linker (T:M:X) in the MIP using a three-level full factorial design. The chemometrics design required the synthesis of a small number of MIPs with different T:M:X ratios and used the results of their testing to determine the polymer composition with optimum recognition characteristics. The polymer with the optimized ratio of 1:10:55 was compared to MIPs, prepared with more commonly used molar ratios such as 1:4:20 [180] and 1:8:40 [181]. The experiment proved that the predicted ratio generated a polymer with superior binding properties.

In another interesting study, Kempe et al. [123] employed multivariate data analysis (Modde 6.0 software, Umetrics, Umea, Sweden) for the optimization of monomer and cross-linker ratios in the design of polymers specific for propranolol.

Mijangos et al. [182] used chemometrics (MODDE 6.0 software, Umetrics, Sweden) to optimize several parameters such as concentration of initiator (1,1′-azobis(cyclohexane-1-carbonitrile) or 2,2-dimethoxy-2-phenylacetophenone) and polymerization time, required for the synthesis of high performance MIPs for ephedrine. A small set of (−) ephedrine imprinted polymers was synthesized and tested by HPLC for their ability to interact with (+) and (−) ephedrine. This chemometric study provided evidence that in order to achieve high performance in chiral separation MIPs should be synthesized for long reaction times using low concentrations of initiator and at low temperatures.

2.11 Examples of Neural Network Methods Used in MIP Design

Prachayasittikul's group used data from the literature, such as monomer composition, retention, and imprinting factors of MIP and NIP, mobile phase composition, etc., in calculations aimed at prediction of optimal template–monomer pairs [75]. The imprinting factor was predicted by artificial neural networks as a function of the calculated molecular and mobile phase descriptors. The quantum chemical descriptors were computed using Gaussian 03W. The model confirmed that the stronger template–functional monomer interactions lead to higher imprinting factors.

Most experiments are influenced by a variety of factors, and screening is often the first step in efficient assessment of which factors are important in influencing the desired outcome of the system under study [183, 184]. Factors that depend on monitoring the effects of changing one factor at a time on a response are extremely time consuming producing erroneous optimums in experiments [183]. Chemometrics often combined with artificial network simulations [75] can help significantly in optimization and design of experiments for simultaneous changing factors to predict optimum conditions thus reducing the number of experiments necessary [183, 184]. While chemometrics offers several advantages, complex numerical solutions that are generated by chemometrics approaches could often be misinterpreted unless a proper procedure is implemented. Since problem solving involves multivariate analysis, the interpreter needs to be fairly skilled in terms of analysis and interpretation. This approach also requires a number of experiments to be made, and as such cannot be performed entirely in silico.

3 Conclusion

Various methods such as MM, MD, QC methods, and QM have been clearly shown to be extremely useful for the design of MIPs. Each of these methods should take account of the various complex physical and chemical processes taking place during the formation of monomer–template complexes, as well as processes involved in polymer formation. QC methods in the recent past have been heavily used because of their low computational cost and their ability to provide relatively accurate estimations of the electronic structures of molecules pertaining to non-covalent interactions in pre-polymerization mixtures. Generally in the rational design of MIPs, it is important to note that although formation of monomer–template complexes in solution can be relatively easily modeled, the challenge still lies in modeling different stages of polymerization. The exact mechanism of events related to the incorporation of monomers into the polymer network and their effects on MIP recognition properties need to be more fully understood. This would lead to a greater understanding of binding events and lead to better design of MIPs. Future improvements in computational protocols could also throw light onto the reasons for the discrepancies observed between the predicted and experimental results of the performance of polymers where the recognition process involves hydrophobic interactions. Current and future computational approaches would certainly help in the design of artificial receptors and generate knowledge that could help to build a better understanding of molecular interactions in biological systems. A combination of two or more methods in a single computational platform could help solve several problems in the design of MIPs and further lead to greater understanding of the processes leading to the creation of specific and highly selective binding pockets.

References

1. Naidoo KJ, Chen YJ, Jansson JLM et al (2004) Molecular properties related to the anomalous solubility of beta cyclodextrin. J Phys Chem B 108:4236–4238
2. Chen W, Huang J, Gilson MK (2004) Identification of symmetries in molecules and complexes. J Chem Inf Comput Sci 44:1301–1313
3. Zheng XM, Lu WM, Sun DZ (2001) Enthalpy and entropy criterion for the molecular recognize of some organic compounds with beta cyclodextrin. Acta Phys Chim Sin 17:343–347
4. Rekharsky MV, Inoue Y (1998) Complexation thermodynamics of cyclodextrins. Chem Rev 98:1875–1917
5. Liu L, Guo QX (2002) The driving forces in the inclusion complexation of cyclodextrins. J Incl Phenom Macro 42:1–14
6. Cagın T, Wang G, Martin R et al (2000) Molecular modelling of dendrimers for nanoscale applications. Nanotechnology 11:77–84
7. Cruz-Morales JA, Guadarrama P (2005) Synthesis, characterization and computational modeling of cyclen substituted with dendrimeric branches – dendrimeric and macrocyclic moieties working together in a collective fashion. J Mol Struct 779:1–10
8. Gorman CB (2003) Dendritic encapsulation as probed in redox active core dendrimers. Comptes Rendus Chimie 6:911–918
9. Amatore C, Oleinick A, Svir I (2005) Diffusion within nanometric and micrometric spherical-type domains limited by nanometric ring or pore active interfaces. Part 1: conformal mapping approach. J Electroanal Chem 575:103–123
10. Cagin T, Wang G, Martin R et al (2001) Multiscale modeling and simulation methods with applications to dendritic polymers. Comput Theor Polym Sci 11:345–356
11. Casnati A, Della Ca' Sansone F et al (2004) Enlarging the size of calix[4]arene-crowns-6 to improve Cs^+/K^+ selectivity: a theoretical and experimental study. Tetrahedron 60:7869–7876
12. Majerski KM, Kragol G (2001) Design, synthesis and cation-binding properties of novel adamantane and 2-oxaadamantane-containing crown ethers. Tetrahedron 57:449–457
13. Furer VL, Borisoglebskaya EI, Kovalenko VI (2006) Modelling conformations and IR spectra of p-tert-butylthiacalix[4]arene tetraester using DFT method. J Mol Struct 825:38–44
14. Inese SI, Smithrud DB (2001) Condensation reactions of calix[4]arenes with unprotected hydroxyamines, and their resulting water solubilities. Tetrahedron 57:9555–9561
15. Korochkina M, Fontanella M, Casnati A et al (2005) Synthesis and spectroscopic studies of isosteviol-calix[4]arene and -calix[6]arene conjugates. Tetrahedron 61:5457–5463
16. Chen W, Gilson MK (2007) ConCept: de novo design of synthetic receptors for targeted ligands. J Chem Inf Model 47:425–434
17. Luo R, Gilson MK (2000) Synthetic adenine receptors: direct calculation of binding affinities. J Am Chem Soc 122:2934–2937
18. Gilson MK, Given JA, Head MS (1997) A new class of models for computing receptor-ligand binding affinities. Chem Biol 4:87–92
19. Gilson MK, Given JA, Bush B et al (1997) The statistical-thermodynamic basis for computation of binding affinities. A critical review. Biophys J 72:1047–1069
20. Looger LL, Dwyer MA, Smith JJ et al (2003) Computational design of receptor and sensor proteins with novel functions. Nature 423:185–190

21. Moore G, Levacher V, Bourguignon J et al (2001) Synthesis of a heterocyclic receptor for carboxylic acids. Tetrahedron Lett 42:261–263
22. Jaramillo A, Tortosa P, Rodrigo G et al (2007) Computational design of proteins with new functions. BMC Syst Biol 1:S15. doi:10.1186/1752-0509-1-S1-S15
23. Evans SM, Burrows CJ, Venanzi CA (1995) Design of cholic acid macrocycles as hosts for molecular recognition of monosaccharides. J Mol Struct 334:193–205
24. Cain JP, Mayorov AV, Cai M et al (2000) Design, synthesis, and biological evaluation of a new class of small molecule peptide mimetics targeting the melanocortin receptors. Bioorg Med Chem Lett 16:5462–5467
25. Biagi G, Bianucci AM, Coi A et al (2005) 2,9-Disubstituted-N^6-arylcarbamoyl-8-azaadenines as new selective A_3 adenosine receptor antagonists: synthesis, biochemical and molecular modelling studies. Bioorg Med Chem 13:4679–4693
26. Claramunt RM, Herranz F, María MDS et al (2005) Molecular recognition of biotin, barbital and tolbutamide with new synthetic receptors. Tetrahedron 61:5089–5100
27. Cristalli G, Costanzi S, Lambertucci C et al (2003) Purine and deazapurine nucleosides: synthetic approaches, molecular modelling and biological activity. Farmaco 58:193–204
28. Albert JS, Peczuh MW, Hamilton AD (1997) Design, synthesis and evaluation of synthetic receptors for the recognition of aspartate pairs in an α-helical conformation. Bioorg Med Chem 5:1455–1467
29. Napolitano E, Fiaschi R, Herman LW et al (1996) Synthesis and estrogen receptor binding of 17α,20E- and 17α,20Z-21-phenylthio- and 21-phenylseleno-19-norpregna-1,3,510,20-tetraene-3,17β-diols. Steroids 61:384–389
30. Carrasco MR, Still WC (1995) Engineering of a synthetic receptor to alter peptide binding selectivity. Chem Biol 2:205–212
31. Brizzi A, Cascio MG, Brizzi V et al (2007) Design, synthesis, binding, and molecular modeling studies of new potent ligands of cannabinoid receptors. Bioorg Med Chem 15:5406–5416
32. Menuel S, Joly JP, Courcot B et al (2007) Synthesis and inclusion ability of a bis-β-cyclodextrin pseudo-cryptand towards Busulfan anticancer agent. Tetrahedron 63:1706–1714
33. Cain JP, Mayorov AV, Cai M et al (2006) Design, synthesis, and biological evaluation of a new class of small molecule peptide mimetics targeting the melanocortin receptors. Bioorg Med Chem Lett 16:5462–5467
34. Lassila JK, Privett HK, Allen BD et al (2006) Combinatorial methods for small molecule replacement in computational enzyme design. PNAS 45:16710–16715
35. Vleeschauwer MD, Vaillancourt M, Goudreau N et al (2001) Design and synthesis of a new Sialyl Lewis X Mimetic: how selective are the selectin receptors? Bioorg Med Chem Lett 11:1109–1112
36. Piletsky SA, Turner APF (eds) (2006) A new generation of chemical sensors based on MIPs, in molecular imprinting of polymers. Landes Bioscience, Georgetown, Texas, United States, pp 64–79 (Ch 6)
37. Subrahmanyam S (2003) Design of molecularly imprinted polymers for sensors and solid phase extraction. PhD Thesis, Cranfield University
38. Piletsky SA, Piletska EV, Chen B et al (2000) Chemical grafting of molecularly imprinted homopolymers to the surface of microplates – application of artificial adrenergic receptor in enzyme-linked assay for beta agonists determination. Anal Chem 72:4381–4385
39. Sellergren B (ed) (2001) Molecularly imprinted polymers, man-made mimics of antibodies and their practical application in analytical chemistry. Elsevier, Amsterdam, pp 59–70
40. Wulff G (2002) Molecular imprinting-a way to prepare effective mimics of natural antibodies and enzymes. Stud Surf Sci Catal 141:35–44
41. Shea KJ (1994) Molecular imprinting of synthetic network polymers: the de-novo synthesis of macromolecular binding and catalytic sites. Trends Polym Sci 19:9–14

42. Janiak DS, Kofinas P (2007) Molecular imprinting of peptides and proteins in aqueous media. Anal Bioanal Chem 389:399–404
43. Ansell RJ, Kriz D, Mosbach K (1996) Molecularly imprinted polymers for bioanalysis: chromatography, binding assays and biomimetic sensors. Curr Opin Biotechnol 7:89–94
44. Spivak D, Gilmore MA, Shea KJ (1997) Evaluation of binding and origins of specificity of 9-ethyladenine imprinted polymers. J Am Chem Soc 119:4388–4393
45. Lanza F, Sellergren B (2001) The application of molecular imprinting technology to solid phase extraction. Chromatographia 53:599–611
46. Masque N, Marce RM, Borrull F (2001) Molecularly imprinted polymers: new tailor-made materials for selective solid-phase extraction. Trends Anal Chem 20:477–486
47. Wulff G (2002) Enzyme-like catalysis by molecularly imprinted polymer. Chem Rev 102:1–27
48. Haupt K, Mosbach K (2000) Molecularly imprinted polymers and their use in biomimetic sensors. Chem Rev 100:2495–2504
49. Kriz D, Ramstrom O, Mosbach K (1997) Molecular imprinting-new possibilities for sensor technology. Anal Chem 69:A345–A349
50. Kriz D, Ramström O, Svensson A et al (1995) Introducing biomimetic sensors based on molecularly imprinted polymers as recognition elements. Anal Chem 67:2142–2144
51. Jakusch M, Janotta M, Mizaikoff B et al (1999) Molecularly imprinted polymers and infrared evanescent wave spectroscopy – a chemical sensors approach. Anal Chem 71:4786–4791
52. Greene NT, Shimizu KD (2005) Colorimetric molecularly imprinted polymer sensor array using dye displacement. J Am Chem Soc 127:5695–5700
53. Hennion MC, Pichon V (2003) Immuno-based sample preparation for trace analysis. J Chromatogr A 1000:29–52
54. Wei S, Molinelli A, Mizaikoff B (2006) Molecularly imprinted micro and nanospheres for the selective recognition of 17β-estradiol. Biosens Bioelectron 21:1943–1951
55. Vidyasankar S, Ru M, Arnold FH (1997) Molecularly imprinted ligand-exchange adsorbents for the chiral separation of underivatized amino acids. J Chromatogr A 775:51–63
56. Takeuchi T, Fakuma D, Matsui J (1999) Combinatorial molecular imprinting: an approach to synthetic polymer receptors. Anal Chem 71:285–290
57. Takeuchi T, Dobashi A, Kimura K (2000) Molecular imprinting of biotin derivatives and its application to competitive binding assay using nonisotopic labeled ligands. Anal Chem 72:2418–2422
58. Takeuchi T, Fukuma D, Matsui J et al (2001) Combinatorial molecular imprinting for formation of atrazine decomposing polymers. Chem Lett 1:530–531
59. Lanza F, Sellergren B (1999) Method for synthesis and screening of large groups of molecularly imprinted polymers. Anal Chem 71:2092–2096
60. Lanza F, Hall AJ, Sellergren B et al (2001) Development of a semiautomated procedure for the synthesis and evaluation of molecularly imprinted polymers applied to the search for functional monomers for phenytoin and nifedipine. Anal Chim Acta 435:91–106
61. Chen B, Day RM Subrahmanyam S et al (2001) Molecularly imprinted polymer. WO0155235 A1, 2 Aug 2001
62. Piletsky S, Karim K, Piletska EV et al (2001) Recognition of ephedrine enantiomers by molecularly imprinted polymers designed using a computational approach. Analyst 126:1826–1830
63. Subrahmanyam S, Piletsky S, Piletska E et al (2001) Bite-and-Switch approach using computationally designed molecularly imprinted polymers for sensing of creatinine. Biosens Bioelectron 16:631–637
64. Chianella I, Lotierzo M, Piletsky SA et al (2002) Rational design of a polymer specific for microcystin-LR using a computational approach. Anal Chem 74:1288–1293
65. Pavel D, Lagowski J, Lepage CJ (2006) Computationally designed monomers for molecular imprinting of chemical warfare agents part V. Polymer 47:8389–8399

66. Wu L, Zhu K, Zhao M et al (2005) Theoretical and experimental study of nicotinamide molecularly imprinted polymers with different porogens. Anal Chim Acta 549:39–44
67. Dong W, Yan M, Zhang M et al (2005) A computational and experimental investigation of the interaction between the template molecule and the functional monomer used in the molecularly imprinted polymer. Anal Chim Acta 542:186–192
68. Farrington K, Magner E, Regan F (2006) Predicting the performance of molecularly imprinted polymers, selective extraction of caffeine by molecularly imprinted solid phase extraction. Anal Chim Acta 566:60–68
69. Baggiani C, Baravalle P, Anfossi L et al (2005) Comparison of pyrimethanil-imprinted beads and bulk polymer as stationary phase by non-linear chromatography. Anal Chim Acta 542:125–134
70. Chapuis F, Pichon V, Lanza F et al (2003) Optimization of the class-selective extraction of triazines from aqueous samples using a molecularly imprinted polymer by a comprehensive approach of the retention mechanism. J Chromatogr A 999:23–33
71. Meng Z, Yamazaki T, Sode K (2004) A molecularly imprinted catalyst designed by a computational approach in catalysing a transesterification process. Biosens Bioelectron 20:1068–1075
72. Benito-Peña E, Moreno-Bondi M, Aparicio S et al (2006) Molecular engineering of fluorescent penicillins for molecularly imprinted polymer assays. Anal Chem 78:2019–2027
73. Rathbone DL, Ge Y (2001) Selectivity of response in fluorescent polymers imprinted with N1-benzylidene pyridine-2-carboxamidrazones. Anal Chim Acta 435:129–136
74. Davies MP, De BV, Perrett D (2004) Approaches to the rational design of molecularly imprinted polymers. Anal Chim Acta 504:7–14
75. Nantasenamat C, Naenna T, Ayudhya CIN et al (2005) Quantitative prediction of imprinting factor of molecularly imprinted polymers by artificial neural network. J Comput Aided Mol Des 19:509–524
76. Nantasenamat C, Ayudhya CIN, Naenna T et al (2007) Quantitative structure-imprinting factor relationship of molecularly imprinted polymers. Biosens Bioelectron 22:3309–3317
77. Turner NW, Piletska EV, Karim K et al (2004) Effect of the solvent on recognition properties of molecularly imprinted polymer specific for ochratoxin A. Biosens Bioelectron 20:1060–1067
78. Chianella I, Karim K, Piletska EV et al (2006) Computational design and synthesis of molecularly imprinted polymers with high binding capacity for pharmaceutical applications-model case: adsorbent for abacavir. Anal Chim Acta 559:73–78
79. Piletska EV, Piletsky SA, Karim K et al (2004) Biotin-specific synthetic receptors prepared using molecular imprinting. Anal Chim Acta 504:179–183
80. Barragán IS, Karim K, Fernández JMC et al (2007) A molecularly imprinted polymer for carbaryl determination in water. Sens Actuators B 123:798–804
81. Piletska EV, Romero GM, Chianella I et al (2005) Towards the development of multisensor for drugs of abuse based on molecular imprinted polymers. Anal Chim Acta 542:111–117
82. Piletska EV, Turner NW, Turner APF et al (2005) Controlled release of the herbicide simazine from computationally designed molecularly imprinted polymers. J Control Release 108:132–139
83. Breton F, Rouillon R, Piletska EV et al (2007) Virtual imprinting as a tool to design efficient MIPs for photosynthesis-inhibiting herbicides. Biosens Bioelectron 22:1948–1954
84. Piletsky S, Piletska EV, Karim K et al (2004) Custom synthesis of molecular imprinted polymers for biotechnological application. Preparation of a polymer selective for tylosin. Anal Chim Acta 504:123–130
85. Cavaleiro E, Chianella I, Whitcombe MJ et al (2011) Synthetic polymers capable of suppressing virulence of bacterial fish pathogens. Curr Opin Biotechnol 22:S79
86. Piletska E, Karim K, Coker R et al (2010) Development of the custom polymeric materials specific for aflatoxin B1 and ochratoxin A for application with the ToxiQuant T1 sensor tool. J Chromatogr A 16:2543–2547

87. Guerreiro A, Soares A, Piletska E et al (2008) Preliminary evaluation of new polymer matrix for solid-phase extraction of nonylphenol from water samples. Anal Chim Acta 612:99–104
88. Shiigi H, Kijima D, Ikenaga Y et al (2005) Molecular recognition for bile acids using a molecularly imprinted overoxidized polypyrrole film. J Electrochem Soc 152:H129–H134
89. Pascale M, Annalisa DG, Visconti A et al (2008) Use of itaconic acid-based polymers for solid-phase extraction of deoxynivalenol and application to pasta analysis. Anal Chim Acta 609:131–138
90. Muhammad T, Cui L, Jide W et al (2012) Rational design and synthesis of water-compatible molecularly imprinted polymers for selective solid phase extraction of amiodarone. Anal Chim Acta 709:98–104
91. Monti S, Cappelli C, Bronco S et al (2006) Towards the design of highly selective recognition sites into molecular imprinting polymers: a computational approach. Biosens Bioelectron 22:153–163
92. Yoshida M, Hatate Y, Uezu K et al (2000) Chiral-recognition polymer prepared by surface molecular imprinting technique. Colloid Surf A 169:259–269
93. Toorisaka E, Uezua K, Goto M et al (2003) A molecularly imprinted polymer that shows enzymatic activity. Biochem Eng J 14:85–91
94. Farrington K, Regan F (2007) Investigation of the nature of MIP recognition: the development and characterisation of a MIP for Ibuprofen. Biosens Bioelectron 22:1138–1146
95. Liu Y, Wang F, Tan T et al (2007) Study of the properties of molecularly imprinted polymers by computational and conformational analysis. Anal Chim Acta 581:137–146
96. Kubo T, Matsumoto H, Shiraishi F et al (2007) Selective separation of hydroxy polychlorinated biphenyls HO-PCBs by the structural recognition on the molecularly imprinted polymers: direct separation of the thyroid hormone active analogues from mixtures. Anal Chim Acta 589:180–185
97. Baggiani C, Anfossi L, Baravalle P et al (2005) Selectivity features of molecularly imprinted polymers recognising the carbamate group. Anal Chim Acta 531:199–207
98. Pavel D, Lagowski J (2005) Computationally designed monomers and polymers for molecular imprinting of theophylline part II. Polymer 46:7543–7556
99. Dong C, Li X, Guo Z et al (2009) Development of a model for the rational design of molecular imprinted polymer: computational approach for combined molecular dynamics/quantum mechanics calculations. Anal Chim Acta 647:117–124
100. Li Y, Li X, Li Y et al (2009) Selective recognition of veterinary drugs residues by artificial antibodies designed using a computational approach. Biomaterials 18:3205–3211
101. Lv Y, Lin Z, Tan T et al (2008) Application of molecular dynamics modeling for the prediction of selective adsorption properties of dimethoate imprinting polymer. Sens Actuators B Chem 28:15–23
102. Babakhani A, Talley TT, Taylor P et al (2009) A virtual screening study of the acetylcholine binding protein using a relaxed – complex approach. Comput Biol Chem 33:160–170
103. Toorisaka E, Uezu E, Goto M et al (2003) A molecularly imprinted polymer that shows enzymatic activity. Biochem Eng J 14:85–91
104. Henthorn DB, Peppas NA (2007) Molecular simulations of recognitive behavior of molecularly imprinted intelligent polymeric networks. Ind Eng Chem Res 46:6084–6091
105. Molinelli A, O'Mahony J, Nolan K et al (2005) Analyzing the mechanisms of selectivity in biomimetic self-assemblies via IR and NMR spectroscopy of prepolymerization solutions and molecular dynamics simulations. Anal Chem 77:5196–5204
106. O'Mahony J, Karlsson BCG, Mizaikoff B et al (2007) Correlated theoretical, spectroscopic and X-ray crystallographic studies of a non-covalent molecularly imprinted polymerisation system. Analyst 11:1161–1168
107. Chapuis F, Pichon V, Lanza F et al (2004) Retention mechanism of analytes in the solid-phase extraction process using molecularly imprinted polymers: application to the extraction of triazines from complex matrices. J Chromatogr B 804:93–101

108. Diñeiro Y, Menéndez MI, Blanco-López MC et al (2006) Computational predictions and experimental affinity distributions for a homovanillic acid molecularly imprinted polymer. Biosens Bioelectron 22:364–371
109. Kowalska A, Stobiecka A, Wysocki S (2009) A computational investigation of the interactions between harmane and the functional monomers commonly used in molecular imprinting. J Mol Struct (THEOCHEM) 901:88–95
110. Saloni J, Lipkowski P, Dasary SSR (2011) Theoretical study of molecular interactions of TNT, acrylic acid, and ethylene glycol dimethacrylate – elements of molecularly imprinted polymer modeling process. Polymer 52:1206–1216
111. Fu Q, Sanbe H, Kagawa C et al (2003) Uniformly sized molecularly imprinted polymer for S-nilvadipine-comparison of chiral recognition ability with HPLC chiral stationary phases based on a protein. Anal Chem 75:191–198
112. Holdsworth CI, Bowyer MC, Lennard C et al (2005) Formulation of cocaine-imprinted polymers utilizing molecular modelling and NMR analysis. Aust J Chem 58:315–320
113. Ogawa T, Hoshina K, Haginaka J et al (2005) Screening of bitterness-suppressing agents for quinine: the use of molecularly imprinted polymers. J Pharm Sci 94:353–362
114. Lai EPC, Feng SY (2003) Molecularly imprinted solid phase extraction for rapid screening of metformin. Microchem J 75:159–168
115. Wu L, Li Y (2004) Study on the recognition of templates and their analogues on molecularly imprinted polymer using computational and conformational analysis approaches. J Mol Recognit 17:567–574
116. Wang J, Guo R, Chen J et al (2005) Phenylurea herbicides-selective polymer prepared by molecular imprinting using N-4-isopropylphenyl-N′-butyleneurea as dummy template. Anal Chim Acta 540:307–315
117. Schwarz L, Holdsworth CI, McCluskey A et al (2004) Synthesis and evaluation of a molecularly imprinted polymer selective to 2,4,6-trichlorophenol. Aust J Chem 57:759–764
118. Li P, Rong F, Xie Y et al (2004) Study on the binding characteristic of S-naproxen imprinted polymer and the interactions between templates and monomers. J Anal Chem 59:939–944
119. Wu L, Li Y (2003) Separation procedures for the pharmacologically active components of rhubarb. Anal Chim Acta 482:175–181
120. Mukawa T, Goto T, Nariai H et al (2003) Novel strategy for molecular imprinting of phenolic compounds utilizing disulfide templates. J Pharm Biomed Anal 30:1943–1947
121. Tada M, Sasaki T, Iwasawa Y (2004) Design of a novel molecular-imprinted Rh–Amine complex on SiO_2 and its shape-selective catalysis for α-methylstyrene. J Phys Chem B Hydrogenation 108:2918–2930
122. Ren Y, Ma W, Ma J (2012) Synthesis and properties of bisphenol A molecular imprinted particle for selective recognition of BPA from water. J Colloid Interface Sci 367:355–361
123. Kempe H, Kempe M (2004) Novel methods for the synthesis of molecularly imprinted polymer bead libraries. Macromol Rapid Commun 25:315–320
124. Karim K, Breton F, Rouillon R et al (2005) How to find effective functional monomers for effective molecularly imprinted polymers? Adv Drug Del Rev 57:1795–1808
125. Nicholls IA (1995) Thermodynamic considerations for the design of and ligand recognition by molecularly imprinted polymers. Chem Lett 24:1035–1036
126. Nicholls IA, Andersson HS (2001) Thermodynamic principles underlying molecularly imprinted polymer formulation and ligand recognition. In: Sellergren (ed) Molecularly imprinted polymers, man-made mimics of antibodies and their practical application in analytical chemistry. Elsevier, Amsterdam, pp 59–70
127. Nicholls IA, Adbo K, Andersson HS et al (2001) Can we rationally design molecularly imprinted polymers? Anal Chim Acta 435:9–18
128. Andrews PR, Craik DJ, Martin JL (1984) Functional group contributions to drug-receptor interactions. J Med Chem 27:1648–1657
129. Nicholls IA (1998) Towards the rational design of molecularly imprinted polymers. J Mol Recognit 11:79–82

130. Dixon S, Blaney J, Weininger D (1993) Characterizing and satisfying the steric and chemical restraints of binding sites. Presented at the Third York Meeting
131. Payne AWR, Glen RC (1993) Molecular recognition using a binary genetic search algorithm. J Mol Graph 11:74–91
132. Nishibata Y, Itai A (1991) Automatic creation of dug candidate structures based on receptor structure – starting point for artificial lead generation. Tetrahedron 47:8985–8990
133. Böhm HJ (1992) The computer program LUDI: a new simple method for the de-novo design of enzyme inhibitors. J Comput Aided Mol Des 6:61–78
134. Gillett VJ, Myatt G, Zsoldos Z et al (1995) SPROUT, HIPPO and CAESA: tools for de novo structure generation and estimation of synthetic accessibility. Perspect Drug Discov Des 3:34–50
135. Eisen MB, Wiley DC, Karplus M et al (1994) HOOK: a program for finding novel molecular architectures that satisfy the chemical and steric requirements of a macromolecule binding site. Proteins 19:199–221
136. Clark DE, Frenkel D, Levy SA et al (1995) PRO LIGAND: an approach to de novo molecular design – application to the design of organic molecules. J Comput Aided Mol Des 9:13–32
137. Luthra R (2006) Computer based de novo design of drug like molecules, Lecture Seminar Report Depart Chem, University of Alabama, US
138. Goodsell DS, Olson AJ (1990) Automated docking of substrates to proteins by simulated annealing. Proteins 8:195–202
139. Pattabiraman N, Levitt M, Ferrin TE et al (1985) Computer graphics in real-time docking with energy calculation and minimization. J Comput Chem 6:432–436
140. Piletska EV, Piletsky SA, Subrahmanyam S et al (2000) A new reactive polymer suitable for covalent immobilization and monitoring of the primary amines. Polymer 42:3603–3608
141. EPA (1990) Fed Regist 35:5991
142. Gould H, Tobochnik J (1988) An introduction to computer simulation methods: applications to physical systems parts 1 and 2. Addison-Wesley, Reading, MA
143. Case DA, Pearlman DA, Caldwell JW et al (2002) AMBER 7. University of California, San Francisco
144. Bronco S, Cappelli C, Monti S (2004) Understanding the interaction between collagen and modifying agents: a theoretical perspective. J Phys Chem B 108:10101–10112
145. Monti S, Bronco S, Cappelli C (2005) Toward the supramolecular structure of collagen: a molecular dynamics approach. J Phys Chem B 109:11389–11398
146. Hobza P, Kabelac M, Sponer J et al (1997) Performance of empirical potentials AMBER, CFF95, CVFF, CHARMM, OPLS, POLTEV, semiempirical quantum chemical methods AMI, MNDO/M, PM3, and ab initio Hartree–Fock method for interaction of DNA bases: comparison with nonempirical beyond Hartree–Fock results. J Comput Chem 18:1136–1150
147. Becke AD (1993) Density functional thermochemistry. III. The role of exact exchange. J Chem Phys 98:5648–5652
148. Koch W, Holthausen MAC (2001) A chemists guide to density functional theory. Wiley-VCH, Berlin
149. Hehre WJ, Radom L, Schleyer PV et al (1986) Ab initio molecular orbital theory. Wiley, New York
150. Cieplak P, Cornell WD, Bayly CI et al (1995) Application of the multi-molecule and multi-conformational RESP methodology to biopolymers: charge derivation for DNA, RNA, and proteins. J Comput Chem 16:1357–1377
151. Bayly CI, Cieplak P, Cornell WD et al (1993) A well-behaved electrostatic potential based method using charge restraints for deriving atomic charges: the RESP model. J Phys Chem 97:10269–10280
152. Berendsen HJC, Postma JPM, VanGunsteren WF et al (1984) Molecular dynamics with coupling to an external bath. J Chem Phys 81:3684–3690
153. Grabuleda X, Jaime C, Kollman PA (2000) Molecular dynamics simulation studies of liquid acetonitrile: new six-site model. J Comput Chem 21:901–908

154. Ryckaert JP, Ciccotti G, Berendsen HJC (1977) Numerical integration of the cartesian equations of motion of a system with constraints: molecular dynamics of n-alkanes. J Comput Phys 23:327–341
155. Ewing TJA, Kuntz ID (1997) Critical evaluation of search algorithms used in automated molecular docking. J Comput Chem 18:1175–1189
156. Kuntz ID, Blaney JM, Oatley SJ et al (1982) A geometric approach to macromolecule–ligand interactions. J Mol Biol 161:269–288
157. Kuntz ID, Brooijmans N (2003) Molecular recognition and docking algorithms. Annu Rev Biophys Biomol Struct 32:335–373
158. Subrahmanyam S, Piletsky SA (2008) Computational design of molecularly imprinted polymers. In: Potyrailo RA, Mirsky VM (eds) Combinatorial methods for chemical and biological sensors. Springer, Germany
159. Schuettelkopf AW, Aalten DMF (2004) PRODRG: a tool for high throughput crystallography of protein–ligand complexes. Acta Cryst 60:1355
160. Gromacs User Manual version 3.2, http://www.gromacs.org/
161. Biomolecular Simulation: The GROMOS96 Manual and User Guide
162. Pavel D, Lagowski J (2005) Computationally designed monomers and polymers for molecular imprinting of theophylline and its derivatives part I. Polymer 46:7528–7542
163. Sun HJ (1994) Force field for computation of conformational energies, structures, and vibrational frequencies of aromatic polyesters. Comput Chem 15:752–768
164. Sun H, Mumby SJ, Maple JR et al (1994) An ab initio CFF93 all-atom force field for polycarbonates. J Am Chem Soc 116:2978–2987
165. Brostow W (1979) Science of materials. Wiley, New York
166. Stewart JJP (1989) Optimization of parameters for semiempirical methods II applications. J Comput Chem 10:221–264
167. Woolley RG (1991) Quantum chemistry beyond the Born–Oppenheimer approximation. J Mol Struct (THEOCHEM) 230:17–46
168. Sherwood P (2006) Unpublished data on the website, Daresbury Laboratory, U.K. http://www.docstoc.com/docs/82865507/Computational-Chemistry-at-Daresbury-Laboratory. Accessed on 28 Mar 2012
169. Frisch MJ, Trucks GW, Schlegel HB et al (1998) Gaussian 03, Ver 1997, Gaussian Inc., Pittsburgh
170. Frisch MJ, Trucks GW, Schlegel HB et al (2003) Gaussian 03, Revision B. 05. Gaussian, Inc., Pittsburgh
171. Lee CT, Yang WT, Parr RG (1988) Development of the Colle-Salvetti correlation-energy formula into a functional of the electron density. Phys Rev 37:785–789
172. Alexander C, Davidson L, Hayes W (2003) Imprinted polymers: artificial molecular recognition materials with applications in synthesis and catalysis. Tetrahedron 59:2025–2057
173. Brady L, Brzozowski AM, Derewenda ZS et al (1990) A serine protease triad forms the catalytic centre of a triacylglycerol lipase. Nature 343:767–770
174. Schrag JD, Li YG, Wu S et al (1991) Ser-His-Glu triad forms the catalytic site of the lipase from Geotrichum candidum. Nature 351:761–764
175. Baggiani C, Atanfossi L, Baravelle P et al (2005) Selectivity features of molecularly imprinted polymers recognising the carbamate group. Anal Chim Acta 531:199–207
176. Wu L, Sun B, Li Y et al (2003) Rapid electrophoretic separations in short capillaries using contactless conductivity detection and a sequential injection analysis manifold for hydrodynamic sample loading. Analyst 128:944–949
177. Araujo PW, Brereton RG (1996) Experimental design III. Quantification. Trends Anal Chem 15:156–163
178. Box GEP, Hunter WG, Hunter JS (1978) Statistics for experimenters. Wiley, New York
179. Steinke J, Sherrington DC, Dunkin IR (1995) Imprinting of synthetic polymers using molecular templates, advances in polymer science. Springer, Berlin

180. Koeber R, Fleischer C, Lanza F et al (2001) Evaluation of a multidimensional solid-phase extraction platform for highly selective on-line cleanup and high-throughput LC-MS analysis of triazines in river water samples using molecularly imprinted polymers. Anal Chem 73:2437–2444
181. Molinelli A, Weiss R, Mizaikoff B (2002) Advanced solid phase extraction using molecularly imprinted polymers for the determination of quercetin in red wine. J Agric Food Chem 50:1804–1808
182. Mijangos I, Villoslada FN, Guerreiro A et al (2006) Influence of initiator and different polymerisation conditions on performance of molecularly imprinted polymers. Biosens Bioelectron 22:381–387
183. Araujo PW, Brereton RG (1996) Experimental design I. Screening. Trends Anal Chem 15:26–31
184. Araujo PW, Brereton RG (1996) Experimental design II. Optimization. Trends Anal Chem 15:63–70
185. Castro B, Whitcombe MJ, Vulfson EN et al (2001) Molecular imprinting for the selective adsorption of organosulphur compounds present in fuels. Anal Chim Acta 435:83–90

MIP Sensors on the Way to Real-World Applications

Ghulam Mustafa and Peter A. Lieberzeit

Abstract Molecularly imprinted polymers are mostly confined to laboratories and their standardized environments. Chemical sensors based on MIP are no exception to this; however, there are increasing efforts to span the gap toward technological applications and thus exposing the devices to real-life environments and thereby assessing selectivity, sensitivity, and ruggedness of the respective sensors. In some application areas this has already been successful, namely in detecting volatile organics and their mixtures, sensing pesticides in environmental water samples, in assessing oxidation processes, e.g., in engine oils, and in some applications of bioanalysis targeting both signaling molecules/drugs and whole cells, viruses, or bacteria. Here, we summarize the selected aspects for transferring MIP strategies out from lab-bench conditions and highlight some of the successful examples.

Keywords MIP biosensing, MIP sensor materials, Oxidative degradation, Pesticides, Real-life matrices, Sensor arrays, Volatile organics

Contents

G. Mustafa and P.A. Lieberzeit (✉)
University of Vienna, Department of Analytical Chemistry, Währinger Strasse 38, 1090, Vienna, Austria
e-mail: Peter.Lieberzeit@univie.ac.at

S.A. Piletsky and M.J. Whitcombe (eds.), *Designing Receptors for the Next Generation of Biosensors*, Springer Series on Chemical Sensors and Biosensors (2013) 12: 167–188
DOI 10.1007/5346_2012_21, © Springer-Verlag Berlin Heidelberg 2012,
Published online: 4 July 2012

1 Molecular Imprinting

Life in general is based on a wide range of molecules working in a predefined ordered fashion ensuring survival. For functioning properly, it is imperative that these molecules interact with one another highly selectively in molecular recognition processes. Usually recognition takes place through non-covalent interactions in all inter- and intracellular events, e.g., binding of antibodies to antigens, interaction of receptors with hormones, and enzymatic catalysis. The excellent fit between natural binding partners has triggered the wish in researchers to fabricate artificial recognition materials [1] with bio-analogous binding properties. A both highly feasible and appreciably straightforward synthetic approach to achieve this is given by molecular imprinting. Here substrate-selective binding sites in man-made polymers are generated by means of a template-assisted method. Generally, the strategy can be defined as "the construction of ligand selective recognition sites in synthetic polymers where a template (atom, ion, molecule, complex or a molecular, ionic or macromolecular assembly, including microorganisms) is employed in order to facilitate recognition site formation during the covalent assembly of the bulk phase by a polymerization or polycondensation process, with subsequent removal of some or all of the template being necessary for recognition to occur in the spaces vacated by the templating species" [2]. In the early 1970s, related research approaches were extended from silica-based matrices to synthetic organic polymers. Since then numerous ideas and techniques have been developed in the field. Main credit for this "New Era" goes to three individuals [3], namely Wulff [4, 5], Mosbach [6], and Shea [7, 8] as a consequence of their respective pioneering work. Initially named host–guest polymerization, now the term "molecularly imprinted polymer (MIP)" is common. Interestingly, it was not used until 1993 and Mosbach himself has preferred "molecularly imprinted absorbents (MIA)" [9]. The general synthetic strategy for molecular imprinting is summarized in Fig. 1.

Usually, monomer(s) and a template compound that share complementary functional groups self-organize with one another forming association complexes. After adding suitable cross-linking monomers, the mixture is polymerized to stabilize the spatial arrangement of the preformed aggregates, which rigidly fixes it in the polymeric matrix. Extraction of template species from the polymer leaves behind cavities with complementary size, shape, and chemical functionality. Ideally, these cavities are highly selective and reversibly bind either the template itself or a closely related compound.

Generally speaking, molecular imprinting can be achieved in two ways, namely covalently and non-covalently. However, the non-covalent approach can be further classified into two categories, one based on hydrogen bonds to prearrange monomers and templates, and the other one relying on weaker types of interaction

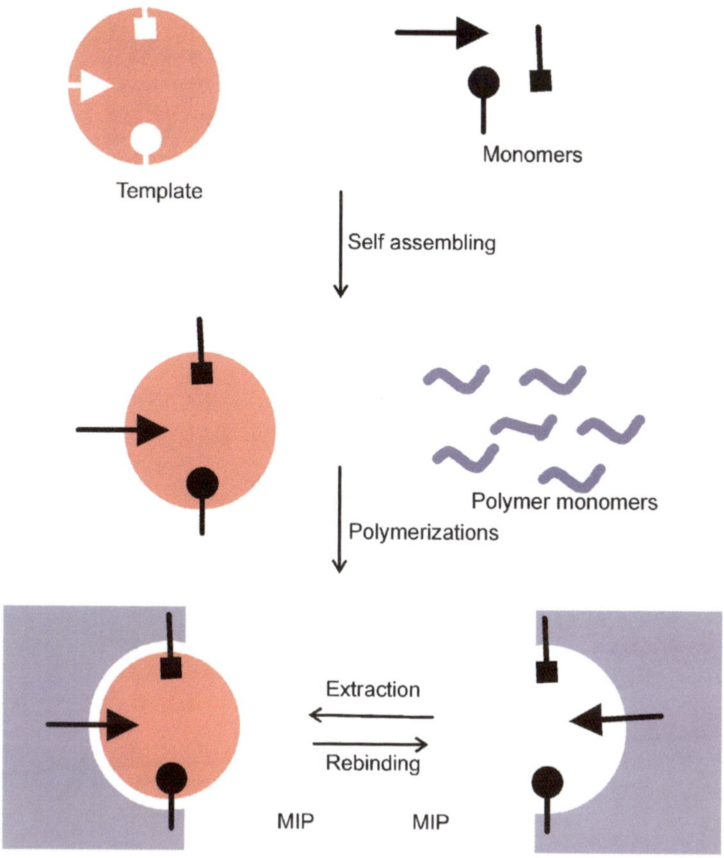

Fig. 1 General procedure for molecular imprinting

including Van der Waals forces. Covalent imprinting on the other hand relies on chemically binding the template to the monomer via a functionality that can easily be cleaved. Non-covalent imprinting is more preferably used for organic molecules having lower molar masses and sometimes even without pronounced functional groups, such as organic solvents or anesthetics, but larger species can also be suitable templates. After polymerization, the template is removed from the polymeric matrix by evaporation or washing and thus leaves behind geometrically adapted cavities and diffusion pathways in the way mentioned before. Surprisingly, the final polymers can reach similar selectivity and sensitivity as covalently imprinted ones [10]. MIPs attract increasing interest, because they combine a very straightforward synthesis method of an inherently technologically processable material with outstanding chemical recognition capabilities [11]. As the analyte directly influences the generation of cavities in the recognition material, the strategy provides a powerful tool to design highly affine sites even for analytes whose chemical composition is either not precisely known or that consist of a complex mixture. Furthermore, it can be extended to larger template species and hence allows for facile synthesis of receptors, e.g., against biopolymers, viruses, or cells. Generally speaking, MIPs combine three highly advantageous features [12]:

- First, they are highly affine and selective and in some cases may even compete with natural receptors in these features.
- Second, their synthesis procedure is very straightforward and usually inexpensive.
- Third, they usually are thermally and mechanically stable and thus have appreciable ruggedness.

Their interactions with the respective analytes are somewhat similar to those observed in natural antibody–antigen systems. However, in contrast to them, MIPs as sensor coatings usually yield reversible responses, which enables researchers to synthesize a new class of materials referred to as plastic antibodies or plastibodies [13–16]. In applications outside the scope of sensors, MIPs can, e.g., be optimized to catalyze a range of chemical reactions thus mimicking naturally occurring enzymes [17, 18]. Therefore it is even possible to synthesize an MIP for a compound or reaction for which no natural antibody or enzyme is known.

2 Molecular Imprinting Strategies

Depending on the nature of the analyte, two different imprinting protocols are employed to generate sensor layers, namely bulk imprinting and surface imprinting. Furthermore, synthesizing MIP nanoparticles is also receiving increasing attention.

2.1 Bulk Imprinting

Bulk imprinting is usually employed for molecules with masses of up to a few hundred atomic mass units, such as volatile organics, polycyclic aromatic hydrocarbons, pharmaceutically active compounds, environmental contaminants, and others. During polymerization, the respective template is directly added to the monomer mixture. Therefore, interaction sites are not only present on the surface of the respective sensor material, but they are also distributed within the whole bulk of the matrix, hence the term "bulk imprinting."

Consequently, the respective sensitive layer can accommodate a comparably large number of interaction sites, even though the majority of them may only be accessible via diffusion pathways within the polymeric matrix. Albeit sometimes leading to increased sensor response times (depending on layer height), in the case of small analyte molecules this guarantees the necessary sensitivity. This is of fundamental interest, since many of these templates are non-immunogenic and therefore not directly accessible for immunoassays due to the lack of suitable antibodies. A schematic protocol for bulk imprinting is shown in Fig. 2.

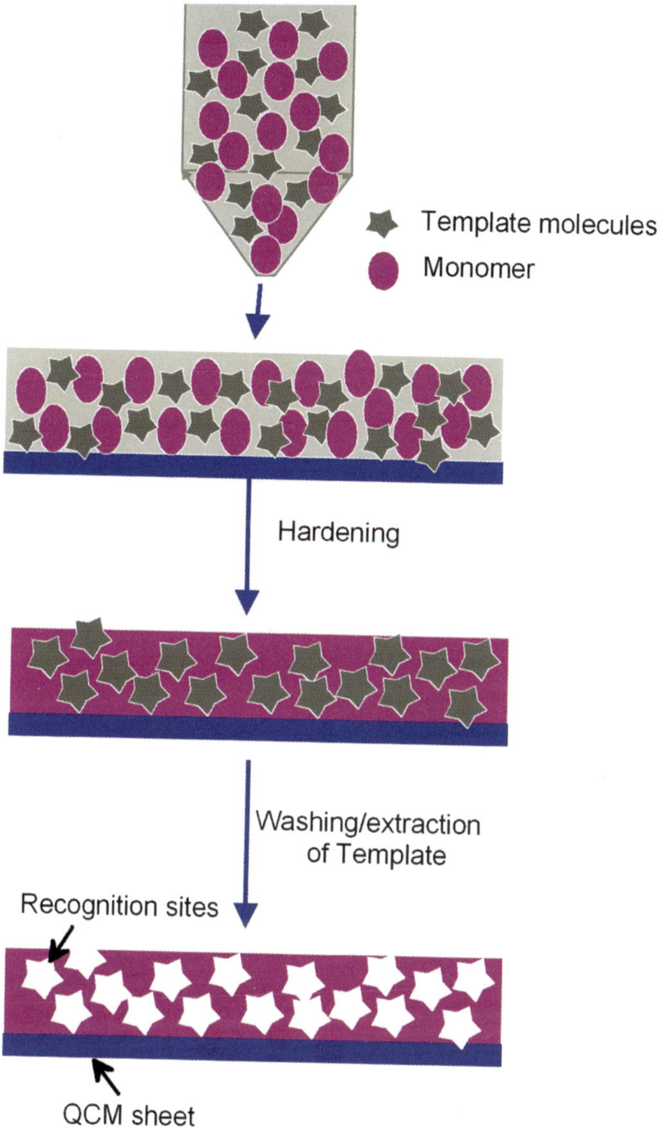

Fig. 2 General procedure for bulk imprinting

2.2 *Surface Imprinting*

When aiming at large templates, i.e., especially biospecies such as proteins, cells, or microorganisms, bulk imprinting fails to generate suitable recognition sites for three reasons: first, the resulting materials would be too thick for most sensor applications. Second, it would be very difficult to entirely remove the template from the matrix. And third, even if that should prove possible, diffusion into cavities lying further below the surface would be substantially sterically hindered.

To overcome these limitations, surface-imprinting approaches have been developed as shown in Fig. 3. In this case the respective template species are self-organized on a stamp surface which is then pressed onto the pre-polymerized oligomeric mixture coated onto the surface of the respective transducer. This again is followed by hardening the material and washing out the template. It has turned out that the selectivity of such "artificial antibodies" is determined by shape, size, and surface chemistry of the respective template species in the same way as for bulk approaches.

2.3 *Imprinted Nanoparticles*

One of the continuous major efforts in chemical or biochemical sensing is to achieve optimum sensitivity and selectivity of a sensor toward its corresponding analyte. The former can be achieved in different ways, e.g., by increasing the fundamental device frequency in the case of mass-sensitive transducers [19]. However, from the material point of view, either increasing the number of interaction sites within the respective recognition layer or increasing their accessibility is another feasible approach. Out of these two possibilities, the latter turned out to be more practicable [20], because increasing the number of recognition sites would require higher amounts of template, which can inhibit polymerization. Improved accessibility can be realized by applying nanoparticles rather than thin films as recognition materials.

Such MIP nanoparticles provide substantially increased surface area for analyte–receptor interactions as compared to thin films. Figure 4a shows a typical AFM image of glass surface coated with imprinted nanoparticles and Fig. 4b represents the mass-sensitive sensor responses of quartz crystal microbalances (QCM) coated with a thin film and MIP nanoparticles, respectively, toward folic acid. Evidently, the nanoparticle-MIP coated surface yields higher response because of shorter diffusion pathways and therefore an increased number of available interaction sites as compared to a thin film consisting of the same amount of selective material.

3 Molecular Imprinting in Chemical Sensing

MIPs are especially suitable for recognition in chemical sensing, because they combine high selectivity with comparably low cost per layer. Among others, this leads to constantly increasing number of publications in this field. Generally speaking, chemical sensors are miniaturized devices capable of continuously

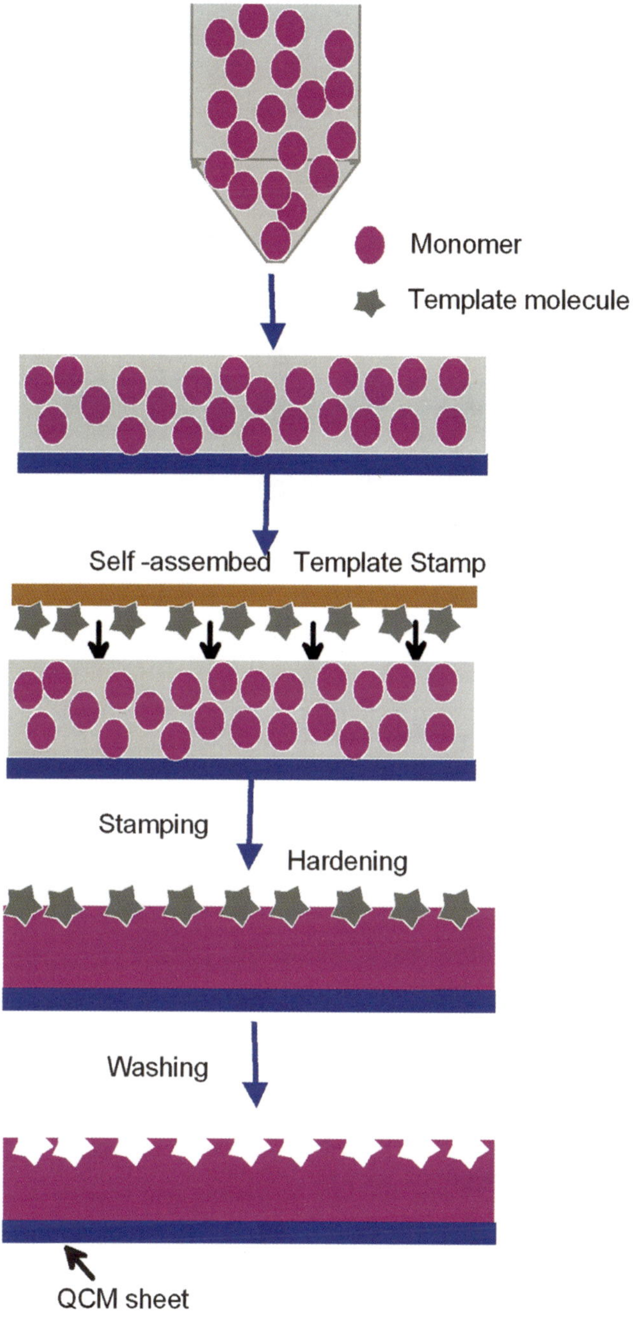

Fig. 3 General procedure for surface imprinting

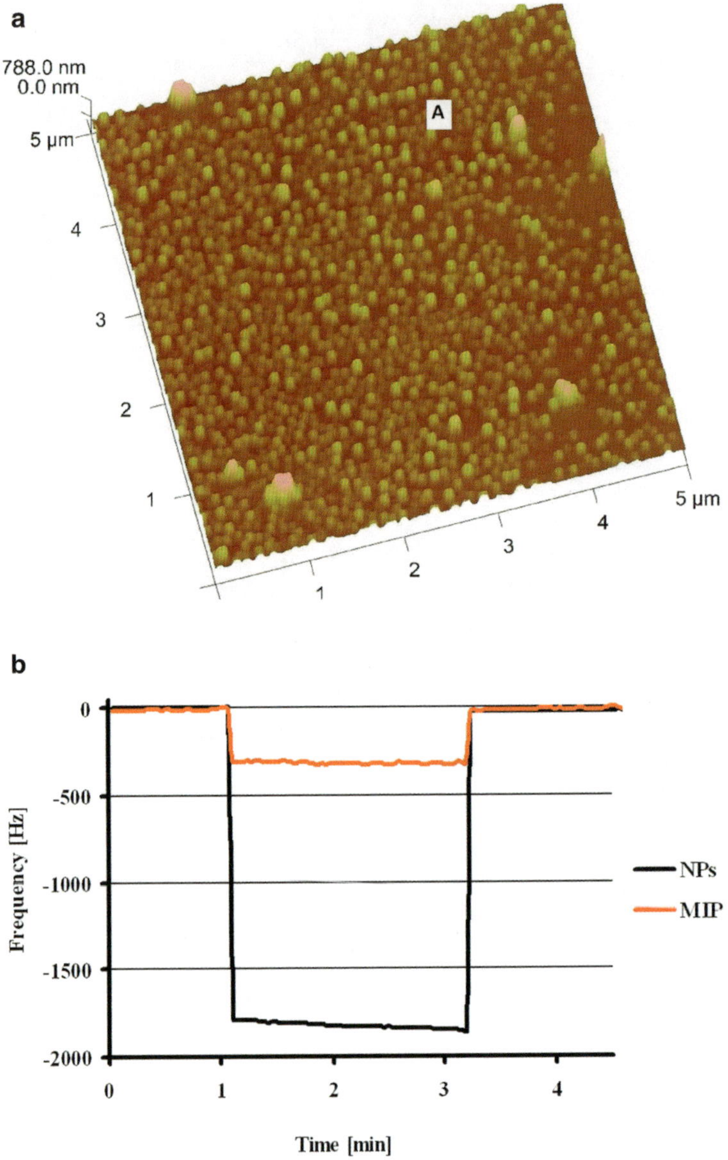

Fig. 4 (**a**) AFM image of molecularly imprinted nanaoparticles; (**b**) shows the comparison of MIP and nanoparticles sensor layers against 500 ppm of folic acid with same polymer system

detecting chemical constituents online in liquids or gases and converting this information in real time into an electrical or optical signal.

They possess many desirable and advantageous features [21, 22], among which are: small size, low cost, ease of operation, and manufacturing by established

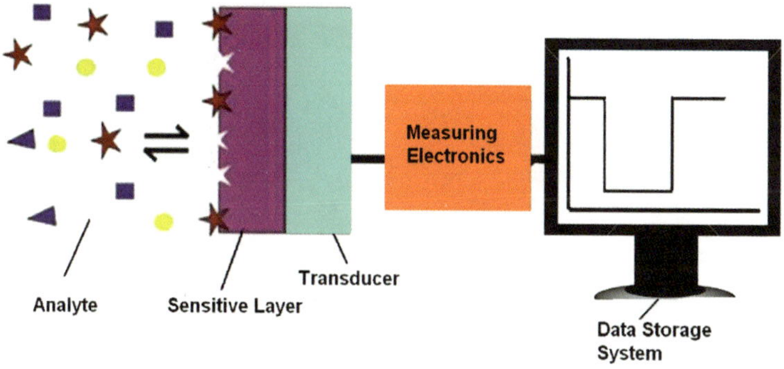

Fig. 5 General diagram of chemical sensor

technological methods. A general schematic diagram of a chemical sensor is shown in Fig. 5.

Generally speaking, a chemical sensor consists of a recognition layer that is sensitive to chemical changes taking place in the surrounding environment. Upon interaction with the target analyte, a change takes place in one of the physical properties of the layer. The transducer detects this change and converts it into a useful analytical signal and thus quantifiable data. The resulting electronic signals are usually amplified and recorded by a data processing unit, often including evaluation. Sensor layers play a crucial role in the efficiency of a chemical sensor, because they partly determine sensitivity and are mainly responsible for selectivity, response time, and lifetime of a chemical sensor. On the transducer side, a wide range of physical parameters can be exploited for generating the sensor signal. Hence, the most common chemical sensors can be classified into electrochemical, optical, piezoelectric, and semiconductor sensors.

Electrochemical sensors measure a change in, e.g., voltage ocurring at the selective layer on analyte exposure against a reference or also amperometric and voltammetric signals. One of the first MIP-based amperometric sensors was developed by Piletsky et al. [23] to determine aniline and phenol. Similarly, Vinokurov reported an early potentiometric MIP sensor [24] based on the electropolymerization of pyrrole, amines, and substituted phenols. In contrast to these, piezoelectric devices detect mass of an analyte and convert it to an electrically measurable quantity, usually a change in frequency.

The most commonly used piezoelectric material in mass-sensitive chemical sensing is AT-cut quartz with circular electrodes patterned on both sides leading to so-called QCM in which thickness shear waves are generated by applying electric potential between the electrodes. Figure 6 shows a typical example of such a device with dual-electrode geometry to ensure compensation of interferences caused, e.g., by changing temperature or viscosity. QCM are useful for achieving a broad range of sensitivity: for example Cabanilla et al. introduced a piezoelectric sensor for the detection of caffeine, reaching a dynamic range from 0.1 to 10 mg mL^{-1} [25]. The sensitivity of piezoelectric sensors is often somewhat lower by comparison with optical and electrochemical sensors; however, it can be

Fig. 6 Mass-sensitive transducer-quartz crystal microbalance (QCM)

improved by increasing the fundamental frequency of the respective transducer [26].

Especially when keeping real-world applications in mind, chemical and biochemical sensors require high sensitivity, selectivity, adsorption capacity, and reversibility. For instance, the selectivity of chemical sensor materials has to be substantially higher than in other applications, such as separation materials: a sensor can only make use of one plate of separation, whereas a chromatographic column can have tens to hundreds of thousands of them. Furthermore, high adsorption capacity of the layer is also of seminal importance, because the volume in which recognition takes place is usually limited. Therefore, the recognition layer also has substantial influence on the overall sensitivity of a sensor. As artificial layers are usually developed for long-term application of the respective system, the layer should show reversible interaction behavior with the analyte combined with low tendency for drift. The main challenge is that these parameters often contradict one another: high selectivity may mean reduced sensitivity and high sensitivity often means partly irreversible binding and therefore limited reusability.

4 Real-Life Applications of MIP

In real-life and online monitoring, the main challenge for an analytical chemist is to deal with very low concentrations of analytes in complex environments, especially under the strong influence of humidity and interfering compounds. MIPs have already received substantial interest by laboratory chemists. However, only a limited number of papers can be found where such materials have indeed been taken from the lab bench into real-life environments. These include virus detection in plant saps [27], monitoring oxidative degradation phenomena in oil matrices [28–30], sensing bacteria [31, 32], yeast cells [33], or volatile organic compounds [11, 34–40]. MIPs have also already been applied in the field of environmental monitoring, especially for sensing pesticides [41–44], endocrine-disrupting compounds, and heavy and toxic metals in soil and water. In the latter case, such materials can in principle also be used for waste water treatment. The application of chemical sensors is not restricted to single analyte detection within a matrix but can also be extended to complex mixtures, such as in engine oil degradation sensing [28–30]. In this case, not only is the application interesting, but it also covers a more unusual aspect of sensing: the overall changes in a rather complex matrix are translated into a single sensor response thus yielding a signal that indeed relies on chemical changes in the matrix, but without actually knowing them in detail. Of course, parallel detection of several compounds is also possible with MIP-based sensors: arrays can be constructed for such multicomponent detection and quantification that contain between two and a few dozen individual sensors. The data obtained from such systems have then to be evaluated by chemometric data analysis, including among other techniques artificial neural networks (ANN) or principal component analysis (PCA). The following paragraphs shall discuss in more detail some examples of the above-mentioned MIP applications.

4.1 MIP Sensors in Multicomponent Environments

Usually the detection of environmentally important contaminants faces the challenges of achieving high sensitivity and selectivity as a consequence of low analyte concentrations on one hand and complex matrices on the other hand. Therefore, there is still only a very limited number of chemosensors published for such applications. These include sensors for: alcohols, esters, ketones, aldehydes, organic acids, and terpenes from different sources. Their quantitative detection in the surrounding atmosphere is of importance for environmental, epidemiological, and toxicological studies [45], but may also be relevant to process control applications. An example for the latter case is given by an approach to determine the lead analytes produced in composting procedures, namely limonene, ethyl acetate, and aliphatic alcohols, by using an array of six MIP-coated QCM as transducer [11]. In such cases relatively high detection limits are an acceptable characteristic. Each analyte group is addressed by a tailor-made polymer, i.e., a polystyrene-based material for the aliphatic terpene, polyurethane for the aliphatic alcohols, and both systems for ethyl acetate.

The respective arrays (for example see Fig. 7) operate in the composter environment for several weeks and continuously monitor the headspace, giving rise to signals that can be validated by GC-MS. MIPs contribute to the success of these measurements in two ways: first, they add selectivity to the array, hence allowing for appreciable detection limits, and second, they do so by providing a matrix that can be exposed to highly humid environments (up to condensing at 40–45°C) over several weeks and then be reused. Such a combination of selectivities generated by both MIPs (that typically yield selectivity factors of 3–10 between closely related compounds) and chemometry allows one to distinguish between different terpenes emanating from a variety of natural sources [34]. The release of terpenes to the atmosphere has the potential to affect the global radiation balance due to their

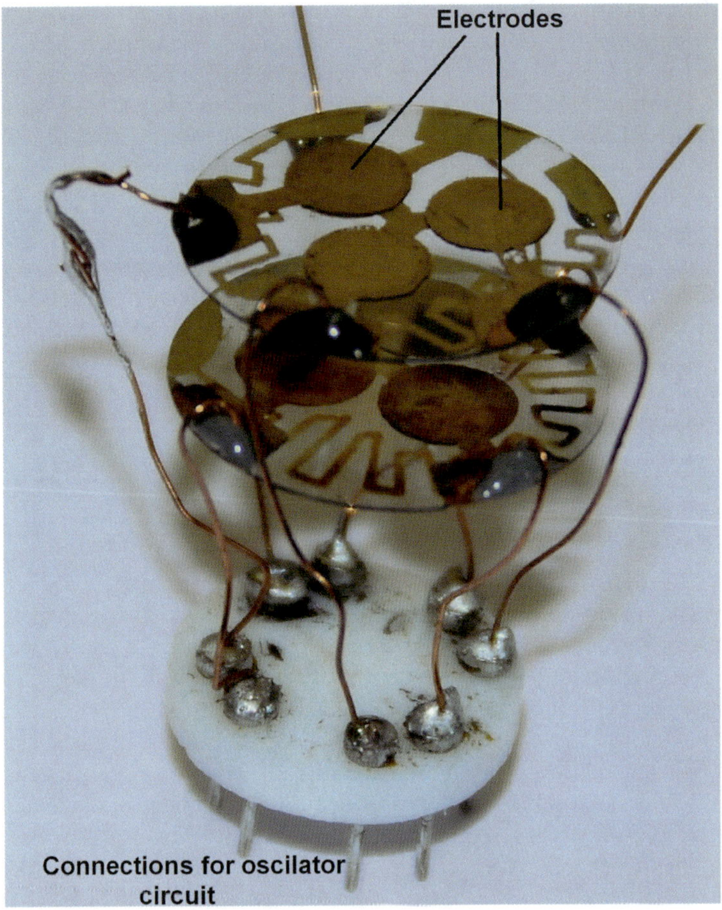

Fig. 7 QCM arrays consisting of six MIP-coated electrodes

vigorous reactions with ozone and OH in the troposphere. The respective MIPs for the analytes are accessible via non-covalent imprinting into polystyrene-divinyl benzene copolymers. The final sensor system exhibits isomeric selectivity and detection limits below 20 ppm in air, as can be seen in Fig. 8. A comparison of the sensor array data with GC-MS results for emanations from fresh basil showed a perfect correlation between the two data sets (Fig. 8).

Another approach toward an MIP-based mass-sensitive sensor for terpenes was reported by Percival et al. [46], who templated poly-(ethylene glycol dimethacrylate)-co-methacrylic acid with D- and L-menthol for detection in the liquid phase. Although probably not of "practical use" in real-life measurements yet, the paper very impressively shows the abilities of MIP: First, the QCM sensor could distinguish between the two enantiomers D-menthol and L-menthol, and secondly it featured a detection limit of 200 ppb for L-menthol in water. One reason

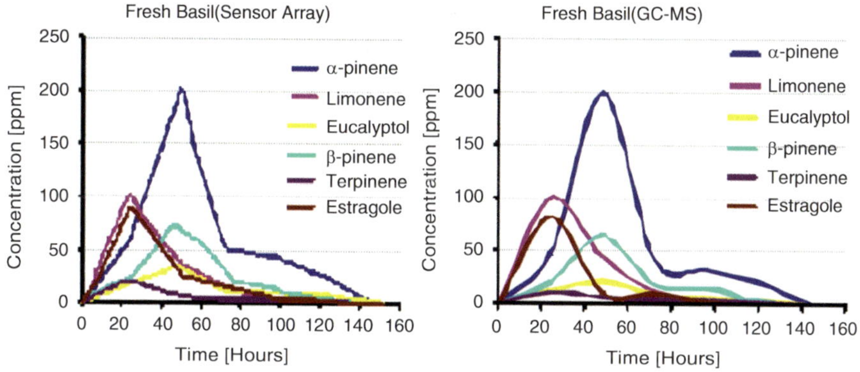

Fig. 8 QCM array data and GC-MS measurements comparison of emanated terpenes from fresh basil

for the comparably better performance of the acrylate-based system here may be the difference in polarity between water and the sensor material which favored extraction of the (rather nonpolar) analyte from its polar matrix.

A further—rather complex—real-life matrix points out some other appreciable aspects of MIP-based chemosensors, namely engine oil degradation sensing. Due to the sheer volume of production, automotive engine oils play a substantial economic and ecological role. The point of main interest of course is to find a technique that actually responds to chemical changes within the matrix rather than more general parameters, such as polarity or viscosity. MIP coated on QCM for instance can serve this need [28]. In this case the template is a model compound for degraded base oil components, namely capric acid. For reasons of ruggedness, the polymers in this case consist of inorganic frameworks—mostly titanate, synthesized by the sol–gel route, resulting in a reincorporation rate of about 70% of the template and thus appreciable sensitivity toward degradation. There is one main difference between this strategy and the sensor arrays introduced previously: while the latter are optimized to selectively detect a range of analytes in parallel, the former aims at translating a huge amount of oxidation reactions and their respective products into a single sensor signal.

4.2 *Pesticide MIP Sensors as an Example of Environmental Sensing*

Mainly as a consequence of sensitivity restrictions, there are hardly any MIPs reported that are employed in outdoor environments. One notable exception deals with an analyte (group) related to the agricultural sector: modern farming widely applies pesticides in the field, aiming to both comply with food regulations and achieve optimal yield and quality of crops. However, this has made them a major source of contamination of water, soil, and as trace contaminants in leafy vegetables and fruits [47]. This has led to attempts to design suitable sensor materials: as early as 1999, Sergeyeva et al. [48] developed a conductometric sensor for the detection of atrazine in aqueous solutions, using MIP membranes prepared by photopolymerization of methacrylic acid as the functional monomer and tri(ethylene glycol) dimethacrylate as the cross-linker. First of all, the imprinted polymer exhibited very high selectivity toward atrazine, with signals an order of magnitude higher than structurally related compounds, such as simazine, prometryn, and others. Furthermore, the sensor response is almost independent of environmental parameters, such as pH or salt load. Therefore the approach, as such, is highly suitable for real-life applications, especially since the lower limit of detection was 5 nM. Similarly Pardieu et al. [49] reported an electrochemical sensor for the detection of atrazine, in which a molecularly imprinted conducting polymer (MICP)—poly (3,4-ethylenedioxythiophene-co-thiophene-acetic acid)—was electrochemically synthesized on a platinum electrode in the presence of atrazine as template. This MICP-based sensor showed selectivity toward the triazine family with a wide dynamic range (10^{-7} mol L^{-1} to 1.5×10^{-2} mol L^{-1} atrazine) and low detection limit (10^{-7} mol L^{-1}). Detection in this case took place via cyclic voltammetry, an approach that had already proven useful in an earlier paper targeting 2,4-dichlorophenoxyacetic acid (2,4-D) [41]; however this did not include assessment of the sensor response in real-life samples.

Substantial efforts have been made for the fabrication of MIP-based sensors for the detection of these pesticides down to ppb levels. All these papers led to highly appreciable sensitivity that at least approaches the levels that are realistically required in environmental sensing. They do have electrochemical detection in common, which of course clearly indicates that the overall sensor properties are not governed only by the MIP, but by the entire sensor setup. Hence, the interplay between layer and device plays a crucial role for these applications.

A different approach to atrazine sensing has been published by Schirhagl et al. [43], who used a double imprinting strategy to generate "artificial antibodies." First, nanoparticles were templated with natural anti-atrazine immunoglobulins. In a second step these nanoparticles were used as a stamp for surface imprinting of polymer layers to produce antibody replicas in a copolymer system based on methacrylic acid, vinylpyrrolidone, and dihydroxyethylene-bisacrylamide (DHEBA). These layers exhibit nearly four times higher sensor response—and hence sensitivity—than their natural counterparts with the limit of detection being 0.04 μg mL^{-1}. Such a comparison is of course of academic interest, but also strongly influences the potential applications, as it allows for assessing the quality of artificial recognition materials and how they compare to the corresponding natural systems. There is one thing to be said about the atrazine-selective polymer: it is not surprising that this analyte yields outstandingly appreciable results. First, a heteroaromatic system that is inherently rigid dominates the structure. Second, triazine compounds have very pronounced functionalities that are able to undergo hydrogen bonding. Most of the polymer systems applied provide such hydrogen bonding between analyte and layer, leading to the appreciable sensor effects mentioned. This is worth mentioning when comparing MIP with antibody–antigen interactions—MIP are sometimes referred to as "artificial antibodies"—because the latter are usually governed by hydrophobic interactions despite the fact that most of the analytes are water soluble, at least to some extent.

4.3 MIP Sensors Applications in Biosensing

Right from the birth of molecular imprinting, the resulting receptor materials have frequently been referred to as "artificial antibodies" [13, 50, 51]. Of course there is substantial interest in MIPs on account of the hope that they can be used as a replacement for biological antibodies [52]. The reasons are twofold: first, antibodies for specific applications can come at a rather high price, and second, replacing biological materials by comparably low cost and mass-producible polymers is tempting, both economically and in terms of manufacturing technology. This "logical" proximity to antibodies led to the development of the so-called "MIA" or molecularly imprinted sorbent assays that have been reviewed by Ansell [51]. One of the earliest approaches was reported in 1993 by Vlatakis et al. [13] already aiming at two compounds that are of clinical interest. The first MIA focused on the detection of diazepam, a sedative, the other on theophylline, which is used as a bronchodilator. Both systems relied on acrylate-based polymers, once more making use of their ability to undergo hydrogen bonding and the facile polymerization properties of (meth)acrylate monomers (which has made acrylic polymers some of the most widely researched matrices in molecular imprinting). Comparing the results of the MIP immunoassays with enzyme-multiplied immunoassays (EMIT) showed that both methods gave the same results when used to determine the serum concentrations of theophylline in 32 patient samples. Similar results were observed in the case of diazepam.

Especially for smaller molecules such as the two above-mentioned ones, MIPs show some advantageous features when compared to their natural counterparts [3]:

- It is easier to produce an MIP against a small molecule, than an antibody.
- Antibodies are only functional in aqueous environments, but MIPs are inherently equally active in aqueous and nonaqueous conditions.
- Antibodies are sensitive and stable only within a specific range of temperature and pH, whereas MIPs are more rugged.

A further strong point of the imprinting technique is the fact that it is not restricted to molecular analytes, but can also be extended to larger structures. This opens up the possibility to address even entire microorganisms. However, for that purpose bulk imprinting is not feasible, because this would on one hand lead to layer heights in the range of some ten of micrometers and on the other hand make diffusion of the prospective analyte to the interaction centers almost impossible. Vulfson et al. [31, 32] were the first to overcome these problems by generating bacteria MIP on a polymer surface via emulsion polymerization. The outcome is MIP "latex" particles comprising surface cavities that reproduce the geometrical features of the template bacteria. The authors also pointed out the parallels between their approach and lithographic techniques to generate surface structures. This groundbreaking work proved the principal feasibility of microorganism imprinting. The first actual sensor measurements, however, were obtained with stamp surface-imprinting strategies of yeast cells into both polyurethane [33] and sol–gel material [53]. By applying SAW devices with a fundamental frequency of 433 MHz it is even possible to sense a recognition event between the MIP and a single yeast cell. Additionally, these MIPs show appreciable selectivity, not only against different microorganisms, such as bacteria, but also between different types of yeast, indicating that during "bioimprinting" not only the morphology of the respective species plays an important role, but also its surface chemistry. This observation is further corroborated by the design of sensor systems for biospecies that share common geometrical features, but differ in surface chemistry. One example for this is erythrocytes (i.e., red blood cells), which hold glycosidic moieties on their respective cell surfaces, which determines blood group. When applied as a template they lead to surface MIP in copolymer systems of acrylic/acrylamide monomers and *N*-vinyl pyrrolidone [54]. Admittedly, the monomer composition in this case is rather complex. However, this allows for tailoring the surface properties of the final polymer to optimally match the template cell. The sensitivity of the red blood cells does not allow a stamping protocol to be used, but sedimentation of erythrocytes in a solution deposited on top of a spin-coated pre-polymer leads to surface cavities having the original "donut" shape of the template cell (see Fig. 9).

Somewhat astonishingly, these cavities proved highly selective: on 10 MHz QCM the MIPs not only allow to distinguish between different blood groups, but even between subgroups that only differ by the amount of glycolipids on the cell surface. It is also worth mentioning that measurements can take place in a diluted whole blood matrix.

From the application point of view, however, viruses are among the most interesting biospecies for chemical sensing, because currently no analysis techniques exist for their rapid detection. The reason for this is their size: it is usually between 10 and 100 nm. Hence virions are not accessible via light scattering or other straightforward optical strategies that dominate in the field of automated bioanalyte detection. Knez et al. published rather early work on the selective nucleation of inorganic matrices on the tobacco mosaic virus (TMV) [27]. TMV is not only a plant pathogen that is very harmful to different crops (obviously tobacco, but also cucumber and pumpkin), but it also provides researchers with a robust template for imprinting. The first actual sensors achieved with TMV-imprinted polyurethanes were reported by Dickert et al. [55]. QCM coated with such MIP can detect TMV highly selectively, even in complex media—namely plant sap—without sample preparation, reaching a limit of detection of 1 mg L^{-1}. Experiments with different serotypes of the human rhinovirus (HRV) show that in this case differences in the surface protein structure also led to selectivity of the respective bio-MIP [56].

Finally, an MIP sensor for the detection of homocysteine in blood [57] highlights a further interesting approach for sensing as well as strategies to utilize molecular imprinting to find novel catalysts. This fluorescent sensor made use of an MIP

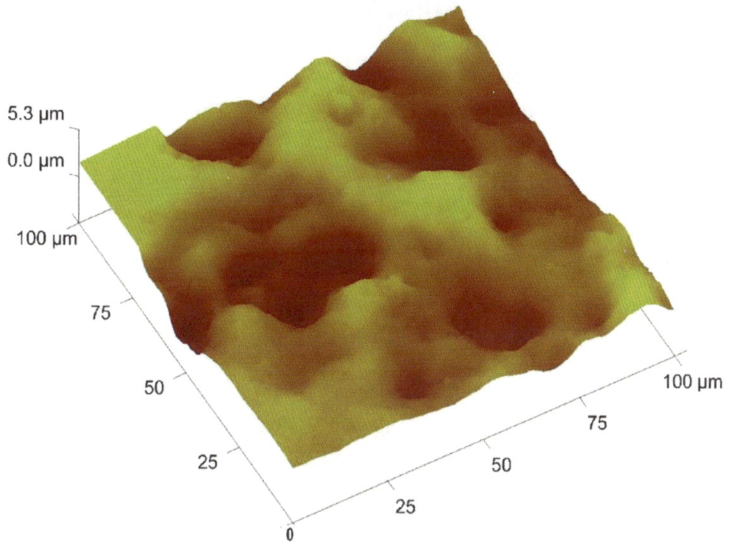

Fig. 9 AFM image of erythrocytes imprinted polymer surface

catalyst to derivatize D/L-homocysteine with *N*-(1-pyrenyl)maleimide. The product of this reaction served as the template during imprinting. The derivatization step introduces a fluorescent group into the homocysteine molecule. Therefore fluorescent measurements directly at the MIP allowed for determination directly in blood plasma.

5 Commercially Available MIPs for Analytical Applications

As already mentioned, MIPs have substantial application possibilities in the fields of separation, catalysis, biomimetic assays, and sensors [12, 58, 59]. However, the number of MIPs in commercial production still lags behind the expectations from laboratory experience. Nonetheless, there are a number of companies that have introduced MIP-based products for a range of applications, including sensing, on a commercial basis [58]. For example MIP Technologies, which is now part of Biotage (http://www.biotage.com), have commercialized products such as "MIP4SPE®" for solid phase extraction, "MIP4LC®" used for HPLC and "MIP4Discovery®" for drug screening. Together with Supelco (http://www.sigmaaldrich.com) they are also selling a product named "SupelMIP™ SPE" for solid phase extraction. Furthermore, Biotage has commercialized a selection of MIP-based materials named "AFFINILUTE MIP products" used for solid phase extraction (SPE) analysis of chloramphenicol, tobacco-specific nitrosamine derivatives, NNAL, triazines, clenbuterol, β-agonists, and β-blockers. Chrysalis Scientific Technologies Inc. is marketing MIPs for solid phase extraction (MIP[4]

SPF) of particular analytes: MIP[4]SPF products are available that target amphetamines (class specific), β-agonist drugs (class specific), β-blocker drugs (class specific), β-receptor (β-agonists + β-blockers), chloramphenicol, clenbuterol, fluoroquinolones (class specific), nitroimidazoles (class specific), NNAL (carcinogenic tobacco-specific nitrosamine), nonsteroidal anti-inflammatory drugs (NSAIDs), polycyclic aromatic hydrocarbons (PAHs), riboflavin (Vitamin B_2), tobacco-specific nitrosamines (NNN, NNK, NAB, and NAT), and triazine herbicides. Other companies dealing with commercial scale applications of MIPs are Oxonon (http://www.oxonon.net) working on MIP synthesis and characterization, Semorex (http://www.semorex.com) developing sensors for detection and diagnostics, Imego (http://www.imego.com) dealing with fabrication of MIP-based sensors for environmental pollutant, drugs and warfare detection. Generally, one can see that currently most commercially available matrices are targeting extraction rather than sensing. One reason for this may be the fact that batch-to-batch reproducibility of the materials is still an issue that has to be overcome to avoid the requirement for each sensor to be calibrated individually.

6 Summary and Outlook

Sensor technology based on MIPs has achieved substantial interest during the last few decades triggering a wide range of applications. However, to date only a few of them have been applied to “real-life” matrices, especially in the fields of detecting microorganisms or volatile organic compounds including complex environments. Future challenges in this field are still the achievement of the sensitivity, selectivity, and reproducibility of material mandatory for industrial requirements. Furthermore, systems will have to undergo the transition from laboratory bench to mass manufacturing of thin films and (preferably) nanoparticles in order to meet the current key challenge for commercialization.

References

1. Komiyama M, Takeuchi T, Mukawa T et al (2003) Molecular imprinting: from fundamentals to applications. Wiley-VCH, Weinheim
2. Alexander C, Andersson HS, Andersson LI et al (2006) Molecular imprinting science and technology: a survey of the literature for the years up to and including 2003. J Mol Recognit 19:106–180
3. Yan M, Ramström O (2005) Molecularly imprinted material-science and technology. Marcel Dekker, New York
4. Wulff G, Sarhan A (1972) Use of polymers with enzyme-analogous structures for the resolution of racemates. Angew Chem Int Ed Engl 11:341
5. Wulff G, Sarhan A, Zabrocki K et al (1973) Enzyme-analogue built polymers and their use for the resolution of racemates. Tetrahedron Lett 14:4329–4332

6. Arshady R, Mosbach K (1981) Synthesis of substrate-selective polymers by host–guest polymerization. Makromol Chem 182:687–692
7. Shea KJ, Dougherty TK (1986) Molecular recognition on synthetic amorphous surfaces: the influence of functional group positioning on the effectiveness of molecular recognition. J Am Chem Soc 108:1091–1093
8. Shea KJ, Sasaki DY (1989) On the control of microenvironment shape of functionalized network polymers prepared by template polymerization. J Am Chem Soc 111:3442–3444
9. Ramström O, Andersson LI, Mosbach K (1993) Recognition sites incorporating both pyridinyl and carboxy functionalities prepared by molecular imprinting. J Org Chem 58:7562–7564
10. Dickert FL, Hayden O (1999) Molecular imprinting in chemical sensing. Trends Anal Chem 18:192–198
11. Dickert FL, Lieberzeit PA, Achatz P et al (2004) QCM array for on-line-monitoring of composting procedures. Analyst 129:432–437
12. Piletsky SA, Alcock S, Turner AF (2001) Molecular imprinting: at the edge of the third millennium. Trends Biotechnol 19:9–12
13. Vlatakis G, Andersson LI, Muller R et al (1993) Drug assay using antibody mimics made by molecular imprinting. Nature 361:645–647
14. Haupt K, Mosbach K (1998) Plastic antibodies: developments and applications. Trends Biotechnol 16:468–478
15. Haupt K, Mosbach K (1999) Molecularly imprinted polymers in chemical and biological sensing. Biochem Soc Trans 27:344–350
16. Haupt K, Mosbach K (2000) Molecularly imprinted polymers and their use in biomimetic sensors. Chem Rev 100:2495–2504
17. Ramström O, Mosbach K (1999) Synthesis and catalysis by molecularly imprinted materials. Curr Opin Chem Biol 3:759–764
18. Wulff G (2002) Enzyme-like catalysis by molecularly imprinted polymers. Chem Rev 102:1–27
19. Dickert FL, Forth P, Fischerauer G et al (1998) SAW devices-sensitivity enhancement in going from 80 MHz to 1 GHz. Sens Actuators B Chem 46:120–125
20. di Natale C, Paolesse R, Macagnano A et al (2004) Sensitivity-selectivity balance in mass sensors: the case of metalloporphyrins. J Mater Chem 14:1281–1287
21. Dickert FL, Lieberzeit PA (2000) Solid-state sensors for field measurements of gases and vapours. In: Meyers RA (ed) Encyclopaedia of analytical chemistry. Wiley, Chichester, pp 3831–3855
22. Janata J (2010) Principles of chemical sensors. Springer, Dordrecht
23. Piletsky SA, Kurys YI, Rachkov AE et al (1994) Formation of matrix polymers sensitive to aniline and phenol. Russ J Electrochem 30:990–992
24. Vinokurov IA (1992) A new kind of redox sensor based on conducting polymer films. Sens Actuators B 10:31–35
25. Cabanilla S, Ebarvia BS, Sevilla F (2003) Piezoelectric biomimetic sensor for caffeine based on electrosynthesized polypyrrole. AsiaSENSE SENSOR 105–109
26. Jakoby B, Ismail GM, Byfield MP et al (1999) A novel molecularly imprinted thin film applied to a love wave gas sensor. Sens Actuators A 76:93–97
27. Knez M, Sumser M, Bittner AM et al (2004) Spatially selective nucleation of metal clusters on the tobacco mosaic virus. Adv Funct Mater 14:116–124
28. Lieberzeit PA, Glanznig G, Leidl A et al (2006) Ceramic materials for mass-sensitive sensors-detection of VOCs and monitoring oil degradation. Adv Sci Technol 45:1799–1802
29. Dickert FL, Greibl W, Rohrer A et al (2001) Sol–gel-coated quartz crystal microbalances for monitoring automotive oil degradation. Adv Mater 13:1327–1330
30. Dickert FL, Forth P, Lieberzeit PA et al (2000) Quality control of automotive engine oils with mass-sensitive chemical sensors-QCMs and molecularly imprinted polymers. Fresenius J Anal Chem 366:802–806
31. Aherne A, Alexander C, Payne MJ et al (1996) Bacteria-mediated lithography of polymer surfaces. J Am Chem Soc 118:8771–8772

32. Alexander C, Vulfson EN (1997) Imprinted polymers as protecting groups for regioselective modification of polyfunctional substrates. Adv Mater 9:751–755
33. Hayden O, Dickert FL (2001) Selective microorganism detection with cell surface imprinted polymers. Adv Mater 13:1480–1483
34. Iqbal N, Mustafa G, Rehman A et al (2010) QCM-arrays for sensing terpenes in fresh and dried herbs via bio-mimetic MIP layers. Sensors 10:6361–6376
35. Lieberzeit PA, Rehman A, Yaqub S et al (2008) Nanostructured particles and layers for sensing contaminants in air and water. Nano 3:205–208
36. Stetter JR, Jurs PC, Rose SL (1986) Detection of hazardous gases and vapors: pattern recognition analysis of data from an electrochemical sensor array. Anal Chem 58:860–866
37. Ji HS (1999) Selective piezoelectric odor sensing using molecularly imprinted polymers. Anal Chim Acta 390:93–100
38. Krupadam RJ (2011) An efficient fluorescent polymer sensing material for detection of traces of benzo[a]pyrene in environmental samples. Environ Chem Lett 9:389–395
39. Dickert FL, Forth P, Lieberzeit P et al (1998) Molecular imprinting in chemical sensing-detection of aromatic and halogenated hydrocarbons as well as polar solvent vapors. Fresenius J Anal Chem 360:759–762
40. Dickert FL, Thierer S (1996) Molecularly imprinted polymers for optochemical sensors. Adv Mater 8:987–990
41. Kröger S, Turner AP, Mosbach K et al (1999) Imprinted polymer based sensor system for herbicides using differential-pulse voltammetry on screen printed electrodes. Anal Chem 71:3698–3702
42. Xie C, Gao S, Zhou H et al (2011) Chemiluminescence sensor for sulfonylurea herbicide using molecular imprinted microspheres as recognition element. Luminescence 26:271–279
43. Schirhagl R, Latif U, Dickert FL (2011) Atrazine detection based on antibody replicas. J Mater Chem 21:14594–14598
44. Yaqub S, Latif U, Dickert FL (2011) Plastic antibodies as chemical sensor material for atrazine detection. Sens Actuators B 160:227–233
45. Iglesias RA, Tsow F, Wang R et al (2009) Hybrid separation and detection device for analysis of benzene, toluene, ethylbenzene, and xylenes in complex samples. Anal Chem 81:8930–8935
46. Percival CJ, Stanley S, Galle TM (2001) Crystal microbalances for the detection of terpenes. Anal Chem 73:4225–4228
47. González-Rodríguez RM, Rial-Otero R, Cancho-Grande B et al (2008) Occurrence of fungicide and insecticide residues in trade samples of leafy vegetables. Food Chem 107:1342–1347
48. Sergeyeva TA, Piletsky SA, Brovko AA et al (1999) Selective recognition of atrazine by molecularly imprinted polymer membranes. Development of conductometric sensor for herbicides detection. Anal Chim Acta 392:105–111
49. Pardieu E, Cheap H, Vedrine C et al (2009) Molecularly imprinted conducting polymer based electrochemical sensor for detection of atrazine. Anal Chim Acta 649:236–245
50. Ye L, Haupt K (2004) Molecularly imprinted polymers as antibody and receptor mimics for assays, sensors and drug discovery. Anal Bioanal Chem 378:1887–1897
51. Ansell RJ (2004) Molecularly imprinted polymers in pseudoimmunoassay. J Chromatogr B 804:151–165
52. Chen L, Xu S, Li J (2011) Recent advances in molecular imprinting technology: current status, challenges and highlighted applications. Chem Soc Rev 40:2922–2942
53. Dickert FL, Hayden O (2002) Bioimprinting of polymers and sol–gel-phases. Selective detection of yeasts with imprinted polymers. Anal Chem 74:1302–1306
54. Seifner A, Lieberzeit P, Jungbauer C et al (2009) Synthetic receptors for selectively detecting erythrocyte ABO subgroups. Anal Chim Acta 651:215–219
55. Dickert FL, Hayden O, Bindeus R et al (2004) Bioimprinted QCM sensors for virus detection-screening of plant sap. Anal Bioanal Chem 378:1929–1934

56. Jenik M, Schirhagl R, Schirk C et al (2009) Sensing picornaviruses using molecular imprinting techniques on a quartz crystal microbalance. Anal Chem 81:6320–6326
57. Chow CF, Lam MHW, Leung MKP (2002) Fluorescent sensing of homocysteine by molecular imprinting. Anal Chim Acta 466:17–30
58. Nostrum CFV (2005) Molecular imprinting: a new tool for drug innovation. Drug Discov Today Technol 1:119–124
59. Moreno-Bondi MC, Navarro-Villoslada F, Benito-Pena E et al (2008) Molecularly imprinted polymers as selective recognition elements in optical sensing. Curr Anal Chem 4:316–340

Molecularly Imprinted Au Nanoparticle Composites for Selective Sensing Applications

Ran Tel-Vered and Itamar Willner

Abstract In this study, a new method to imprint molecular recognition sites into Au nanoparticles (NPs) composites is described. The method includes the electropolymerization of thioaniline-functionalized Au NPs in the presence of imprint substrates that exhibit affinity interactions with the thioaniline-functionalized Au NPs or with a co-added ligand associated with the electropolymerizable NPs. Exclusion of the imprint substrate from the composite leads to the formation of selective imprinted sites in the Au NPs matrices. The imprinted matrices are implemented for the sensing of explosives, herbicides, saccharides, and ions. π-Donor–acceptor interactions, ionic interactions and H-bonds, or ligand–substrate interactions are used to generate the imprinted sites. The coupling between the localized plasmon of the NPs and the surface plasmon wave of the support is used to amplify the dielectric changes occurring in the NPs matrices upon the binding of the analytes to the imprinted sites, thus enabling the surface plasmon resonance (SPR) transduction of the sensing events. The imprinted Au NPs matrices demonstrate highly selective, stereoselective, and chiroselective sensing performance.

Keywords Electropolymerization, Molecular imprinting, Nanoparticles, Sensors, Surface plasmon resonance

Contents

R. Tel-Vered and I. Willner (✉)
The Institute of Chemistry, The Hebrew University of Jerusalem, Jerusalem 91904, Israel
e-mail: willnea@vms.huji.ac.il

S.A. Piletsky and M.J. Whitcombe (eds.), *Designing Receptors for the Next Generation of Biosensors*, Springer Series on Chemical Sensors and Biosensors (2013) 12: 189–212
DOI 10.1007/5346_2012_18,
Published online: 27 June 2012

1 Introduction

During the past several decades, and in accord with the rapidly growing field of nanotechnology, a considerable amount of scientific efforts has been directed toward research that includes metal nanoparticles (NPs). Due to the quantum-size dimensions, metallic NPs offer unique electrical [1–3], optical [4], and catalytic functions [5], which allow their diverse applications in many different fields. For example, metallic NPs were recently reported as promising building blocks of enhanced sensing platforms [6], or nanoscale devices [7], or as catalysts for a variety of chemical reactions [8]. Among the different metallic NPs, Au NPs are most extensively investigated. The aggregation of Au NPs as a result of sensing events, and the subsequent formation of interparticle coupled plasmon excitons were used for the development of colorimetric sensors [9, 10]. For example, the red-to-blue color changes arising from the aggregation of Au NPs were used to detect phosphatase activity [11], polynucleotides [12], or alkali ions [13]. Also, the occurrence of shifts in the plasmon absorption bands of Au NPs upon binding of different analytes, and due to the consequent emergence of dielectric changes on the surface, was used in the development of optical sensors, such as for dopamine [14], adrenaline [15], cholesterol [16], pH [17] and biosensors that probed DNA hybridization [18], or the formation of aptamer-substrate [19] or antigen–antibody [20, 21] complexes.

In recent years electropolymerized films on electrodes have played a key role as functional materials for sensing [22–25], in the actuation of a mechanical motion in microdevices [26, 27], in the fabrication of solar cells [28, 29] or light-emitting diodes [30, 31], and for the bioelectrocatalytic activation of enzymes [32–34]. The methodology of imprinting organic or inorganic polymer matrices is known since the twentieth century, and has been extensively developed by Mosbach [35–37] and Wulff [38–40]. This technology still constitutes a simple, yet a solid means, for generating specific molecular recognition sites in polymer systems [41–43]. Over the years, a broad range of applications based on molecularly imprinted polymers have been demonstrated, including their use as separation and controlled release matrices [44], as catalytic supports [45, 46], or as sensing materials [47]. In this context, particularly interesting is the application of imprinted polymers immobilized on electrodes for sensing. Despite the inherent advantages provided by the imprinted sensing electrodes, which are primarily associated with the formation of molecular templates for the selective binding of analytes, several fundamental limitations related to these electrodes are often encountered; (1) The density of the imprinted sites is typically low, and thus relatively thick polymer interfaces are required. This results in a slow diffusion of the analyte into the thick sensing matrix, leading to long response times for the sensor devices. (2) The need for a thick sensing layer often perturbs the communication between the bound analyte and the electrode transducer, causing a substantial attenuation to the readout signal associated with the recognition events.

Au NPs possess several attractive characteristics in serving as building blocks for molecularly imprinted sensing platforms on electrodes: (1) The nanometric dimensions of the particles and the well-developed, yet simple, surface chemistry associated with the functionalization of Au surfaces, imply that the particles may be chemically modified and aggregated by different means to yield a high surface area, porous, Au NPs assemblies. Using the appropriate synthetic cross-linking conditions, and introducing a target analyte-molecule into the reaction chamber, imprinted sites corresponding to the molecular contours of the target analyte can, then, be generated within the cross-linked Au NPs composite. The high porosity associated with the imprinted Au NPs composite is expected to facilitate the diffusion of the analyte within the layer, and, thus, the association of the analyte with the recognition sites becomes rapid, leading to fast response times of the sensing system. (2) The chemical modification of the exterior capping layer of the Au NPs by functional molecules, such as electropolymerizable and/or ligand units, yields a simple and controllable route for an electrochemical synthesis of Au NPs arrays, acting as receptors for specific target analytes. Also, the versatility of the available methods for the size- and shape- controlled synthesis of the Au NPs is expected to affect the resulting sensing performance by the imprinted matrices and to play a major role in the optimization of the sensors. (3) Several intrinsic features of Au NPs, such as the occurrence of a localized plasmon upon their excitation by light, provide unique optical enhancement phenomena in close proximity to the interface of the NPs, and can be, thus, used for the transduction and amplification of recognition events. Specifically, when functionalized Au NPs are electropolymerized on a Au-coated surface, the coupling between the localized surface plasmon of the metal NPs and the surface plasmon wave associated with the thin metal film leads to a shift in the surface plasmon resonance (SPR) energy. The effects of this coupling were theoretically addressed [48], and the changes in the SPR curves, resulting from the variations in the dielectric properties at the vicinity of the Au surface, were discussed.

2 Molecularly Imprinted Bis-Aniline-Cross-linked Au NPs Matrices for Sensing

Recently, we developed a generic method to electropolymerize assemblies of bis-aniline-cross-linked Au NPs on Au electrodes (Fig. 1a). The procedure involves the synthesis of Au NPs (ca. 4 nm diameter) modified with a protective capping layer consisting of electropolymerizable thioaniline groups (**1**), and mercaptoethane sulfonic acid (**2**) stabilizing units. The subsequent electropolymerization of the functionalized Au NPs is carried out by the application of a fixed number of potential cycles, between −0.35 and 0.8 V vs. a Ag quasi reference electrode (Ag QRE), on a

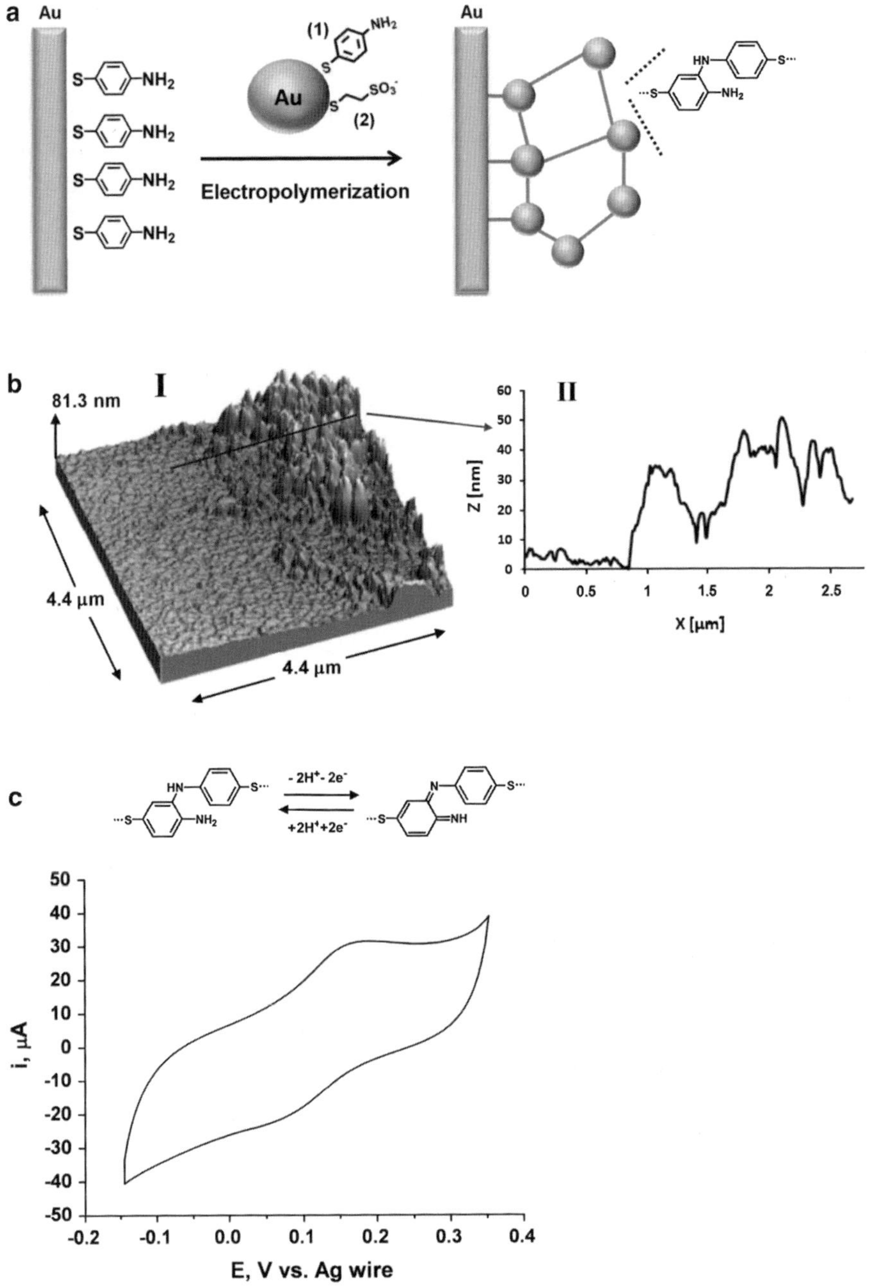

Fig. 1 (**a**) Schematic presentation for the electrochemical synthesis of a bis-aniline-cross-linked Au NPs composite-modified Au electrode. (**b**) I: An AFM image demonstrating the morphology of the Au NPs composite at the border region between the electropolymerized film and the bare Au surface. II: Cross-sectional analysis of the electropolymerized composite. (**c**) Cyclic voltammogram

thioaniline monolayer-modified Au-coated glass surface. Due to the electronic conductivity of the Au NPs, the electropolymerization process yields a three-dimensional film that is being gradually deposited on the surface of the electrode. An AFM analysis corresponding to the border region between the electropolymerized surface and the bare Au support reveals a complex morphology (Fig. 1b). Evidently, the composite consists of aggregated domains exhibiting an average thickness of ca. 30 nm and a nonhomogenous topography. A cyclic voltammetry analysis of the bis-aniline-cross-linked Au NPs deposit at pH = 7.2 reveals a quasi-reversible redox wave around 0.1 V vs. Ag QRE (Fig. 1c) consistent with the electrochemical transformation of the bridging units between the reduced bis-aniline and the oxidized quinoid states.

The simplicity associated with the one-step electrochemical synthesis of the bis-aniline-cross-linked Au NPs composite, the ability to control its electrochemical preparative conditions, and the diversity in the functionalization of the Au NPs building blocks suggested the application of the electrodes as attractive host materials for molecular imprinting. Upon their synthesis, the imprinted electrodes were applied in different fields, including the sensitive and selective detection of different target analytes [49–56], as photochemically and/or electrochemically triggered "sponges" [57, 58], the signal-controlled wettability of surfaces [59, 60], and for the generation of enhanced photoelectrochemical cells [61].

By the incorporation of a substrate that demonstrates high structural and/or chemical analogy to the selected target analyte in the electropolymerization mixture, and by designing high affinity interactions between the substrate and the bis-aniline-bridging units (or the capping of the Au NPs, for further information *vide infra*), specific imprinted sites for the target analyte in the Au NPs composite can be generated (Fig. 2). Different host–guest interactions between the matrix and the template molecule facilitate the imprinting process, and these include donor–acceptor, ionic (acid–base) interactions, hydrogen bonds, and ligand complexation. It should be noted that the analog imprint molecules are often significantly more soluble in the electropolymerization medium in comparison to the target analytes themselves. This enhanced solubility is essential for increasing the density of imprinted sites in the electro-synthesized matrices. Following the entrapment of the imprint analog molecule in the matrix, it can be removed away by weakening the associative interactions. This process is typically carried out by rinsing off the composite for an appropriate time-interval with the pure background buffer solution, by using a chemical agent, or by the application of an external potential on the composite-modified electrode, allowing the transformation of the bridging units between their oxidized and reduced states. The extraction process yields a matrix containing high affinity imprinted sites with a remarkable similarity

Fig. 1 (continued) corresponding to the bis-aniline-cross-linked Au NPs composite, recorded in a 0.1 M, pH = 7.2 HEPES buffer at 100 mV s^{-1}. Part B – Reprinted with permission from ref. 59. Copyright 2011, American Chemical Society. Part C – Reprinted with permission from ref. 57. Copyright 2010, American Chemical Society

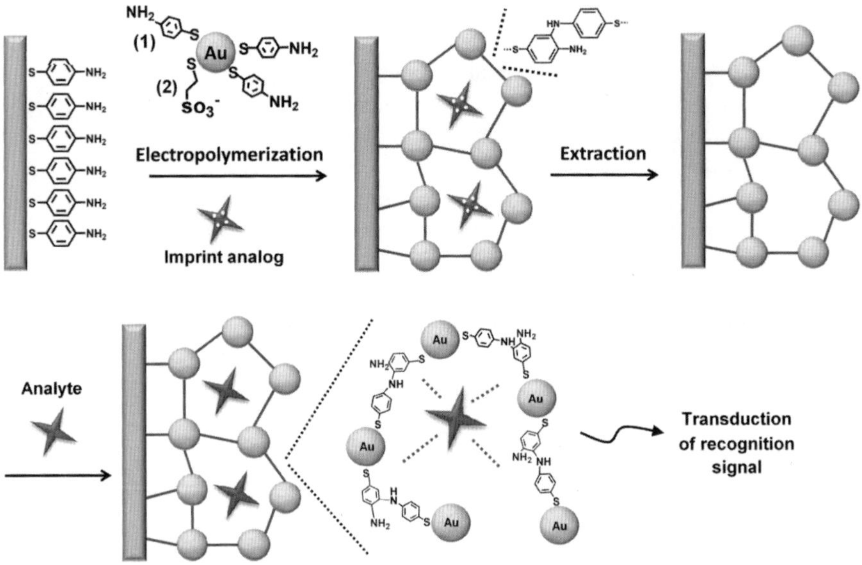

Fig. 2 Schematic presentation for the imprinting of molecular recognition sites into a bis-aniline-cross-linked Au NPs matrix for the specific sensing of a target molecule. Adapted with permission from ref. 52. Copyright 2011 American Chemical Society

to the target analyte structure. Upon reaction of the vacant imprinted matrix with the analyte, the affinity interactions between the analyte and the matrix facilitate its concentration within the imprinted sites. These recognition events are, then, transduced by a physical signal proportional to the concentration of the analyte.

The high affinity of the target analytes to the imprinted sites in the bis-aniline-cross-linked Au NPs composites is reflected by high association constants, typically ranging between 10^7 M^{-1} and 10^{12} M^{-1}, resulting in low detection limits (lowest detectable concentrations) for the sensing of different imprinted substrates by these matrices (for further information, *vide infra*). Furthermore, due to the delicate process of imprinting, the imprinted matrices demonstrate an impressive selectivity in the analysis of different substrates, and as will be shown, in addition to stereoselectivity, in some cases chiroselectivity can also observed (e.g., see Sect. 2.3).

An important issue relates to the selection of the readout technique. The appropriate choice of an analytical methodology that is used for the transduction of the recognition signals, upon the binding of the analyte to the matrix, depends substantially on the chemical and physical properties of the target analyte. For example, the association of nitro-aromatic explosive compounds with the bis-aniline-cross-linked Au NPs composite by π-donor–acceptor interactions can be followed by the electrochemical reduction of the nitrobenzene units associated with these explosives to hydroxylamine groups [62, 63]. Using the inherent electrochemical activity of nitro-aromatic explosives and the three-dimensional electronic conductivity by the Au NPs in the bis-aniline-cross-linked matrix, an electrochemical sensor for explosives, such as trinitrotoluene (TNT), dinitrotoluene (DNT), and mononitrotoluene (MNT) was demonstrated [49]. Whereas this methodology was proved to be efficient, facilitating a detection limit of 200 pM for TNT with an impressive selectivity between the structurally related nitoaromatic explosives, a major drawback of the electrochemical detection scheme is in the necessity for the target analyte to be redox active (the redox activity must also be outside the potential region associated with the broad quasi-reversible electrical response of the bis-aniline bridges). To overcome these obstacles, other readout techniques, such as surface plasmon resonance, SPR, were further implemented to follow the sensing capabilities of the bis-aniline-cross-linked Au NPs matrices.

SPR is a common method to probe refractive index changes occurring on thin metal films as a result of recognition events or chemical transformations [64, 65]. Due to its high sensitivity, SPR spectroscopy has been widely used to develop optical sensors and biosensors. For example, protein–protein interactions [66, 67] and the formation of antigen–antibody complexes [68, 69] were probed using SPR. The method is, however, limited to large molecules, such as proteins, or low-molecular-weight substrates which change their color upon various chemical transformations leading to substantial refractive index changes (that are translated to shifts in the SPR spectrum). Nonetheless, for systems with low coverage of the surface with the analytes, or when only small refractive index changes are induced by the binding of low-molecular-weight analytes, amplification of the SPR signal is required. This limitation can be resolved by the conjugation of labels that intensify the refractive index changes. For example, the conjugation of latex particles [70], liposomes [71], or secondary proteins [72] was used to amplify small refractive index changes, therefore leading to observable SPR shifts upon the recognition processes.

Metal NPs, such as Au or Ag NPs, exhibiting a localized plasmon, were extensively used to amplify SPR signals originating from recognition events. The coupling of the localized plasmon of the NPs with the surface plasmon wave was found to affect the SPR energy and, thus, to enhance the SPR shifts [48]. For example, the formation of immuno-complexes [20], DNA hybridization [18], and biocatalytic processes [73] were followed by the amplification of the SPR signals using the Au NPs as labels. The following sections will describe the SPR sensing of a variety of compounds on the respective molecularly imprinted bis-aniline-cross-linked Au NPs matrices. The different target analytes are associated with the matrices through different physical and chemical interactions that will be addressed in detail.

2.1 Molecular Imprinting of Recognition Sites in Au NPs Composites Through Donor–Acceptor Interactions

The electropolymerization of thioaniline-functionalized Au NPs onto a thioaniline-modified Au surface yields a thin film containing layers of Au NPs. The redox state of the bridging units which cross-link the NPs can be altered upon the application of an external potential on the composite-modified electrode. Specifically, by the application of a reducing potential, $E < 0.1$ V vs. Ag QRE, the bridging units undergo an electrochemical transition to a bis-aniline reduced state which exhibits π-donor characteristics, whereas at potentials higher than $E > 0.1$ V, the entire population of the bridges exist in an oxidized quinoid state of a π-acceptor nature. Accordingly, the introduction of an acceptor substrate into the electropolymerization solution results in its association with the π-donor thioaniline residues that cap the Au NPs, by π-donor–acceptor interactions, and upon electropolymerization, the acceptor substrate also binds to the newly formed π-donor bis-aniline bridging units. To exemplify this process, a high concentration, 1 mg mL^{-1}, of picric acid, (**3**), was introduced into the electropolymerization solution [50]. Picric acid exhibits a molecular structure that is analogous to TNT (**4**) (Fig. 3) with only a minor size difference between the hydroxyl and methyl substituents associated with the molecules. Furthermore, both compounds demonstrate comparable electron acceptor characteristics, suggesting that the picric acid might be a good analog for the imprinting of TNT recognition sites. The use of picric acid as imprint analog is favored over the direct implementation of TNT due to its enhanced solubility in the aqueous electropolymerization medium, suggesting the generation of a matrix with a higher density of imprinted sites for the association of TNT. The subsequent removal of the picric acid from the matrix, by a prolonged rinsing of the surface with a buffer, renders the surface with imprinted sites exhibiting high affinity toward TNT. Similarly, a non-imprinted matrix was prepared by the exclusion of the imprint analog (**3**) from the electropolymerization process. It should be noted that both the imprinting and extraction stages were monitored in situ, by following the changes in the SPR spectra corresponding to these processes. Figure 4A shows the SPR curves obtained for the (**3**)-imprinted bis-aniline-cross-linked Au NPs composite before, curve (a), and after, curve (b), the interaction of the matrix with (**4**), 1 pM. A clear shift of the resonance angle is observed, consistent with the association of (**4**) to the imprinted sites results in dielectric changes at the Au NPs matrix. While the change observed in the minimum refractive angle, θ_{min}, is significant, and may be used to provide a quantitative measure for (**4**), an alternative, more convenient sensing mode is facilitated by fixing the incident light angle, θ, while monitoring the time-dependent changes of the reflectance, ΔR, to yield a sensogram. Figure 4B depicts two sensograms corresponding to the analysis of (**4**) by the non-imprinted, left Fig. 4B-I, and the (**3**)-imprinted, right Fig. 4B-II, Au NPs composites. In these figures, the injections of the TNT samples, with increasing concentrations, are marked with arrows. Evidently, whereas the imprinted composite enables the detection of TNT with an ultra-sensitivity of 10 fM, the non-imprinted matrix shows a 1×10^3-fold higher detection limit (10 pM). The enhanced performance by the imprinted Au NPs matrix is attributed to the formation of the high affinity imprinted sites for (**4**), allowing its efficient binding to the matrix. Using the resulting calibration curves (Fig. 4C) and employing a Langmuir-type fitting, the association constants of (**4**) to the imprinted and non-imprinted matrices are estimated to be $K_a^I = 6.4 \times 10^{12}$ M^{-1} and $K_a^{NI} = 3.9 \times 10^9 M^{-1}$, respectively. Interestingly, the two step reflectance

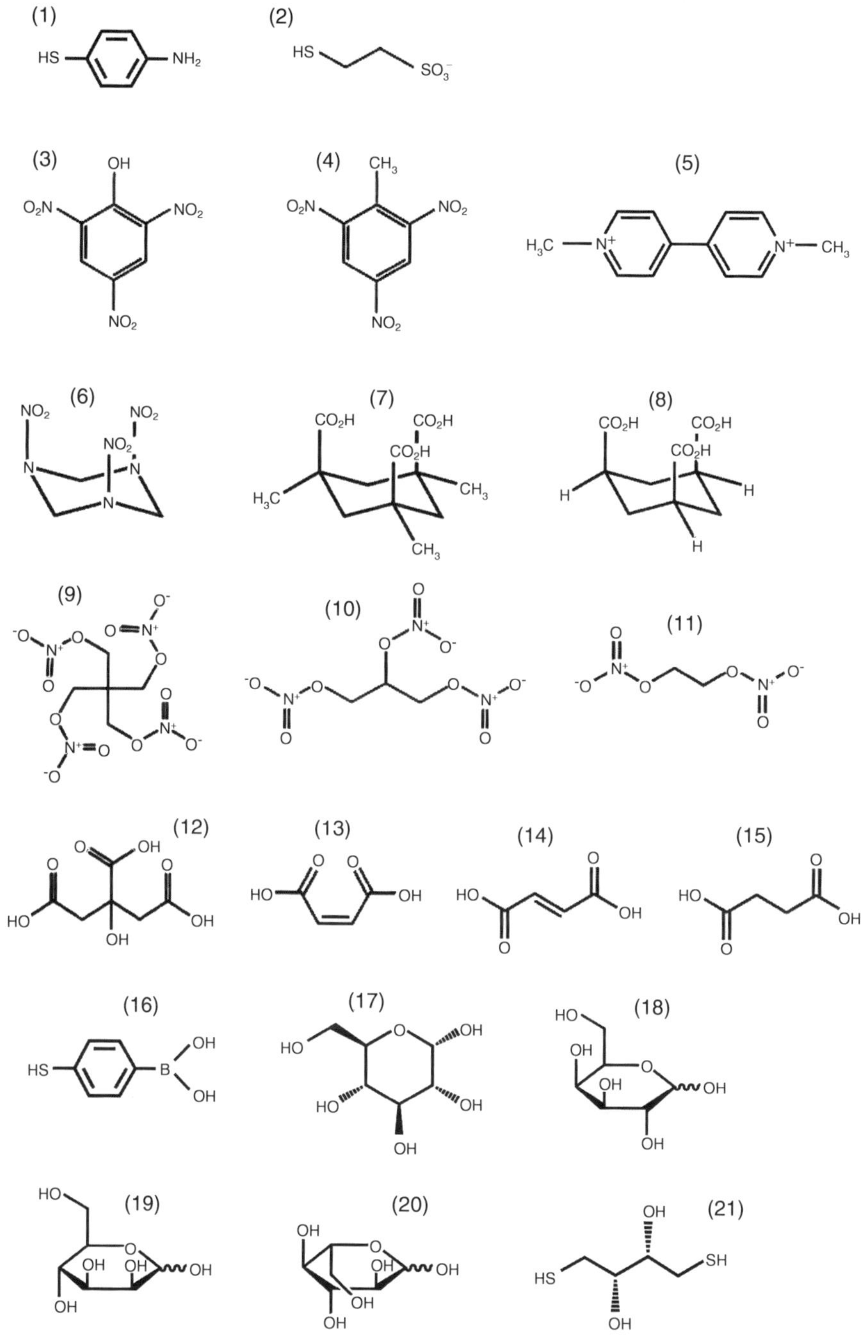

Fig. 3 List of compounds used in the study

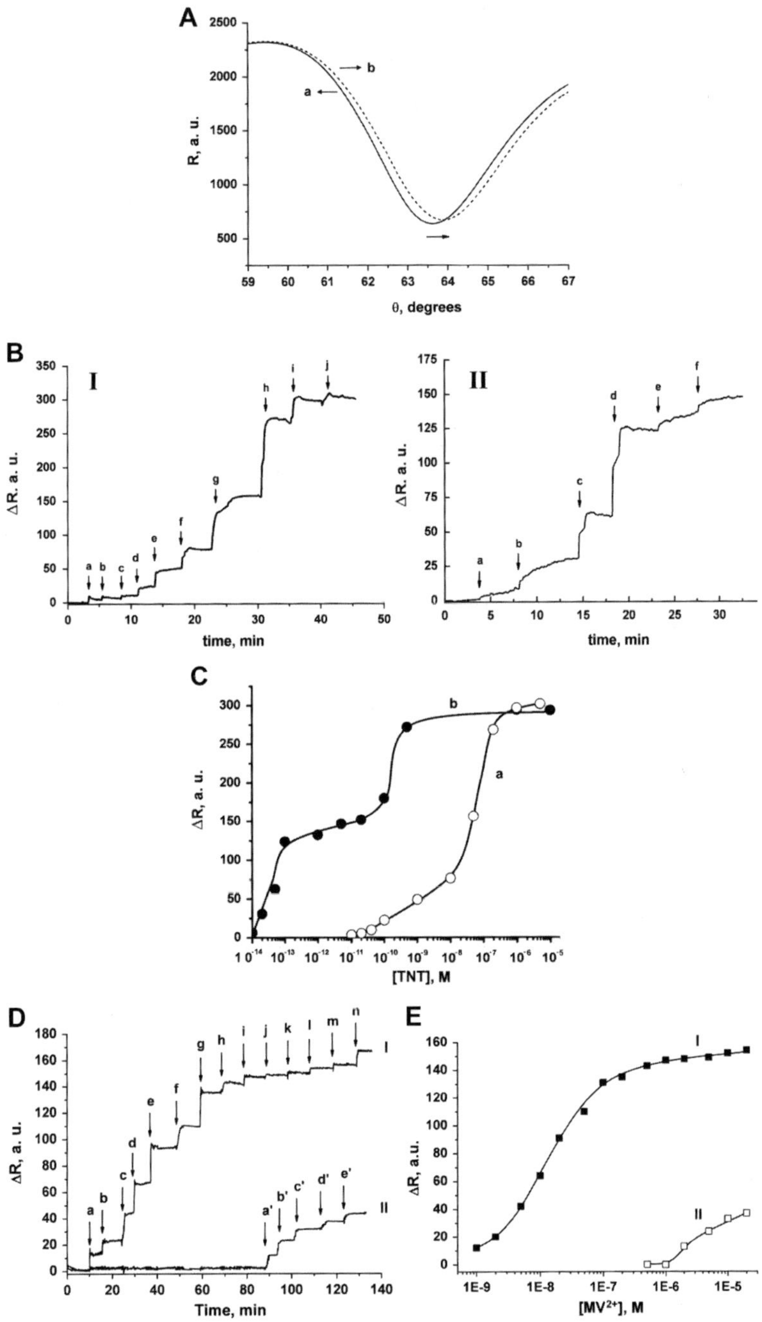

Fig. 4 (**A**) SPR curves obtained for TNT analysis on the picric acid-imprinted bis-aniline-cross-linked Au NPs composite. TNT concentrations: (*a*) 0 M, (*b*) 1 pM. (**B**) Sensograms, at $\theta = 62.4°$, showing the reflectance changes obtained for: I—injection of a series of increasing concentrations of TNT

regions observed in the calibration curve of the imprinted composite, Fig. 4C, curve (b), are attributed to the existence of two types of TNT-binding sites. Whereas the reflectance changes at the lower TNT concentrations of 10 fM < [TNT] < 5 pM correspond to the association of (**4**) to the imprinted π-donor sites of the matrix, the reflectance changes observed at the higher TNT concentrations, 100 pM < [TNT] < 10 μM, which are also observed for the non-imprinted matrix, curve (b), are attributed to the association of TNT to the non-imprinted bis-aniline π-donor bridges as well as to the excess of thioaniline π-donor units modifying the Au NPs in the composites. The calibration curves also reveal a difference of approximately two orders of magnitude in the inflection points between the imprinted and non-imprinted Au NPs matrices. The lower inflection point associated with the imprinted matrix originates from the existence of nonoptimal, partially fitting binding sites for TNT molecules, which are presumably formed during the imprinting process.

To further exemplify the use of π-donor–acceptor interactions for imprinting the bis-aniline-cross-linked Au NPs matrices, the herbicide, *N*,*N*′-dimethyl-4,4′-bipyridinium dichloride, methyl viologen (MV^{2+}), (**5**), was imprinted, and sensed, by similar methodology [57]. In this case, the electropolymerization process involved the direct imprint of MV^{2+} in the composite due to the highly solubility of (**5**) itself. Figure 4D depicts the sensograms obtained for analyzing (**5**) on the (**5**)-imprinted and non-imprinted matrices, and the respective calibration curves are presented in Fig. 4E. A striking difference in the performance of the two matrices is observed, with a 10^3-fold lower detection limit by the imprinted matrix, 1 nM, as compared to the non-imprinted composite, 1 μM, highlighting, once again, the important contribution of the imprinting process to the sensing paradigm.

2.2 *Ionic Interactions in Molecularly Imprinted Au NPs Matrices for Sensing*

A second class of sensing platforms based on bis-aniline-cross-linked Au NPs composites implementing ionic interactions between the analyte (or its imprint analog) and the bis-aniline bridging units as an associative driving force for imprinting was developed. Using this methodology, sensitive sensors for a variety of nitro

Fig. 4 (continued) (a–j: 10 pM–5 μM TNT) on the non-imprinted Au NPs matrix. II—injection of increasing concentrations of TNT (*a*–*f*: 10 fM–5 pM TNT) on the picric acid-imprinted Au NPs matrix. (**C**) The calibration curves derived for the measurements in (**B**). (**D**) Sensograms, at $\theta = 62.0°$, corresponding to MV^{2+} analysis on: I—the MV^{2+}-imprinted Au NPs matrix (*a*–*n*: 1 nM–20 μM MV^{2+}), and II—the non-imprinted matrix (*a*′–*e*′: 1–20 μM MV^{2+}). (**E**) The calibration curves derived for the measurements in (**D**). All measurements were performed in a 0.1 M HEPES buffer solution (pH = 7.2). Parts A-C – Adapted with permission from ref. 50. Copyright 2009, American Chemical Society. Parts D, E – reprinted with permission from ref. 57. Copyright 2010 American Chemical Society

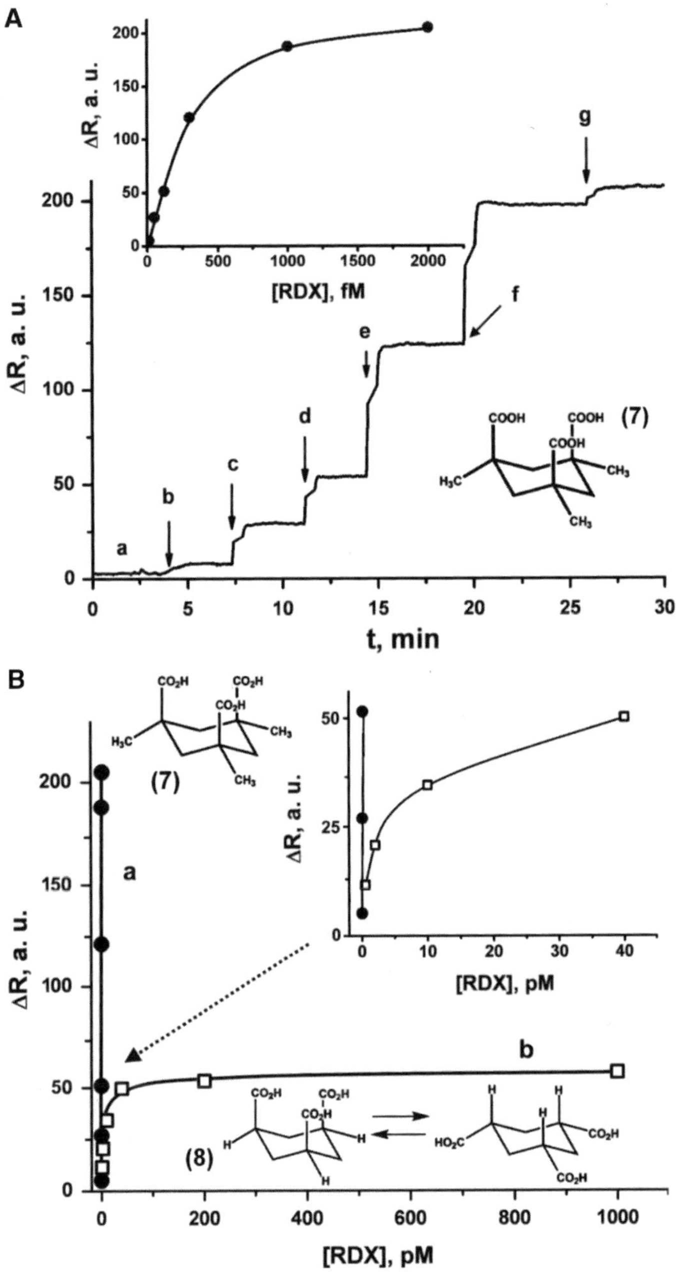

Fig. 5 (**A**) Sensogram, at $\theta = 63.3°$, showing the reflectance changes obtained upon to the analysis of RDX on the Kemp's acid-imprinted bis-aniline-cross-linked Au NPs matrix. RDX concentrations analyzed: $a = 0$ M and b–f: 12 fM–2 pM. The *inset* presents the derived calibration curve for the different RDX concentrations. (**B**) Comparison between the calibration curves obtained for the analysis of variable concentrations of RDX on: (*a*) The Kemp's acid-imprinted Au NPs composite, and (*b*) the cyclohexane tricarboxylic acid-imprinted Au NPs composite. The *inset* shows the lower concentration region of the calibration curves. All measurements were performed in a 0.1 M HEPES buffer solution (pH = 7.2). Reproduced with pemission from [51]. Copyright Willey-VCH Verlag GmbH & Co. KGaA

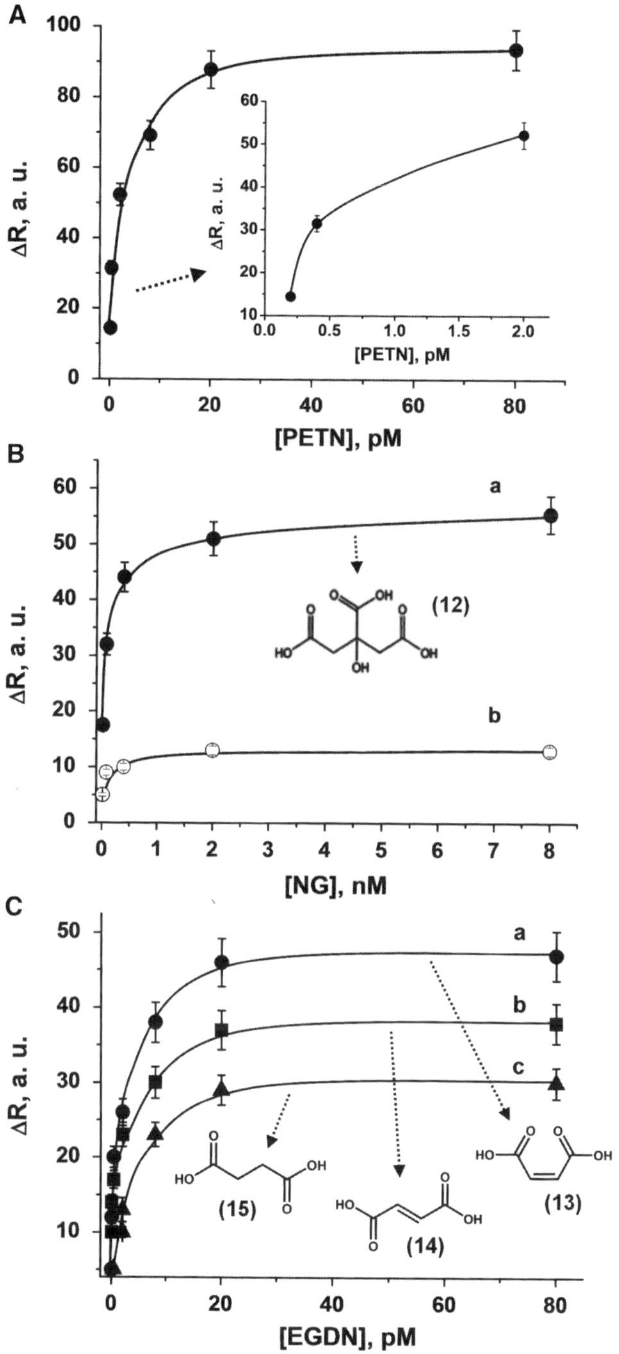

Fig. 6 (A) Calibration curve, corresponding to changes in the reflectance intensities upon the analysis of variable concentrations of PETN on the citric acid-imprinted Au NPs composite.

($–NO_2$ or $–ONO_2$)-substituted explosives were prepared, and the fundamental characteristics of the respective matrices were studied (Figs. 5 and 6). Due to the close dimensional similarity between nitro and carboxylic groups, and due to the expected neutralization of the carboxylic groups by the amine functionalities associated with the thioaniline and the bis-aniline bridging units, several carboxylic acid-functionalized substances were suggested as appropriate analogs for the nitro-substituted explosives. Namely, the ionic interactions formed between the carboxylic acid-modified analogs and the bis-aniline-functionalized Au NPs matrix are anticipated to yield high affinity recognition sites for the sensing of the explosives.

Whereas the detection of the TNT explosive has gained a substantial scientific focus, sensors for more hazardous explosives, such as hexahydro-1,3,5-trinitro-1,3,5-triazine, RDX, (**6**), remained relatively less explored and require the development of new technologies to achieve improved detection sensitivities. The three quasi-axial nitro substituents associated with the RDX explosive [74, 75] (Fig. 3) can be mimicked by using the imprint analog 1,3,5-trimethyl-1,3,5-tricarboxycyclohexane, Kemp's acid, (**7**), which exhibits three equatorial methyl substituents [51]. The methyl groups in (**7**) enable the molecule to adopt a relative rigid "chair-like" structure in which the three carboxylic acid substituents are fixed in quasi-axial positions, thus providing a unique analog to RDX. Figure 5A demonstrates the analysis of RDX on the Kemp's acid-imprinted Au NPs matrix. Evidently, the imprinted matrix reveals a high response in the fM region and a detection limit as low as 12 fM. The sensing of the ultra-low concentrations by the imprinted Au NPs composite is facilitated by the dielectric changes that occur during the high affinity binding of the analyte molecules to the imprinted sites. The intrinsic amplification mechanism, due to the coupling between the localized plasmon of the Au NPs and the Au surface plasmon wave, acts in amplifying the SPR shifts and allows the detection of the ultra-low concentrations of the analyte. Using the calibration curve in Fig. 5A inset, the association constant of RDX to the imprinted matrix was estimated to be $K_a^I = 1.9 \times 10^{12} M^{-1}$. Comparative measurements performed on the non-imprinted Au NPs composite revealed significantly higher detection limit, 4 nM, and a lower association constant, $K_a^{NI} = 3.4 \times 10^7 M^{-1}$. This highlighted the need for the generation of the imprinted sites for the efficient binding and sensing of RDX. Furthermore, the high compatibility of Kemp's acid as an RDX analog is demonstrated by sensing RDX on a matrix imprinted with the

---

Fig. 6 (continued) The *inset* shows the lower concentration region of the calibration curve. (**B**) Calibration curves, corresponding to the analysis of variable concentrations of nitroglycerine on: (*a*) the citric acid-imprinted Au NPs composite, and (*b*) the non-imprinted Au NPs composite. (**C**) Calibration curves, corresponding to the analyses of variable concentrations of EGDN on: (*a*) the maleic acid-imprinted, (*b*) the fumaric acid-imprinted, and (*c*) the succinic acid-imprinted bis-aniline-cross-linked Au NPs matrices. *Error bars* correspond to a set of $N = 5$ measurements. All measurements were performed in ethanol. Reprinted with permission from ref. 52. Copyright 2011 American Chemical Society

structurally related molecule cyclohexane tricarboxylic acid, (**8**). The steric flexibility of (**8**) yields an equilibrium mixture between the axial–equatorial conformations of the carboxylic acid functionalities (Fig. 5B) resulting in the generation of imperfect imprinted recognition sites, exhibiting lower affinity for the binding of RDX. Evidently, a decreased sensing performance for the detection of RDX by the (**8**)-imprinted matrix was observed, as reflected by more than a 40-fold higher detection limit compared to the (**7**)-imprinted matrix.

The successful implementation of a carboxylic acid-functionalized molecule as an analog for imprinting recognition sites for RDX was further expanded for the sensing of other important explosives, such as pentaerythritol tetranitrate, PETN, (**9**), nitroglycerin, NG, (**10**), and ethylene glycol dinitrate, EGDN, (**11**) (Fig. 6) [52]. Figure 6A depicts the calibration curve obtained upon the sensing of PETN on a citric acid, (**12**)-imprinted bis-aniline-cross-linked Au NPs composite. The ionic and/or H-bonds between the carboxylate and the anilinium residues provide strong affinity binding sites for (**9**), leading to a detection limit as low as 200 fM. Under these conditions, a high association constant between the matrix and the explosive, $K_a^I = 9.5 \times 10^{11} M^{-1}$, was evaluated. Citric acid, (**12**), was, similarly, implemented as imprinting substrate to generate imprinted sites in the Au NPs matrix for nitroglycerine, (**10**) (Fig. 6B). Evidently, significant reflectance changes are observed upon the interaction of pM concentrations of (**10**) with the (**12**)-imprinted matrix, curve (a). Similar SPR responses are, as expected, absent in the analysis of (**10**) by the non-imprinted matrix, curve (b), which reconfirms the important role of the imprinting in the detection paradigm. The lower detection limit observed for the association of PETN with the (**12**)-imprinted sites, as compared to the NG, may be attributed to the presence of the OH group in (**12**) (Fig. 3). The hydroxyl group, in addition to the three carboxylic acid residues, mimic the four –ONO_2 functionalities of PETN, whereas in the NG explosive (that includes only three –ONO_2 functionalities), no contribution is made by the additional OH group associated with (**12**). It is, thus, clear that (**12**) is structurally optimized to accommodate the PETN substrate containing the four –ONO_2 functionalities.

Another important explosive, yet lacking significant scientific attention, is ethylene glycol dinitrate, EGDN, (**11**). The chemical structure of (**11**) (Fig. 3) suggested that the imprinting analogs that may be adapted to yield imprinted sites for (**11**) in the Au NPs matrices via electrostatic/ionic interactions are: maleic acid (**13**), fumaric acid (**14**), and succinic acid (**15**). The three calibration curves presented in Fig. 6C show the reflectance changes obtained by the interaction of variable concentrations of EGDN with the (**13**)-, (**14**)-, or (**15**)-imprinted Au NPs composites. The results reveal a comparable sensing performance for all matrices (Fig. 6C). The superior performance by the (**13**)-imprinted matrix may be, however, explained by the favorable existence of (**11**) in a state where the nitro substituents adopt a "gauche" conformation [76]. Thus, the *cis* configuration associated with the carboxylic acid functionalities of (**13**) allows the molecule to act as an effective analog for EGDN, resulting in an enhanced sensing performance by the (**13**)-imprinted matrix.

The imprinting of cross-linked Au NPs composites through ionic interactions provides a general paradigm which is not limited for the sensing of aromatic explosives. For example, using cysteine-modified Au NPs, imprinted matrices were generated for a series of amino acids [53]. The matrices showed high sensitivities, in the nM concentration range, toward the detection of the substrate analytes, and exhibited remarkable stereoselectivity and chiroselectivity during the analysis of different amino acids.

2.3 Imprinted Electropolymerized Au NPs Composites Based on Ligand–Analyte Complexation Processes for Sensing

An alternative approach for imprinting recognition sites in cross-linked Au NPs matrices involves the primary modification of the NPs with target-specific ligand units. To this end, the ligand units are incorporated as a co-additive capping agent in the synthesis of the Au NPs. Figure 7A schematically presents the electropolymerization of Au NPs, modified with a mixed monolayer consisting of thioaniline (**1**) electropolymerizable units, thioethane sulfonate (**2**) stabilizing units, and thiophenyl boronic acid (**16**) ligation units, in the presence of the saccharide D-glucose, (**17**), analyte. The interaction between the ligand (**16**) and the vicinal diols of the sugar (**17**), under basic conditions, results in the formation of a boronic-ester bond, leading to the formation of a complex between the saccharide and the modified Au NP. The (**17**)-complexated Au NPs were, then, electropolymerized onto a thioaniline monolayer-modified Au surface in the presence of an excess saccharide, and under basic conditions, to yield the (**17**)-imprinted bis-aniline-cross-linked Au NPs electrode. The subsequent removal of (**17**) from the matrix was carried out by acidifying the electrolyte to pH = 1.3, which induces the dissociation of the boronate-ester bond, thus releasing the sugar units. Following this stage, a vacant (**17**)-imprinted Au NPs composite is obtained. Figure 7B depicts the analysis of D-glucose (**17**) by the resulting imprinted matrix. The time-dependent reflectance changes correlate with the concentrations of (**17**). The detection limit of (**17**) is 40 pM, Fig. 7C, and an association constant corresponding to $K_a^I = 8.3 \times 10^9 M^{-1}$ between (**17**) and the matrix was estimated. Also, only minute reflectance changes are observed upon challenging the (**17**)-imprinted matrix with D-mannose, (**18**), or D-galactose, (**19**), consistent with the impressive stereoselectivity provided by the (**17**)-imprinted sites.

In addition to the stereoselectivity for sensing the imprinted saccharides, the imprinting methodology leads, also, to chiroselective recognition of sugars. Figure 7D depicts the sensograms obtained upon the analyses of D-glucose, (**17**), and L-glucose, (**20**), on the D-glucose (**17**)-imprinted Au NPs matrix. Whereas significant reflectance responses are observed at D-glucose (**17**) concentrations as low as 10 pM on the (**17**)-imprinted matrix, no substantial reflectance changes are observed, up to a

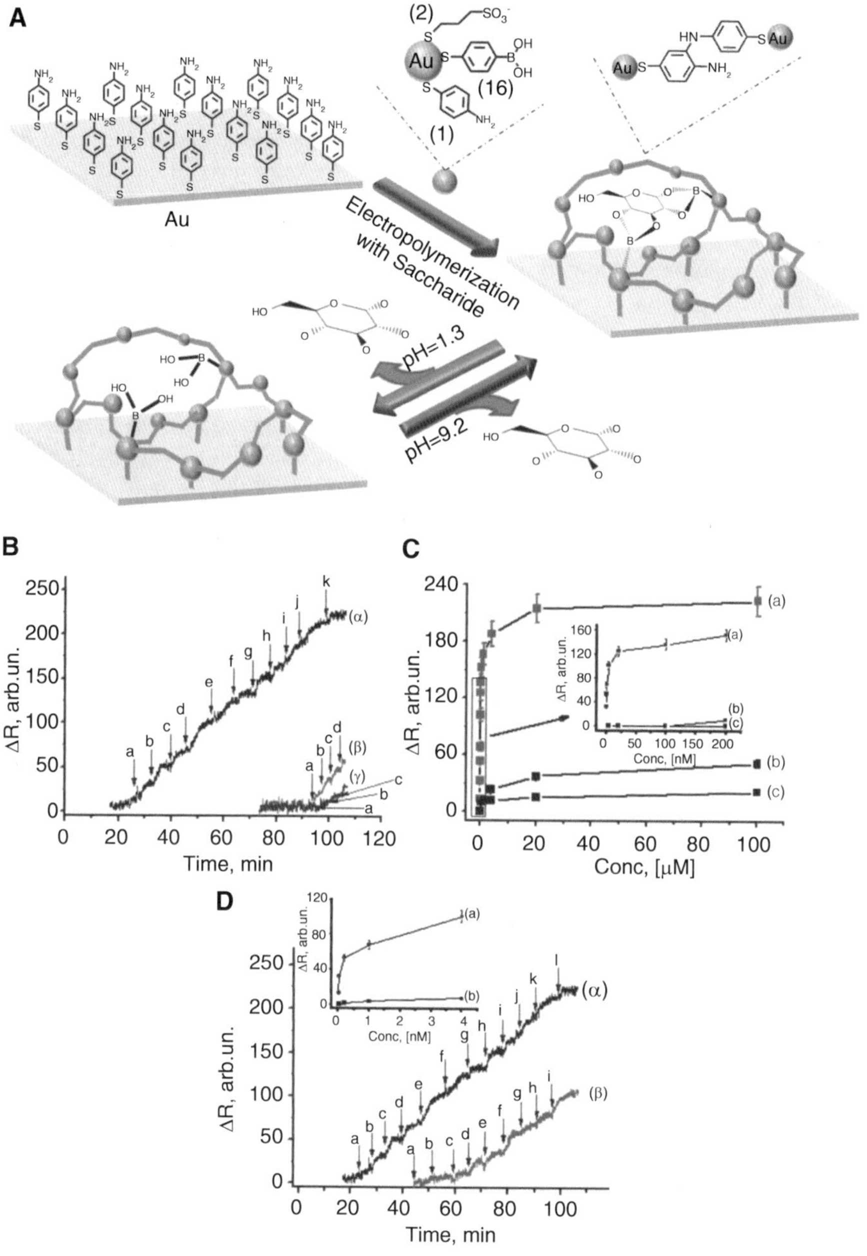

Fig. 7 (**A**) Schematic presentation for the imprinting of saccharides into a bis-aniline-cross-linked Au NPs composite and their association/dissociation to and from the matrix. (**B**) Sensograms, at $\theta = 63.5°$, showing the reflectance changes obtained upon to the analysis of: (α) D-glucose (concentrations *a*–*k*: 40 pM–100 μM), (β) D-mannose (concentrations *a*–*d*: 4–200 μM), and (γ)

concentration of 1 nM, for the analysis of the L-glucose substrate (**20**) on the same matrix. Also, ca. 2×10^3-fold lower association constant was derived for the binding of (**20**) to the (**17**)-imprinted matrix. The results were also reversed upon the imprinting of L-glucose sites in the Au NPs composite, and lower detection limit and enhanced sensing performance were observed in this case for the analysis of (**20**) by the (**20**)-imprinted Au NPs composite [54]. The results suggest that the molecular contours generated in the matrix during the imprinting process provide a unique steric template for the imprinted substrate, and can be used to distinguish between diastereoisomers or enantiomers. It should be noted that in another study, an analogous system also employing boronic acid-functionalized Au NPs was used for the imprinting of different diol-functionalized antibiotics, including Kanamycin, Neomycin, and Streptomycin [55]. The imprinted matrices revealed impressive detection limits for the respectively imprinted substrates, which were several orders of magnitude lower than the maximum allowed residue limits (MRLs). Furthermore, the antibiotic-imprinted Au NPs matrices also allowed the sensitive and selective analysis of the target substances in real food samples, such as in milk [55].

Ligand–analyte interactions which constitute the grounds for imprinting via complexation in electropolymerized Au NPs composites may also be used for the detection of metal ions [56]. Au NPs were co-modified with the complexating dithiothreitol (**21**) ligand and reacted with an alkaline-earth-metal ion, such as Sr^{2+}, Mg^{2+}, Ca^{2+}, or Ba^{2+}. Following the primary complexation of the ion with the ligand-modified particles, the Au NPs were electropolymerized onto a thioaniline-modified Au surface, to yield an imprinted composite, which is schematically exemplified in Fig. 8A. The removal of the metal ions from the respective matrices by exposing them to an acidic, pH = 1.5, solution, resulted in vacant imprinted templates. This process presumably protonated the dithiotreitol ligands, thus releasing the metal ions to the solution.

Figure 8B depicts the sensograms obtained upon the interaction of the Mg^{2+}, Ca^{2+}, Sr^{2+}, and Ba^{2+} ions with the Mg^{2+}-imprinted bis-aniline-cross-linked Au NPs composite. Evidently, the Mg^{2+}-imprinted matrix is most sensitive to the presence of Mg^{2+} ions, exhibiting a low detection limit of 20 fM and a linear dynamic sensing range in the femtomolar region (Fig. 8C). The high sensitivity is attributed to the dielectric changes induced upon the binding of the Mg^{2+} ions to the matrix, and/or due to possible changes in the local plasmon intensity, caused by variations in the inter-Au NPs distances, as a result of the association of the ions with the

Fig. 7 (continued) d-galactose (concentrations *a–c*: 20–200 μM) on the D-glucose-imprinted bis-aniline-cross-linked Au NPs matrix. (**C**) Calibration curves derived from the sensograms in (**B**). The *inset* shows the lower concentration regions of the curves. (**D**) Sensograms, at $\theta = 63.5°$, showing the reflectance changes obtained upon to the analysis of: (α) D-glucose (concentrations *a–l*: 10 pM–100 μM), and (β) L-glucose (concentrations *a–i*: 1 nM–100 μM) on the D-glucose-imprinted bis-aniline-cross-linked Au NPs matrix. The *inset* shows the respective calibration curves derived from the sensograms. All measurements were performed in a 50 mM HEPES buffer solution (pH = 9.2). Reproduced from ref. 54 by permission of the Royal Society of Chemistry (RSC)

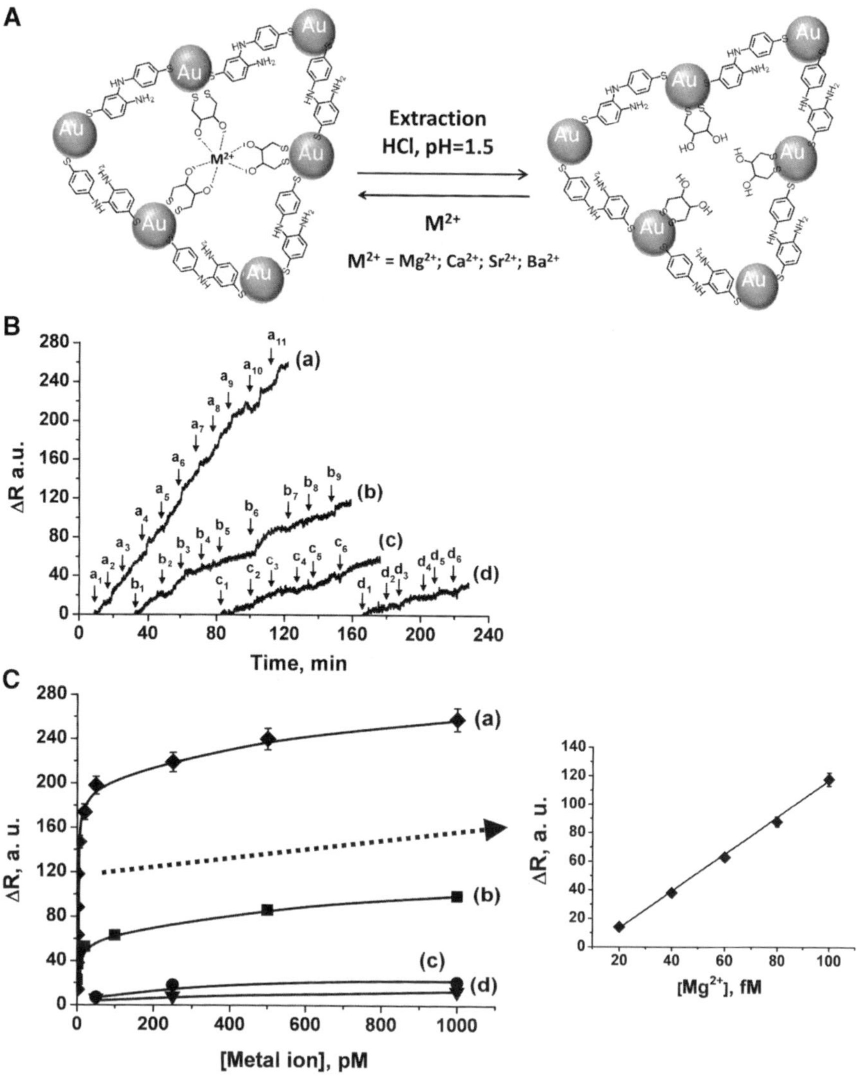

Fig. 8 (**A**) Suggested assembly for the association and dissociation of several earth-metal ions to and from the metal ion-imprinted bis-aniline-cross-linked Au NPs matrix. (**B**) Sensograms, at $\theta = 63.5°$, showing the reflectance changes obtained upon to the analyses of: (*a*) Mg^{2+} (concentrations a_1–a_{11}: 20 fM–250 pM), (*b*) Ca^{2+} (concentrations b_1–b_9: 80 fM–500 pM), (*c*) Sr^{2+} (concentrations c_1–c_6: 50 pM–5 nM), and (*d*) Ba^{2+} (concentrations: d_1–d_6: 50 pM–50 nM), on the Mg^{2+}-imprinted bis-aniline-cross-linked Au NPs matrix. (**C**) Calibration curves derived from the sensograms in (**B**). The *inset* shows the lower concentration region of the Mg^{2+} calibration curve. All measurements were performed in triply-deionized water. *Error bars* correspond to a set of $N = 5$ measurements. Reproduced from ref. 56 by permission of the Royal Society of Chemistry (RSC)

three-dimensional electropolymerized assembly. The high selectivity of the imprinted matrices, evident in Fig. 8B, C and reflected by the lower reflectance responses for the Ca^{2+}, Sr^{2+}, and Ba^{2+} analyses on the Mg^{2+}-imprinted matrix, was further reconfirmed by the synthesis of imprinted composites for each of the earth-metal ions. As expected, in all cases the specific ion-imprinted matrices revealed the highest performance for the sensing of the respective metal ion analytes, and sensitive detection was also demonstrated in the presence of solutions containing a mixture of all relevant ions [56].

3 Conclusions and Perspectives

The chapter introduced the electrochemical assembly and the application of Au electrodes modified with molecularly imprinted bis-aniline-cross-linked Au NPs matrices for sensitive and selective detection of a broad range of analytes, including explosives, herbicides, sugars, earth-metal ions, antibiotics, and amino acids. Three main interactions between the analytes and the matrices, facilitating the imprinting process, were discussed: (1) donor–acceptor association, (2) ionic/H-bonds, and (3) ligand–analyte complexation. The development of sensors, based on those interactions, involved the careful selection of the imprint molecule, a process which proved to be of a primary importance to the performance of the resulting sensors, as well as the nano-engineering of the matrix by the incorporation of receptor elements, such as specific ligands for the target molecules or ions.

The imprinted bis-aniline-cross-linked Au NPs matrices exhibit remarkable and often unprecedented detection capabilities. The Au NPs matrices offer a dynamic analyte sensing range that is often in the fM–pM regime, high stereoselectivity (and/or chiroselectivity) toward the analysis of chemical interferants, a good reproducibility and stability, and a relatively easy and cost-effective preparation. Albeit these advances, many challenging topics associated with the imprinted Au NPs matrices still need to be addressed. Whereas in most of the studies presented in this chapter, the SPR technique has been used as a readout method for the sensing events, other analytical techniques, such as surface enhanced Raman spectroscopy (SERS), might be successfully employed for following the recognition processes. With the advantage of detecting the spectroscopic fingerprints of the analytes, such technique may be well suited for both quantitative and qualitative multiplexed analyses of the various substrates in different environments.

Further challenging topics may involve the investigation of the correlation between the morphology of the NPs, including their size, shape, and composition, and the sensing performance of the imprinted matrices. Also, gas phase detection of volatile explosive analytes and nano-engineering of the matrices for the detection of other important substrates, such as street drugs or toxins, may be envisaged.

Finally, the applications of imprinted bis-aniline-cross-linked Au NPs matrices are not confined to sensing. In fact, the imprinted matrices have already been implemented in several scientific fields, including the generation of electrically and/or photonically triggered "sponges" for the selective uptake of specific pollutants and their concentration for secondary neutralization [57, 58], in the signal-controlled wettability of surfaces [59, 60] and for the generation of enhanced performance photoelectrochemical anodes and cells [61]. This diversity of applications still needs to be broadened, with the aim of exploiting the inherent advantages of the imprinted cross-linked Au NPs composites in other scientific fields as well.

References

1. Mulvaney P (1996) Surface plasmon spectroscopy of nanosized metal particles. Langmuir 12:788–800
2. Wilcoxon JP, Abrams BL (2006) Synthesis, structure and properties of metal nanoclusters. Chem Soc Rev 35:1162–1194
3. Kamat PV (2002) Synthesis, structure and properties of metal nanoclusters. J Phys Chem B 106:7729–7744
4. Murphy CJ, Sau TK, Gole AM et al (2005) Anisotropic metal nanoparticles: synthesis, assembly, and optical applications. Phys Chem B 109:13857–13870
5. Zayats M, Baron R, Popov I et al (2005) Biocatalytic growth of Au nanoparticles: from mechanistic aspects to biosensors design. Nano Lett 5:21–25
6. Daniel MC, Astruc D (2004) Gold nanoparticles: assembly, supramolecular chemistry, quantum-size-related properties, and applications toward biology, catalysis, and nanotechnology. Chem Rev 104:293–346
7. Feldheim DL, Keating CD (1998) Self-assembly of single electron transistors and related devices. Chem Soc Rev 27:1–12
8. Lewis LN (1993) Chemical catalysis by colloids and clusters. Chem Rev 93:2693–2730
9. Wang L, Song S, Pan D et al (2010) Gold nanoparticle-based sensing strategies for biomolecular detection. Pure Appl Chem 82:81–89
10. Weisbecker CS, Merritt MV, Whitesides GM (1996) Molecular self-assembly of aliphatic thiols on gold colloids. Langmuir 12:3763–3772
11. Choi Y, Ho NH, Tung CH (2007) Sensing phosphatase activity by using gold nanoparticles. Angew Chem Int Ed 46:707–709
12. Storhoff JJ, Elghanian R, Mucic RC et al (1998) One-pot colorimetric differentiation of polynucleotides with single base imperfections using gold nanoparticle probes. J Am Chem Soc 120:1959–1964
13. Obare SO, Hollowell RE, Murphy CJ (2002) Sensing strategy for lithium ion based on gold nanoparticles. Langmuir 18:10407–10410
14. Matsui J, Akamatsu K, Hara N et al (2005) SPR sensor chip for detection of small molecules using molecularly imprinted polymer with embedded gold nanoparticles. Anal Chem 77:4282–4285
15. Matsui J, Akamatsu K, Nishiguchi S et al (2004) Composite of Au nanoparticles and molecularly imprinted polymer as a sensing material. Anal Chem 76:1310–1315
16. Tokareva I, Tokarev I, Minko S et al (2006) Ultrathin molecularly imprinted polymer sensors employing enhanced transmission surface plasmon resonance spectroscopy. Chem Commun 3343–3345
17. Jiang G, Baba A, Ikarashi H et al (2007) Signal enhancement and tuning of surface plasmon resonance in Au nanoparticle/polyelectrolyte ultrathin films. Phys Chem C 111:18687–18694
18. He L, Musick MD, Nicewarner SR et al (2000) Colloidal Au-enhanced surface plasmon resonance for ultrasensitive detection of DNA hybridization. J Am Chem Soc 122:9071–9077

19. Golub E, Pelossof G, Freeman R et al (2009) Electrochemical, photoelectrochemical, and surface plasmon resonance detection of cocaine using supramolecular aptamer complexes and metallic or semiconductor nanoparticles. Anal Chem 81:9291–9298
20. Lyon LA, Musick MD, Natan MJ (1998) Colloidal Au-enhanced surface plasmon resonance immunosensing. Anal Chem 70:5177–5183
21. Mauriz E, Calle A, Lechuga LM et al (2006) Real-time detection of chlorpyrifos at part per trillion levels in ground, surface and drinking water samples by a portable surface plasmon resonance immunosensor. Anal Chim Acta 561:40–47
22. Gu H, Su X, Loh KP (2005) Electrochemical impedance sensing of DNA hybridization on conducting polymer film-modified diamond. J Phys Chem B 109:13611–13618
23. Inoue T, Kirchhoff JR (2000) Electrochemical detection of thiols with a coenzyme pyrroloquinoline quinone modified electrode. Anal Chem 72:5755–5760
24. Malitesta C, Palmisano F, Torsi L et al (1990) Glucose fast-response amperometric sensor based on glucose-oxidase immobilized in an electropolymerized poly(ortho-phenylenediamine) film. Anal Chem 62:2735–2740
25. Cosnier S (2005) Affinity biosensors based on electropolymerized films. Electroanalysis 17:1701–1715
26. Jager E, Inganas O, Lundstrom I (2000) Microrobots for micrometer-size objects in aqueous media: potential tools for single-cell manipulation. Science 288:2335–2338
27. Jager E, Smela E, Inganas O (2000) Microfabricating conjugated polymer actuators. Science 290:1540–1545
28. Campos LM, Tontcheva A, Gunes S et al (2005) Extended photocurrent spectrum of a low band gap polymer in a bulk heterojunction solar cell. Chem Mater 17:4031–4033
29. Senadeera R, Fukuri N, Saito Y et al (2005) Volatile solvent-free solid-state polymer-sensitized TiO_2 solar cells with poly(3,4-ethylenedioxythiophene) as a hole-transporting medium. Chem Commun 2259–2261
30. Li M, Tang S, Shen F et al (2006) Highly luminescent network films from electrochemical deposition of peripheral carbazole functionalized fluorene oligomer and their applications for light-emitting diodes. Chem Commun 3393–3395
31. Maness KM, Terrill RH, Meyer TJ et al (1996) Solid-state diode-like chemiluminescence based on serial, immobilized concentration gradients in mixed-valent poly$[Ru(vbpy)_3](PF_6)_2$ films. J Am Chem Soc 118:10609–10616
32. Granot E, Katz E, Basnar B et al (2005) Enhanced bioelectrocatalysis using Au-nanoparticle/polyaniline hybrid systems in thin films and microstructured rods assembled on electrodes. Chem Mater 17:4600–4609
33. Raitman OA, Katz E, Bückmann AF et al (2002) Integration of polyaniline/poly(acrylic acid) films and redox enzymes on electrode supports: An in situ electrochemical/surface plasmon resonance study of the bioelectrocatalyzed oxidation of glucose or lactate in the integrated bioelectrocatalytic systems. J Am Chem Soc 124:6487–6496
34. Morozov SV, Karyakina EE, Zorin NA et al (2002) Direct and electrically wired bioelectrocatalysis by hydrogenase from Thiocapsa roseopersicina. Bioelectrochemistry 55:169–171
35. Kriz D, Ramstrom O, Mosbach K (1997) Molecular imprinting – new possibilities for sensor technology. Anal Chem 69:345A–349A
36. Mosbach K (1994) Molecular imprinting. Trends Biochem Sci 19:9–14
37. Haupt K, Mosbach K (2000) Molecularly imprinted polymers and their use in biomimetic sensors. Chem Rev 100:2495–2504
38. Wulff G (1995) Molecular imprinting in cross-linked materials with the aid of molecular templates – a way towards artificial antibodies. Angew Chem Int Ed 34:1812–1832
39. Liu JQ, Wulff G (2004) Molecularly imprinted polymers with strong carboxypeptidase A-like activity: combination of an amidinium function with a zinc-ion binding site in transition-state imprinted cavities. Angew Chem Int Ed 43:1287
40. Liu J, Wulff G (2004) Functional mimicry of the active site of carboxypeptidase A by a molecular imprinting strategy: cooperativity of an amidinium and a copper ion in a transition-state imprinted cavity giving rise to high catalytic activity. J Am Chem Soc 126:7452–7453

41. Bossi A, Bonini F, Turner APF et al (2007) Molecularly imprinted polymers for the recognition of proteins: the state of the art. Biosens Bioelectron 22:1131–1137
42. Haupt K (2003) Imprinted polymers – tailor-made mimics of antibodies and receptors. Chem Commun 171–178
43. Martin-Esteban A (2001) Molecularly imprinted polymers: new molecular recognition materials for selective solid-phase extraction of organic compounds. Fresenius J Anal Chem 370:795–802
44. Bajpai V, He P, Dai L (2004) Conducting-polymer microcontainers: controlled syntheses and potential applications. Adv Funct Mater 14:145–151
45. Wulff G (2002) Enzyme-like catalysis by molecularly imprinted polymers. Chem Rev 102:1–27
46. Davis ME, Katz A, Ahmad WR (1996) Rational catalyst design via imprinted nanostructured materials. Chem Mater 8:1820–1839
47. Cosnier S (2003) Biosensors based on electropolymerized films: new trends. Anal Bioanal Chem 377:507–520
48. Agarwal GS, Gupta SD (1985) Interaction between surface-plasmons and localized plasmons. Phys Rev B 32:3607–3611
49. Riskin M, Tel-Vered R, Bourenko T et al (2008) Imprinting of molecular recognition sites through electropolymerization of functionalized Au nanoparticles: development of an electrochemical TNT sensor based on pi-donor-acceptor interactions. J Am Chem Soc 130:9726–9733
50. Riskin M, Tel-Vered R, Lioubashevski O et al (2009) Ultrasensitive surface plasmon resonance detection of trinitrotoluene by a bis-aniline-cross-linked Au nanoparticles composite. J Am Chem Soc 131:7368–7378
51. Riskin M, Tel-Vered R, Willner I (2010) Imprinted Au-nanoparticle composites for the ultrasensitive surface plasmon resonance detection of hexahydro-1,3,5-trinitro-1,3,5-triazine (RDX). Adv Mater 22:1387–1391
52. Riskin M, Ben-Amram Y, Tel-Vered R et al (2011) Molecularly Imprinted Au nanoparticles composites on Au surfaces for the surface plasmon resonance detection of pentaerythritol tetranitrate, nitroglycerin, and ethylene glycol dinitrate. Anal Chem 83:3082–3088
53. Riskin M, Tel-Vered R, Frasconi M et al (2010) Stereoselective and chiroselective surface plasmon resonance (SPR) analysis of amino acids by molecularly imprinted Au-nanoparticle composites. Chem Eur J 16:7114–7120
54. Ben-Amram Y, Riskin M, Willner I (2010) Selective and enantioselective analysis of mono- and disaccharides using surface plasmon resonance spectroscopy and imprinted boronic acid-functionalized Au nanoparticle composites. Analyst 135:2952–2959
55. Frasconi M, Tel-Vered R, Riskin M et al (2010) Surface plasmon resonance analysis of antibiotics using imprinted boronic acid-functionalized Au nanoparticle composites. Anal Chem 82:2512–2519
56. Ben-Amram Y, Tel-Vered R, Riskin M et al (2012) Ultrasensitive and selective detection of alkaline-earth metal ions using ion-imprinted Au NPs composites and surface plasmon resonance spectroscopy. Chem Sci 3:162–167
57. Frasconi M, Tel-Vered R, Riskin M et al (2010) Electrified selective "sponges" made of Au nanoparticles. J Am Chem Soc 132:9373–9382
58. Zhang J, Riskin M, Freeman R et al (2011) Electrochemically triggered Au nanoparticles "sponges" for the controlled uptake and release of a photoisomerizable dithienylethene guest substrate. ACS Nano 5:5936–5944
59. Balogh D, Tel-Vered R, Riskin M et al (2011) electrified Au nanoparticle sponges with controlled hydrophilic/hydrophobic properties. ACS Nano 5:299–306
60. Balogh D, Tel-Vered R, Freeman R et al (2011) Photochemically and electrochemically triggered Au nanoparticles "sponges". J Am Chem Soc 133:6533–6536
61. Yildiz HB, Tel-Vered R, Willner I (2008) Solar cells with enhanced photocurrent efficiencies using oligoaniline-crosslinked Au/CdS nanoparticles arrays on electrodes. Adv Funct Mater 18:3497–3505

62. Bratin K, Kissinger PT (1981) Determination of nitro aromatic, nitramine, and nitrate ester explosive compounds in explosive mixtures and gunshot residue by liquid-chromatography and reductive electrochemical detection. Anal Chim Acta 130:295–311
63. Zuman P, Fijalek ZJ (1990) Reaction of electrogenerated arylhydroxylamines and nitrosobenzene in the course of reduction of nitrobenzene under conditions of cyclic voltammetry. Electroanal Chem 296:589–593
64. Knoll W (1998) Interfaces and thin films as seen by bound electromagnetic waves. Annu Rev Phys Chem 49:569–638
65. Phillips KS, Cheng Q (2007) Recent advances in surface plasmon resonance based techniques for bioanalysis. Anal Bioanal Chem 387:1831–1840
66. Schuck P (1997) Use of surface plasmon resonance to probe the equilibrium and dynamic aspects of interactions between biological macromolecules. Annu Rev Biophys Biomol Struct 26:541–566
67. Garland PB (1996) Optical evanescent wave methods for the study of biomolecular interactions. Q Rev Biophys 29:91–117
68. Fivash M, Towler EM, Fisher RJ (1998) BIAcore for macromolecular interaction. Curr Opin Biotechnol 9:97–101
69. Cooper MA (2002) Optical biosensors in drug discovery. Nat Rev Drug Discov 1:515–528
70. Kubitschko S, Spinke J, Bruckner T et al (1997) Sensitivity enhancement of optical immunosensors with nanoparticles. Anal Biochem 253:112–122
71. Wink T, van Zuilen SJ, Bult A et al (1998) Liposome-mediated enhancement of the sensitivity in immunoassays of proteins and peptides in surface plasmon resonance spectrometry. Anal Chem 70:827–832
72. Zayats M, Raitman OA, Chegel VI et al (2002) Probing antigen-antibody binding processes by impedance measurements on ion-sensitive field-effect transistor devices and complementary surface plasmon resonance analyses: development of cholera toxin sensors. Anal Chem 74:4763–4773
73. Zayats M, Pogorelova SP, Kharitonov AB et al (2003) Au nanoparticle-enhanced surface plasmon resonance sensing of biocatalytic transformations. Chem Eur J 9:6108–6114
74. Rice BM, Chabalowski CF (1997) Ab initio and nonlocal density functional study of 1,3,5-trinitro-s-triazine (RDX) conformers. J Phys Chem A 101:8720–8726
75. Politzer P, Ma Y (2004) Noncovalent intermolecular energetics: RDX crystal. Int J Quantum Chem 100:733–739
76. Beresneva GA, Khristenko LV, Krasnoshchekov SV et al (1988) Vibrational spectra and conformational composition of ethylene glycol dinitrate in solid phases. J Appl Spectrosc 48:614–619

Design and Development of In Vivo Sensor Systems: The Long and Tortured Road to a Self-Contained, Implantable Glucose Sensor for Diabetes Management

Christina Thomas, Rachel Weller Roska, and Robert E. Carlson

Abstract We have successfully completed the development of a glucose sensing system, which is the mission-critical component of an implantable glucose sensor for use by diabetic patients. This proof-of-principle demonstration showed that the closed-cycle, self-contained glucose sensing system can produce a consistent, measurable response to physiologically relevant levels of glucose while functioning under biologically relevant conditions. The sensing system requires the interaction of two components: (1) a competitive agent/signaling component, which is a dendrimer-boronic acid (DBA) construct and (2) a glucose-competitive DBA binding environment, which is an immobilized monosaccharide mimic (iDIOL). The demonstrated sensing system meets our primary stability, sensitivity, and specificity criteria. These results, accompanied by our library of synthetic materials and binding affinity database, provide a firm foundation upon which to optimize the glucose sensing system and incorporate it into the implantable sensor device.

Keywords Artificial receptor, Binding affinity, Boronic acid, Closed-cycle, Competitive binding assay, Continuous, Dendrimer-boronic acid, Glucose sensing system, Glucose-competitive binding environment, Implantable, Noninvasive, Saccharide

Contents

C. Thomas, R.W. Roska, and R.E. Carlson (✉)
RECEPTORS LLC, Suite 510/MD 57, 1107 Hazeltine Boulevard, Chaska, MN 55318-2533, USA
e-mail: bc@receptorsllc.com

S.A. Piletsky and M.J. Whitcombe (eds.), *Designing Receptors for the Next Generation of Biosensors*, Springer Series on Chemical Sensors and Biosensors (2013) 12: 213–238
DOI 10.1007/5346_2012_23,
Published online: 28 July 2012

Abbreviations

ARS	Alizarin Red S; also known as 3,4-dihydroxy-9,10-dioxo-2-anthracene-sulfonic acid sodium salt
DBA	Dendrimer-boronic acid
IC_{50}	Half Maximal Inhibitory Concentration
iDIOL	Immobilized diol/saccharide analogue
Keq	Equilibrium constant
MIP	Moleculary imprinted polymer
NIR	Near-infrared
PET	Photoinduced electron transfer
pK_a	The negative logarithm of the dissociation constant
RFID	Radio frequency identification

Definitions

DBA	A dendrimer construct that is functionalized with boronic acid receptor ligands and a fluorescent reporter moiety. The DBA is the sensing system signaling component that can competitively bind to the glucose analyte and to the 1,2- and 1,3-dihydroxy motif(s) of iDIOLs.
diol	A saccharide analogue moiety that typically contains a 1,2- or 1,3-dihydroxy motif and a functional group that can be used for covalent immobilization of the diol on a support to create an iDIOL.
iDIOL	An immobilized diol/saccharide analogue that contains a 1,2- or 1,3-dihydroxy moiety that is covalently attached to a support. The iDIOL competes with glucose for DBA binding, which produces a bound versus free sensing system signal.
DBA:glucose:iDIOL	Designation of the three component competitive system where glucose and the iDIOL compete for DBA binding to produce a signal response that is proportional to glucose concentration.

1 Diabetes and Blood Glucose Management

Diabetes is one of the most significant global health challenges of the twenty-first century [1]. It remains one of the leading causes of death and is a major contributor to cardiovascular disease and is the leading cause of kidney failure, non-traumatic lower-limb amputation, and new cases of blindness in the USA [2]. Worldwide, the predominance and occurrence of diabetes has reached epidemic proportions and is expected to grow to 438 million by 2030 [3, 4]. Currently, diabetes is not curable but can be controlled through proper management, which includes accurate monitoring of blood glucose levels, in order to improve lifestyle and lifespan. Effective and consistent measurement of glucose, which is essential for accurate monitoring, remains a barrier to proper control of this disease due to the invasive and costly nature of currently available monitoring devices and resulting poor patient compliance.

Currently, the self-monitoring blood glucose test is the cornerstone of self-management for patients with diabetes. Unfortunately, this test requires that the patient extract a small drop of blood through an inconvenient and painful finger or torso pricking method three to four times daily for type I diabetes, according to the American Diabetes Association. In addition to this motivational barrier, high out-of-pocket expenditures for device test materials are also cited for noncompliant testing [5]. Over time, suboptimal testing frequency leads to out-of-range blood glucose levels and potential health complications.

Positive societal and economic impact can be achieved with the development of an easy-to-use, implantable glucose monitoring system. An implantable device is beneficial to patients because it provides real-time continuous information regarding glucose levels. Early detection of rapidly changing glucose levels is especially important for patients with type I diabetes when the onset of hypoglycemia can come without warning and can incur potentially dangerous consequences [6, 7]. An implanted, RFID-enabled device would minimize the continual cost, pain, and complications of current diagnostic systems. In terms of limiting expense and increasing comfort and testing compliance, diabetic patients would benefit from the long operational life of a one-time invasive, implanted device. An implantable glucose monitoring device is superior to other systems because, although initially more invasive upon implantation, ultimately and for the long term it is noninvasive on a daily basis. Ease of use makes patient monitoring and compliance a relative nonissue compared to the requirements of sampling blood daily or using an invasive, transdermal cannula. Additionally, glucose fluctuation data can be gathered electronically and stored for observation in real-time with no input from the patient. This chapter gives an overview of the design and proof-of-concept development of a self-contained and closed-cycle, stable glucose sensing system as the integral component of an implantable device for real-time in vivo glucose measurement and diabetes management.

Since the advent of the first commercial glucose testing devices in the 1970s, there has been progress toward the development of glucose detection techniques designed for noninvasive systems [8]. The three most studied techniques include enzyme [9–13], fluorescent [14–18], and NIR spectroscopy [19–23]. Despite various attempts, successful development of a fully functional implantable, noninvasive continuous monitoring device has remained elusive due to critical deficiencies of these detection techniques. Each method has physical and/or chemical limitations that make them impractical for use in a long-term, implantable device. Enzyme-based techniques function on reagents that are consumed and require a continuous reagent supply during the process of detection. The by-products of the reagent reactions are undesirable and cause detection interference. In addition, enzyme-based detection techniques experience reagent degradation and inactivation over the long term, eventually causing inaccurate readings and sensor drift [24–26]. Similar to problems with enzyme-based techniques, there are also reagent limitations for long-term fluorescence-based systems. Current fluorescence-based sensors cannot remain at an implantation site and respond to blood glucose concentrations over an extended period of time [27]. Over the lifetime of the sensor, denaturation, relaxation, or poisoning of the fluorescent molecular recognition element occurs [28]. Gradual deterioration of signaling reagents results in sensitivity and signal shifts that subsequently require continual readjustment and calibration in order to achieve accurate measurement [7]. Using NIR spectroscopy to decipher glucose levels by way of absorption measurements through or at tissues, however conceptually simple, is equally impractical. This approach is currently not acceptable for clinical use due to the fact that a number of factors such as tissue hydration, blood flow, temperature, light scattering, and overlapping absorption by non-glucose molecules cause read-out precision errors [7, 29]. It is no surprise that the search for the ideal glucose detection system continues to motivate the scientific community. However, past efforts in designing an implantable and self-contained glucose sensing system have not been successful because developers have given only partial consideration to the long-term impact and limitations of the in vivo environment.

A technically and commercially successful implantable glucose sensor requires the integrated design and development of several critical components (Fig. 1). The mission-critical self-contained and closed-cycle sensing component must be designed to interface with an appropriate signal transduction/signal processing device that, in turn, is coupled to the sensor's electronics and communication function. Further, the entire device must be enclosed in a porous, biostable, and biocompatible material that simultaneously prevents biofouling of the device and allows biotransport of the glucose analyte in and out of the device. Failure to integrate any of these components into the implantable device invariably leads to product development failure. Our integrated design for the implantable device, as illustrated in Fig. 2, envisions signal transduction using a MEMS cantilever [30–33] that will respond to bound/unbound mass changes of the reporter construct with subsequent processing of the resulting signal on a device-specific ASIC chip [34–37]. Signal export to the external environment will be via RFID communication [38–40] with signal processing to provide the diabetic patient and their medical team with glucose concentration and rate-of-change information both onboard the RFID reader module and wirelessly exported to an external database. Additionally, the biocompatible/biotransport membrane [41, 42] will: (1) protect the device from

BIOSTABLE INTERFACE	CLOSED-CYCLE SENSING	SIGNAL PROCESSING	RFID ENABLED COMMUNICATION

Fig. 1 Critical components schematic. The key components of our approach are the bioselective interface between the in vivo environment and the sensing system, the closed-cycle glucose sensing system and a mass-sensitive signal transduction interface that is coupled to the RFID-enabled data communication component

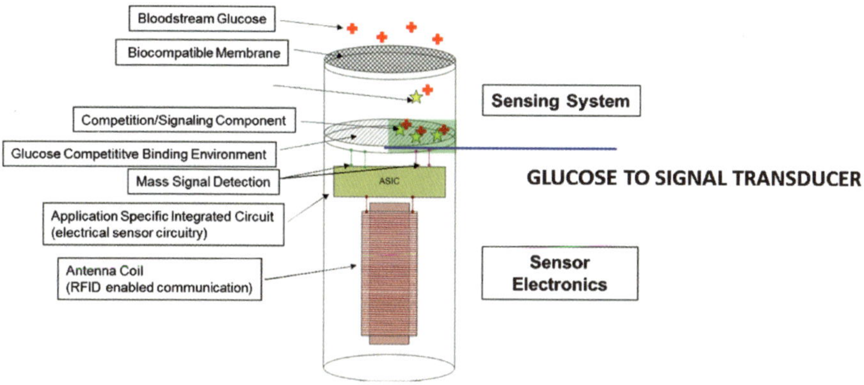

Fig. 2 Integrated sensor device design. The prototype glucose sensor will integrate the glucose sensing system with a mass-sensitive signal transduction mechanism coupled to the RFID-enabled communication electronics, all enclosed in a millimeter scale, implantable package

encapsulation and (2) facilitate the size-selective transport of the low molecular weight fraction of the in vivo fluid matrix in and out of the device, while also containing the mobile sensing system reagents (Fig. 3; see www.receptorsllc.com for an animation of the implantable device). While each of these component pieces is integral to the success of the device, the sensing system is the mission-critical component.

2 Artificial Receptor Technology and the Development of a Sensing System for an Implantable Glucose Sensor

Increasing demand for the detection of bioanalytes has triggered the development of rapid assay techniques in the form of sensor technologies [17, 43–45]. The need for more robust sensors that transcend the cost and stability limitations of current detection systems that require consumable biochemical reagents, such as enzymes and antibodies, has fueled the trend toward the design and development of sensing systems that are based on synthetic components like aptamers [46–48], MIPs [49–51] and receptor constructs [52, 53]. In addition, detection of bioanalytes

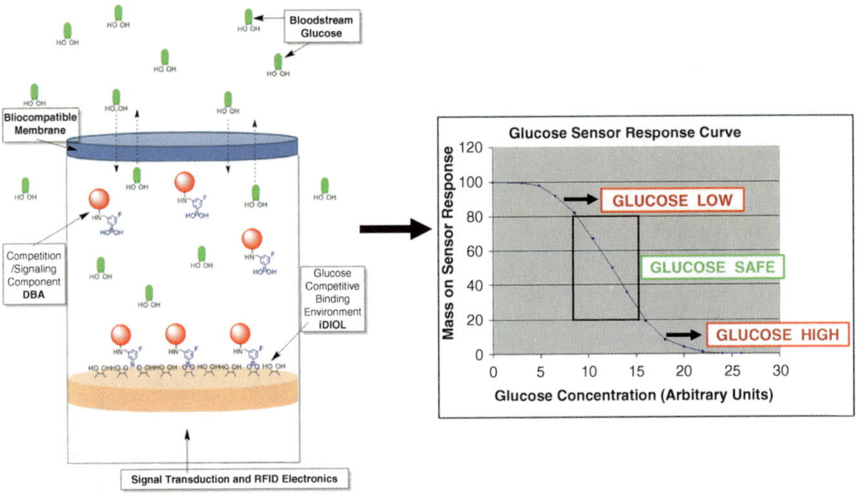

Fig. 3 Sensing system components. The glucose sensing system includes the glucose-competitive DBA binding environment (iDIOL) and the competition/signaling component (DBA). The system will produce a response proportionate to the bloodstream glucose levels. The readout will be one of three messages: LOW, SAFE, HIGH, which correspond to adult glucose levels of below 70 mg/dL, between 70 and 130 mg/dL and above 130 mg/dL, respectively, as they pertain to fasting glucose levels (American Diabetes Association 2010)

may require more advanced sensing component materials in order to substantially increase sensitivity and selectivity due to the complexity of the sample matrix and the inherently low analyte concentration that can exist in a physiological system [43]. The approach of utilizing synthetic materials for the construction of chemical recognition systems provides the structural and functional materials required for effective and robust sensing/receptor function [54]. Developing synthetic recognition materials with known physical and chemical properties provides the advantage of flexibility in selecting compatible sensing system reagents that meet the design criteria for operation within a physiological environment. It is critical that the reagents simultaneously function in complex, aqueous media while maintaining performance integrity under physiological pH and temperature [54]. It is also imperative that the materials not only preserve sensitivity and selectivity within complicated matrices of potentially competing analytes, but also retain sensitivity for a particular moiety whose physiological concentration may be low. The design challenges of an in vivo sensing system can be overcome using synthetically optimized recognition materials.

The applicability of artificial receptor materials to the development of saccharide sensors, especially as it relates to glucose detection, has attracted a great deal of interest [55–58]. Efforts to improve signaling technology continue to make headway because materials with enhanced biocompatibility and superior sensitivity and selectivity toward glucose are fundamental requirements for monitoring glucose levels in an implantable device. Our group has developed a sensing system technology for in vivo glucose analysis that utilizes synthetically optimized materials to fulfill the reagent requirements of a self-contained and closed-cycle, stable glucose sensing system. We have successfully developed components that can detect biologically relevant levels of glucose with the required sensitivity and selectivity in a physiologically relevant matrix solution. The materials are physically and chemically stable in aqueous media at physiological pH. Scalability is also an advantage of these reagents in that they are reproducible on a large scale with the capability to meet commercial demand.

The novel approach to glucose sensor design devised by our group involves two main components: a synthetically optimized boronic acid terminated dendrimer scaffold and a surface immobilized monosaccharide mimic. When these components are exposed to glucose, they competitively interact to produce a detectable and reproducible signal that is responsive to fluctuating levels of glucose. The magnitude of sensitivity and selectivity is tunable through the use of appropriate boronic acid and dihydroxy (diol) analogues and the degree of sensitivity and selectivity can be optimized based on a system specific binding affinity model and database. Reported herein is an overview of the development of our synthetic glucose sensing system. This description includes a discussion of our strategy, along with an overview of the in-depth considerations we used to select system components for optimal detection performance in a physiologically relevant environment.

3 Self-Contained and Closed-Cycle Stable Glucose Sensing System

The design of our device is based on the creation of an integrated, self-contained sensing system that produces an RFID readout, which provides two pieces of information: milligrams per deciliter (mg/dL) glucose values and an indication of whether the physiological glucose concentration is increasing or decreasing. This combination of information can be used by the diabetic patient to determine whether their glucose levels are currently low, safe, or high (Fig. 3). Demonstration of the closed-cycle chemical sensing system required the interaction of two components. These components are (1) the competitive agent/signaling component, which is based on a dendrimer-boronic acid (DBA) construct (Fig. 3) and (2) the glucose-competitive DBA binding environment, which consists of an immobilized monosaccharide mimic (iDIOL, Fig. 3). Our unique detection approach functions through reversible competitive binding between glucose and the iDIOL for the DBA. The amount of DBA that is bound to the iDIOL binding environment on the mass-sensitive transduction interface fluctuates in response to changing levels of glucose. The change in free versus bound DBA is measured via a change in the resonance frequency of the MEMS microcantilever. This signal transduction event gives a measurement of glucose concentration that can be calibrated to bloodstream glucose levels (Fig. 3). The function of this type of sensor relies on the relative affinity of glucose and the iDIOL for the DBA. Consequently, optimization of the glucose sensing system was based on our evaluation of the binding affinities of the DBA for both glucose and the iDIOL. More broadly, our approach for constructing and optimizing component materials was also based on an in-depth consideration of how these materials related to the sensing system and the device as a whole.

3.1 Competition/Signaling Component (DBA)

Macromolecular DBA constructs have been used for the first time by our group as the glucose recognition and signaling agent in a *competitive* binding assay that will ultimately be incorporated as a mass-sensitive detection method for the in vivo determination of glucose concentration. A DBA construct has been described for use in an in vitro saccharide sensor by James et al. In this example, anthracene units are used as the dye indicator that correlates fluorescence intensity changes with saccharide binding [59]. Although useful for detection of saccharides in an in vitro environment, this type of detection technique is not applicable to an implantable device for multiple reasons. The dendrimer constructs have limited aqueous solubility due to the highly insoluble anthracene moiety [59]. More generally, the use of anthracene as a candidate for in vivo applications is unfavorable due to sensitivity issues, toxicity concerns, and lack of metabolic stability [60, 61]. The viability of this type of sensor in a physiological matrix would be compromised, as the material would continue to lose sensitivity over time due to diminishing fluorescence resulting from denaturation, photodegradation, and/or indicator poisoning [28]. This would, in turn, require that the device be continually calibrated and frequently recharged with fresh reagents [7]. In addition, there is not a well-established method for exciting the fluorophore and taking measurements from an implanted fluorescence-based device without inserting an invasive probe into the subcutaneous tissue [7]. The design of our DBA constructs remedies these obstacles to functional implantation.

The first critical step required for demonstration of the glucose sensing system is the construction of the DBA competitive agent/signaling component. The selection of materials for the DBA component was dictated by the need to build synthetic receptor moieties that would respond with optimal binding sensitivity and selectivity for glucose in a complex aqueous matrix of potentially competing analytes. In addition, as deemed essential for extended function in a closed-cycle, long-term implantable device that is continuously exposed to the lytic nature of physiological fluid, the synthetic materials used to synthesize the DBAs must be stable and able to perform without diminished capacity over the lifetime of the sensor. Separately, but equally important, the materials must not be consumed during the detection process or require external reagents. For these reasons, our work focused on the development of a synthetic saccharide sensor that has the capacity to selectively detect glucose with long-term integrity in a physiological system.

3.1.1 Boronic Acids as Recognition Elements for Glucose Detection

Boronic acid analogues as synthetic receptor ligands have been extensively evaluated for their use as the molecular recognition component in the construction of saccharide sensors [62–71]. It is known that boronic acids rapidly and reversibly bind diols in aqueous environments, through boronate ester formation [72–74]. This favorable interaction occurs with high affinity and selectivity [62–64, 69, 70]. As a result, boronic acid-based detection systems find application where quantitative detection of saccharides is critically linked to disease therapies, such as diabetes management [75–77]. The use of appropriately designed boronic acids as molecular recognition units provides the ability to both selectively recognize and signal analytes, such as glucose, at low concentrations and in real time [78–80].

Numerous advances have been made in understanding how the electronic, geometric, and polar properties of functional groups on boronic acid analogues affect the mechanism and process of reversible diol complexation [73, 74, 76, 81–84]. Several groups have demonstrated that saccharide selectivity and binding properties are affected by the location and type of substituents about the aromatic boronic acid substructure [81, 85]. It has also been reported that, in general, aryl boronic acids with lower pK_as tend to have higher binding affinities for diols near neutral pH, although optimal binding depends not only on the pK_a of the boronic acid but also on the structure and properties of the diol in question, as well as the pH and ionic strength of the binding environment [73, 74, 76, 81, 82, 86]. Boronic acid pK_as are tunable by altering the substituents [75]. For example, Badugu et al. [75] have shown that the pK_a of phenylboronic acid can be decreased by adding electron withdrawing groups, while adding electron donating groups increases the pK_a. Alternatively, there is evidence that a neighboring nitrogen can enhance the formation of boronate esters under neutral pH conditions by coordinating intramolecularly with boron to create a more electron deficient atomic center, resulting in a reduction in the apparent pK_a of the boronic acid [66, 87–89]. In our efforts to design a boronic acid-based receptor and signaling component, we exploited the physical and chemical influence of substituent type and location to improve the binding affinity and selectivity of DBAs for glucose and iDIOLs.

The ability of boronic acid-based sensors to function efficiently in a physiological system is reflected by their selective interaction with saccharides. For saccharide recognition to proceed, cyclic boronate ester formation must occur upon binding of a boronic acid to, preferably, a 1,2- or 1,3- diol to form a five- or six-membered cyclic ester [72, 90]. It is possible for boronate esters to form under aqueous conditions, but at neutral pH binding affinity is low [84]. Greater binding affinity can be obtained under elevated pH conditions (pH 10), where the more favorable tetrahedral boronate form dominates [66, 73, 87, 88]. Designing a boronic acid-based sensor component that has greater binding affinity in a neutral physiological system can be achieved by (1) strategically outfitting the phenylboronic acid substructure with electron withdrawing groups in the *meta-* or *para-* position in order to stabilize the boronate form of the acid and lower the pK_a value [73–76, 81, 82, 86] and/or (2) introduce an *ortho*-amino methyl substituent to facilitate boronate ester formation at neutral pH through donation of the nitrogen lone pairs into the empty boron p-orbital [66, 87–89]. Strategic selection of boronic acid receptor molecules containing substituent(s) that have the greatest potential to initiate boronate ester formation was key in designing a signaling component that would perform with the desired glucose binding characteristics.

Our preliminary efforts in designing a boronic acid sensing system focused on selecting commercially available boronic acids with a diversity of substituent(s) about the phenylboronic acid substructure and a reactive group that could be used for coupling the boronic acid to a carrier scaffold. We selected phenylboronic acid molecules whose substituent type(s) and location(s) would increase the electrophilicity of the boronic acid group, reducing its pK_a and ultimately, increasing the binding affinity at neutral pH. The resulting boronic acid constructs, each of which possessed unique functionalities and enabled a diversity of saccharide binding sensitivities and selectivities, formed the basis of our library of candidate DBA signaling component materials.

3.1.2 Dendrimers as Synthetic Receptor Scaffold Materials

Three main considerations influenced the design of the DBA scaffold. These included: (1) selection of the appropriate scaffold to arrange the recognition motif in the correct orientation to support binding affinity and specificity, (2) selection of a scaffold with a mass sufficient to create a differential with glucose in order to generate a detectable signal, and (3) selection of a construct of appropriate size to prevent the signaling/competition component from diffusing out of the sensing system compartment.

Owing to their physical and chemical properties, dendrimers are advantageous for the construction of synthetic receptor materials and stable sensing applications [91]. Dendrimers have a spherical and highly branched 3-D architecture that gives them a well-defined composition and topology [92, 93]. These characteristics, combined with their high-density surface functional group capacity for boronic acid immobilization, give dendrimers desirable physical, chemical, and polyvalency characteristics [91, 94]. Their highly functionalized terminal surfaces also allow for control over the display of surface recognition elements. In addition, dendrimers are frequently exploited in physiological systems because they are water soluble, biocompatible and non-immunogenic [92, 95, 96]. They are commercially available in a number of different generations and have size and mass characteristics that are compatible with our sensing system and implantable device design.

These characteristics make dendrimers ideally suited as scaffolds for the DBA competition/signaling component. They simultaneously provide a water soluble, stable, and polyvalent scaffold that facilitates and stabilizes the conjugation of the otherwise insoluble and unstable boronic acid recognition moieties at the dendrimer surface.

3.1.3 DBA Construction

Boronic acid analogues were selected for inclusion in the DBA library based on prescreening of their interactions with our target analyte (glucose) versus their interactions with our saccharide mimic diol compounds. Utilizing ARS, a diol selective fluorescent dye, we characterized the binding interactions of the initial kit of boronic acids with each diol species [97–100]. Indicator displacement assays, such as the ARS assay, rely on the relative affinity of two competing guests for the receptor host. Specifically, the saccharide or diol-containing species, as the analyte of interest, competes with and preferentially displaces the diol-containing ARS from the boronic acid host. The displacement of the ARS reporter molecule from the boronic acid structure causes a measurable change in fluorescence. The magnitude of the fluorescence change that results from increasing concentrations of analyte provides a straightforward method to determine which boronic acid structures bind competitively with glucose and/or the saccharide mimics under the conditions (e.g., pH, ionic strength) of the assay.

ARS competitive assays were performed in a physiological buffer at pH 7 to confirm that the preselected kit of boronic acid ligands, which were selected to include a range of structural and chemical properties, bound glucose with adequate affinity in an aqueous environment. If the observed ARS fluorescence dropped substantially as the concentration of glucose titrated into the assay solution increased, we could conclude that glucose was competitive with the ARS diol relative to the boronic acid. In that case, the boronic acid was deemed to have passed our screening guidelines. On the other hand, if there was no observed change in fluorescence as increasing amount of glucose was titrated into the assay solution, we could conclude that glucose could not compete for the boronic acid with adequate affinity and, as a result, that particular boronic acid would no longer be considered as a viable candidate.

Response curves from a representative ARS assay experiment are shown in Fig. 4. The observed drop in fluorescence intensity as the concentration of glucose titrated into the solution increased demonstrated that glucose could bind to phenyl boronic acid 1 (Fig. 5a) and compete with ARS. In other words, the affinity of phenyl boronic acid 1 for glucose was greater than the affinity of phenyl boronic acid 1 for ARS, causing phenyl boronic acid 1 to preferentially bind with glucose. In contrast, phenyl boronic acid 2 (Fig. 5b) showed little affinity toward glucose and was not included in construction of the DBA library. As predicted from structure–pK_a relationships, phenyl boronic acid 1 would have greater binding affinity does glucose than does phenyl boronic acid 2. According to Hammet equation predictions, the quantifiable difference between phenyl boronic acid 1 and phenyl boronic acid 2 is the fluoro substituent located in the *para*-position on the phenylboronic acid structure. The electron withdrawing effect of the fluoro substituent in the *para*-position, on phenyl boronic acid 1, combined with a less sterically hindered boronic acid, will cause a drop in pK_a and an increase in binding affinity for glucose.

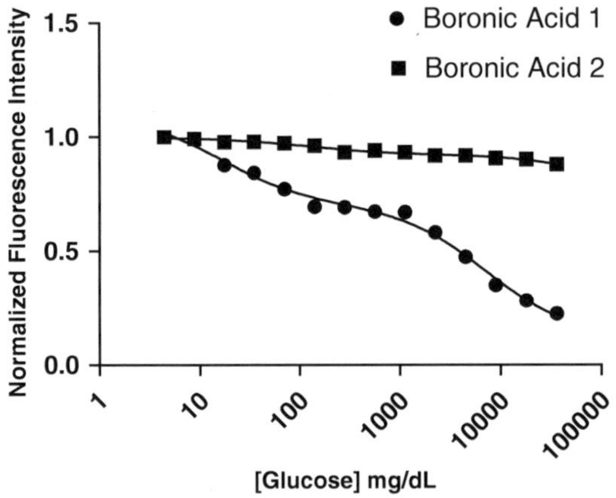

Fig. 4 ARS competition binding assay. Glucose competition curves showing normalized fluorescence intensity versus glucose concentration for boronic acid 1 and boronic acid 2 in a physiological buffer at neutral pH

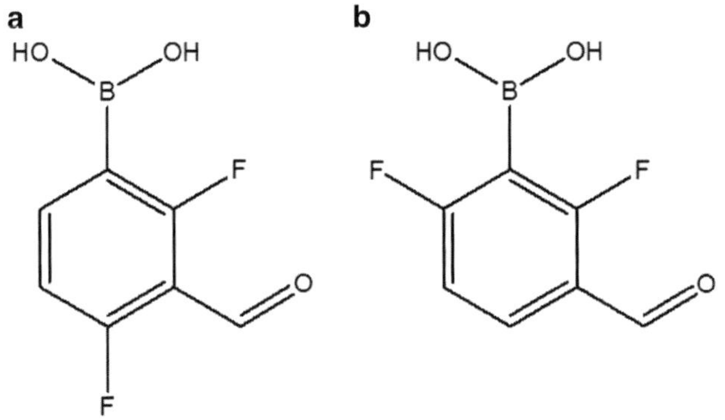

Fig. 5 Structures of boronic acid 1 (**a**) and boronic acid 2 (**b**) evaluated for binding performance in a glucose competition binding assay

Following boronic acid–glucose binding affinity prescreening, we synthesized a library of DBAs using the selected, candidate boronic acids. Each DBA in the library was subsequently screened against each candidate saccharide mimic.

4 Glucose-Competitive DBA Binding Environment (iDIOL)

Diols, in the form of immobilized saccharide mimics (iDIOLs), have been used by our group as a glucose-competitive, DBA binding environment in a competitive binding assay that serves as the prototype for the ultimate mass-sensitive, in vivo glucose sensor. Thus, the second critical step required for the demonstration of the self-contained glucose sensing system was the selection of the glucose-competitive

DBA binding environment (iDIOL). The selection of materials for this component was governed by the need to (1) construct a glucose-competitive binding environment that would form a reversible complex with the DBA signaling component in aqueous media and (2) select commercially available saccharide mimics with a diversity of diol sub-structures and a suitable functional moiety for covalent immobilization to a support.

4.1 iDIOLs as Competitive Binding Environments for Glucose Detection

This iDIOL versus DBA strategy, as discussed in detail earlier, uses the observation that the hydroxyl groups on saccharides, specifically 1,2- or 1,3-diols, are known to competitively bind with boronic acids to form five- or six-membered ring structures [72, 90]. We initially selected diols, which would subsequently be immobilized to produce the required iDIOLs, based on a comparison of their binding affinity to DBAs versus the binding affinity of the respective DBA for glucose. Our diol selection strategy involved exploiting the differential in relative binding affinity that would be created when a DBA is concurrently exposed to an immobilized diol (iDIOL) and a range of glucose concentrations. The objective was to identify DBA: iDIOL pairs that would permit discriminatory binding of the DBA to glucose, due to increased relative affinity over DBA binding to the iDIOL.

Selection of diols for ultimate preparation of iDIOLs, via immobilization of the diol on the sensing system's transduction interface, was based on our evaluation of the interactions between our kit of boronic acid-derived DBAs and various candidate diol species. We again used the ARS assay, as described previously, to characterize the binding of the DBAs with the candidate diols. The magnitude of the change in ARS fluorescence that resulted from increasing the amount of diol titrated into the assay solution provided a straightforward method to determine which diols, and ultimately which iDIOL structures, would competitively interact with the various DBA species.

Response curves from an ARS assay experiment performed in a physiological buffer at neutral pH are shown in Fig. 6. An enhanced response of the DBA 2 (Fig. 7a) for diol 1 (Fig. 7b) versus diol 2 (Fig. 7c) was observed. Phenylboronic acids are known to have different binding affinities for diols depending on the dihedral angle of the diol. Smaller dihedral angles often accompany higher binding constants [72]. Additionally, rigid cyclic *cis* diols tend to form stronger cyclic esters than acyclic diols [72, 90]. Thus, the enhanced binding of diol 1 can be attributed to the improved compatibility of the boronic acid recognition motif on DBA 2 with the dihedral angle of the diol. In contrast, it can be inferred that diol 2 formed a weaker cyclic ester with the same boronic acid of DBA 2 as a result of increased angle strain of the larger dihedral angle structure of the acyclic diol.

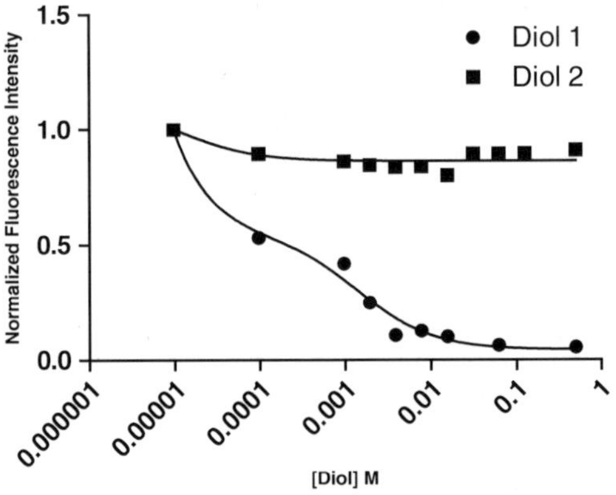

Fig. 6 ARS competition binding assay. Diol competition curves showing normalized fluorescence intensity versus diol 1 and diol 2 concentration for a DBA in a physiological buffer at neutral pH

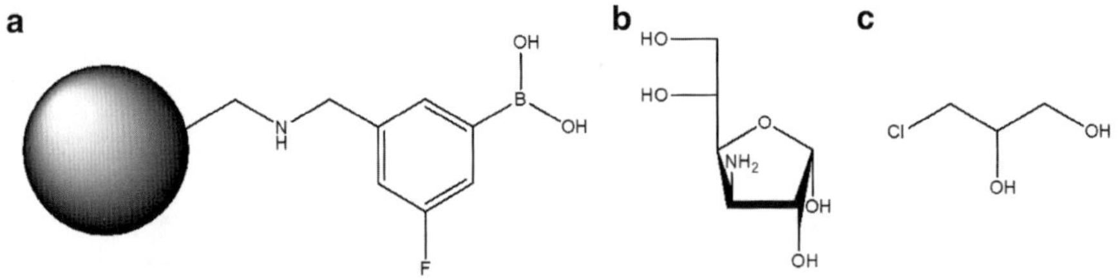

Fig. 7 Structures of DBA 2 (**a**), diol 1 (**b**) and diol 2 (**c**) evaluated for binding performance in a diol competition binding assay

The drop in fluorescence intensity as the concentration of diol titrated into the solution increased demonstrated that diol 1 could bind to the DBA with an affinity sufficient to release the DBA from the DBA:ARS complex. By contrast, the DBA showed little affinity towards diol 2, which was subsequently not considered for immobilization as an iDIOL. Based on the results of this screening process, we generated a library of diols that, when immobilized as iDIOLs, encompassed a range of DBA:diol and DBA:glucose binding affinities. The database of diol/iDIOL chemical and physical properties, as they related to binding affinity, became part of the toolbox that enabled us to screen for the optimal signaling component relative to the desired glucose-competitive DBA binding environment.

5 Coordinated Identification of DBA:iDIOL Lead Pairs

In order to achieve identification of lead DBA:iDIOL pairs for subsequent evaluation as candidate glucose sensing system components, we required a method that would allow us to determine the relative affinities of DBA:glucose versus DBA:

iDIOL. As a consequence, we began systematically evaluating the Keq values of DBA:diol and DBA:glucose candidates using their ARS profiles, over a range of diol/glucose concentrations. Although it could be viewed as necessary to screen every potential candidate DBA:diol combination to determine their response to glucose, even a limited set of boronic acids (e.g. $n = 50$) incorporated into a series of dendrimer generations as DBA constructs (e.g. $n = 5$) and evaluated against iDIOL candidates (e.g., $n = 50$) gives a formidable number (e.g. $n = 50 \times 5 \times 50 = 12{,}500$) of possible combinations. In order to overcome the technical and resource challenges of such a laborious screening process, we built a binding affinity model and database based on a three-component DBA:glucose:diol interaction model [73]. Establishing a foundation based on an affinity model database was critical to furthering our efforts toward designing a system whose function relies on the affinities of the sensing system components. These derived Keq values were used to identify lead DBA and iDIOL candidates. By comparing Keq values, we were able to estimate how sensitively each DBA would respond to glucose and identify components that would best fit a sensing system designed to detect glucose over the physiological range. Not only did this approach significantly limit the number of DBA:diol candidate combinations that would need to be screened, but it also quantified and allowed us to directly compare binding between each DBA: glucose and DBA:diol pair.

Experimental Keq values of DBA:diol and DBA:glucose combinations were generated utilizing the three-component competitive assay developed by Springsteen and Wang [73]. Using ARS as the fluorescent reporter, the association constant between each respective DBA:glucose and DBA:diol pair was determined. Within this system there are two competing equilibria, the first between the candidate DBA and the ARS reporter and the second between the candidate DBA and glucose or saccharide mimic diol. Fluorescence intensity changes, as they relate to the formation and perturbation of each equilibria, were used to calculate the Keq of glucose and the diol relative to the DBA [73]. These data were ranked according to the magnitude of the Keq (Fig. 8) to facilitate selection of DBA(s) for use as competition signaling components and diol(s) for immobilization as iDIOL binding environments.

Keq values of each DBA:diol and DBA:glucose combination were used to generate a scatter plot of the interaction data (Fig. 8), which illustrates the wide range of relative affinities encompassed in our DBA and saccharide mimic libraries. Based on the location of a representative data point on the interaction graph, the relative affinity of glucose versus each diol for that DBA can be easily compared. For example, if a data interaction point is located along the 1:1 line, as depicted in Fig. 8, this indicates that the relative binding affinity of the candidate DBA for glucose is similar to the binding affinity of the same DBA for the diol of the DBA: diol pair. Additionally, if a data interaction point is located along the 2:1 line, the binding strength of the candidate DBA for glucose is approximately twice the binding strength of the same DBA for the diol. This may signify that a data interaction point on the 2:1 line represents a DBA:diol that is more sensitive to glucose than a DBA:diol pair on the 1:1 line. Depending on how the binding affinity

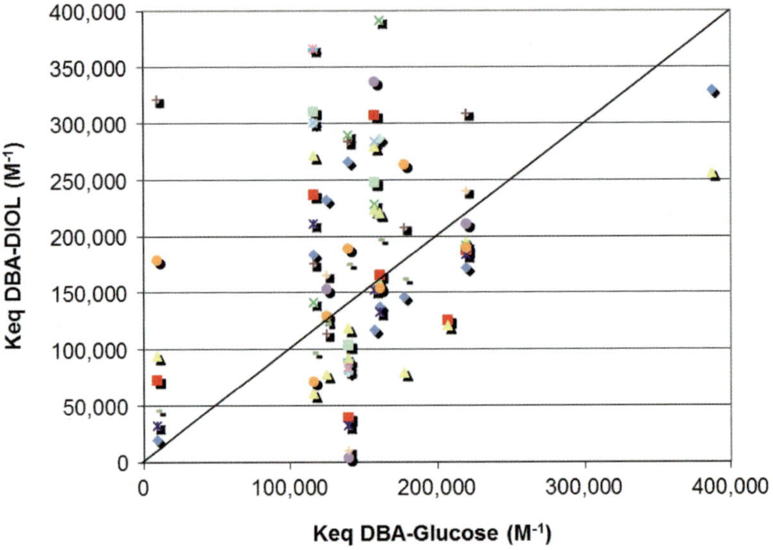

Fig. 8 Keq interaction graph. Comparison of DBA-to-glucose binding affinity (*X*-axis) versus DBA-to-diol binding affinity (*Y*-axis)

values for a DBA:glucose and DBA:diol pair differed in magnitude, that particular DBA:diol pair was either eliminated or included as a lead pair in further glucose competition assay screening experiments.

Our libraries of DBA and diol compounds were systematically evaluated for Keq under conditions (ionic species, pH, etc.) that resembled those of an in vivo environment. This data system was designed as a guide to rapidly compare relative binding affinities of a large number of DBA and diol species before committing to diol immobilization as an iDIOL environment and subsequent DBA:glucose:iDIOL surface competition screening. Significantly, data extrapolated from the Keq interaction graph streamlined our efforts in estimating how each DBA:iDIOL combination would respond to glucose.

6 Glucose Competition Binding Assay

Glucose competition was next assessed using a format more closely related to the format that will eventually be used in the final device. Previously selected diols that demonstrated a range of Keq values with several of the DBAs relative to glucose were covalently immobilized on glass supports as iDIOL environments. A series of mixtures that contained a fixed concentration of fluorescently labeled DBA with varying concentrations of glucose, including the concentration range encompassing physiologically relevant glucose levels (30–300 mg/dL), were incubated with the iDIOL-functionalized surface. Detection of free, labeled DBA indicated loss of fluorescent signal from the iDIOL environment following exposure to glucose, confirming successful competition. A plot of the fluorescence signal in response to increasing glucose concentrations produced a response curve that defined the glucose sensitivity of the candidate DBA relative to the iDIOL. Response profiles of DBAs that showed a significant, competitive response to increasing glucose concentration were considered to have a desirable binding equilibrium between glucose and the iDIOL. On the one hand, the DBA needed to bind to the iDIOL with sufficient affinity to produce a useful signal. On the other hand, the DBA needed to bind to the iDIOL weakly enough relative to the DBA:glucose affinity so that glucose could compete to produce a signal. The slope and IC_{50} values of each response curve were the parameters used to compare the binding sensitivity of each DBA:glucose:iDIOL detection system.

6.1 *Examples of Different Component Sensitivity and Selectivity*

In one representative study, multiple candidate DBAs were used to generate glucose response curves using a reference iDIOL, over a broad glucose concentration range. Figure 9a shows the glucose response curves, which are the inverse of the free solution fluorescence intensity measured during the assay. Upon addition of glucose, the fluorescence intensity of DBA not bound to the iDIOL increased. This was due to the competitive binding of glucose to the boronic acid receptors of the DBA, which prevented the fluorescently labeled DBA from binding to the iDIOL. Previously determined binding constants for DBA:glucose and DBA:diol were correlated with the glucose response curves of each DBA:iDIOL system (Fig. 9b).

These binding curves illustrate that the candidate DBAs (Fig. 10a–c) respond differently to changing levels of glucose when exposed to a particular iDIOL (Fig. 10d), as would be expected from their DBA:diol Keq values. DBA 2 and DBA 3 are on or below the DBA:glucose versus DBA:diol 1:1 line, indicating that glucose has equal or greater affinity for DBA 2 and DBA 3 than the diol. The opposite is true for DBA 1, which has minimal DBA:glucose affinity relative to the DBA:diol. These data correlate with the observed glucose response curves where DBA 1 produced a minimally responsive curve and DBA 2 and DBA 3 showed typical competitive assay curves. Furthermore, the greater I_{50} sensitivity of DBA 2 relative to DBA 3 (Fig. 11) is in agreement with the difference in DBA 2:glucose affinity versus DBA 3:glucose affinity.

In a second representative study, multiple candidate iDIOL conjugates (Fig. 13a–c) were used to generate glucose response curves (Fig. 12a) using a reference DBA (Fig. 13d) over a broad glucose concentration range. As in the previous example, upon addition of glucose, the fluorescence intensity of unbound DBA increased due to the competitive binding of glucose to the boronic acid receptors of the DBA, which prevented further binding of the DBA to the iDIOL

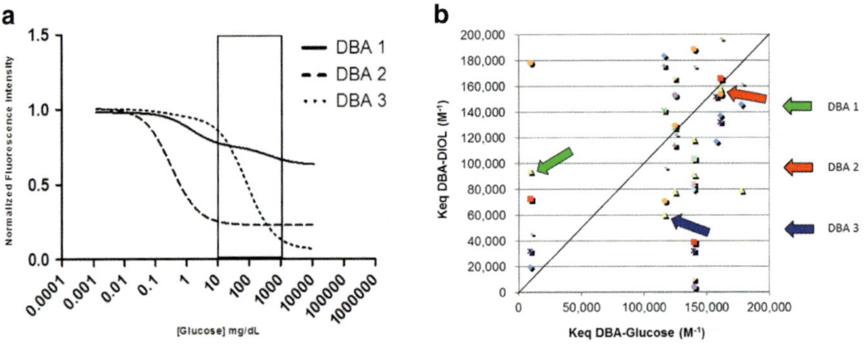

Fig. 9 Glucose competition curves showing the normalized DBA fluorescence intensity versus glucose concentration for DBA 1, DBA 2, and DBA 3 in physiological buffer at neutral pH on an iDIOL 3 surface (**a**). Binding constants, in the Keq interaction graph, for DBA 1, 2, and/or 3: glucose and DBA 1, 2, and/or 3:diol combinations were correlated with the glucose response curves of each DBA:iDIOL system (**b**)

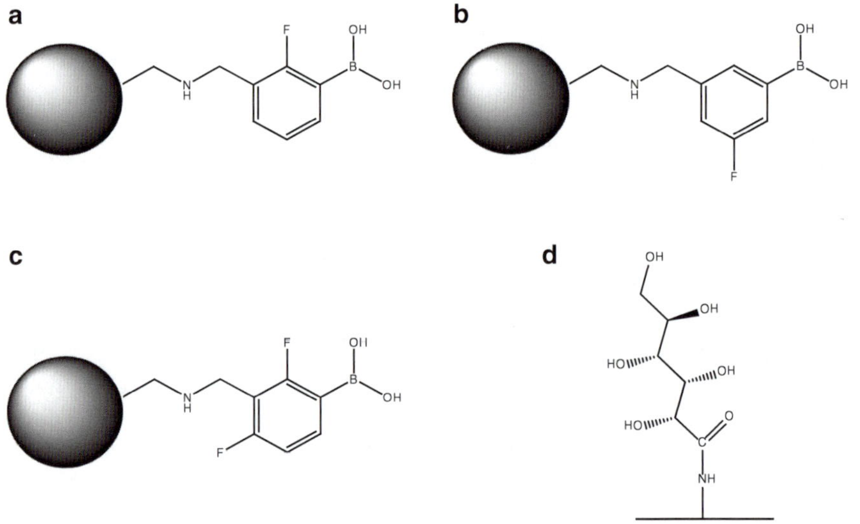

Fig. 10 Structures of DBA 1 (**a**), DBA 2 (**b**), DBA 3 (**c**), and iDIOL 3 (**d**) evaluated for binding performance in a glucose competition binding assay

surface. These glucose competition curves illustrate that the DBA responded, as would be expected from their DBA:diol Keq values, to changing levels of glucose with significant diversity relative to the iDIOLs. Previously determined binding constants for DBA:glucose and DBA:diol combinations were correlated with the glucose response curves of each DBA:iDIOL system (Fig. 12b). Keq vlaues for the diols corresponding to iDIOL 1 and iDIOL 2 are above the DBA:glucose versus DBA:diol 1:1 line, indicating that the DBA has less affinity for glucose than either the

	DBA:iDIOL Combination	IC_{50} (mg/dL)
A)	DBA 1:iDIOL 3	>10,000
	DBA 2:iDIOL 3	0.5
	DBA 3:iDIOL 3	100
B)	DBA 3:iDIOL 1	>10,000
	DBA 3:iDIOL 2	10,000
	DBA 3:iDIOL 3	100

Fig. 11 IC_{50} values from glucose competition response curves of various DBA:iDIOL combinations

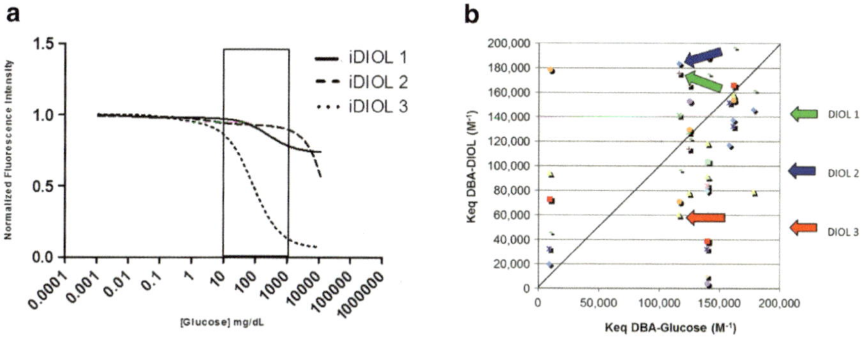

Fig. 12 Glucose competition curves showing the normalized DBA fluorescence intensity versus glucose concentration of DBA 3 in a physiological buffer at neutral pH on an iDIOL 1, iDIOL 2, and iDIOL 3 surface (**a**). Binding constants, in the Keq interaction graph, for DBA 3:glucose and DBA 3:diol 1, 2, and/or 3 combinations were correlated with the glucose response curves of each DBA:iDIOL system (**b**)

diol corresponding to iDIOL. The opposite is true for the diol corresponding to iDIOL 3, whose corresponding diol lies below the 1:1 line. These data correlate with the observed glucose response curves, wherein iDIOL 1 and iDIOL 2 produce minimally responsive curves while iDIOL 3 produced a competitive assay curve.

Although the above studies established the glucose sensitivity of the illustrated DBA:iDIOL systems, it was also critical to determine glucose specificity. In a representative selectivity study, the DBA 3:iDIOL 3 component pair was evaluated for binding response relative to fructose and galactose (Fig. 14), which are present in vivo and could potentially interfere with the glucose response of the system. Measurements were performed over a broad saccharide concentration range. Upon addition of fructose and/or galactose, the DBA fluorescence intensity signal changed very little due to the inability of fructose and/or galactose to bind to the boronic acid receptors of the DBA. Therefore, the binding equilibrium of the DBA with the iDIOL binding environment was undisturbed. These curves show that this DBA:iDIOL pair is minimally cross-reactive with fructose or galactose.

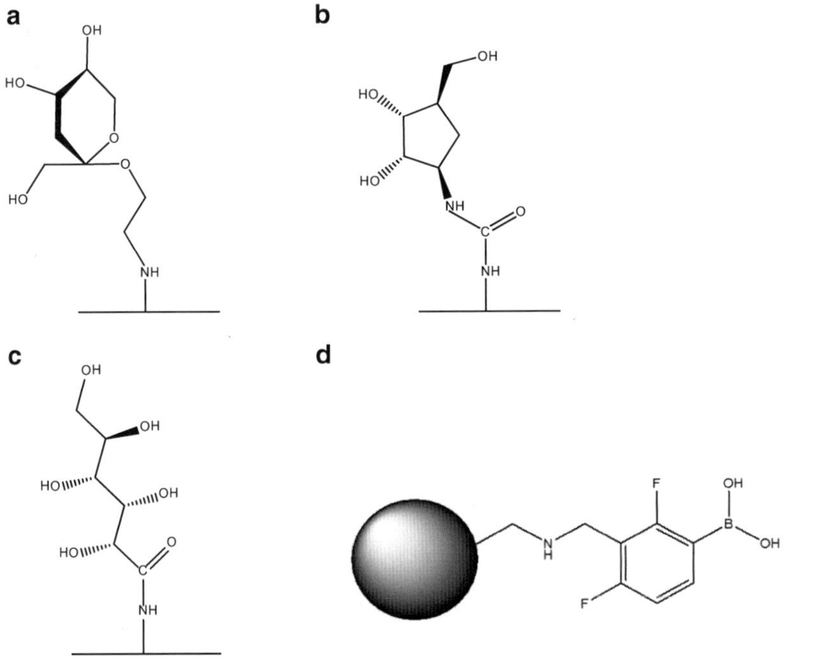

Fig. 13 Chemical structures of iDIOL 1 (**a**), iDIOL 2 (**b**), iDIOL 3 (**c**), and DBA 3 and (**d**) evaluated for binding performance in a glucose competition binding assay

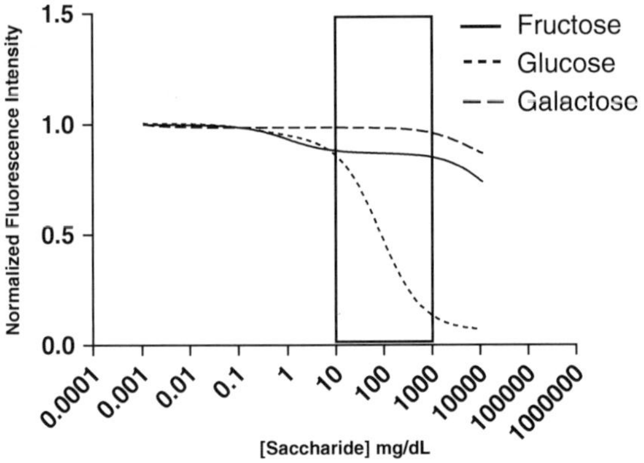

Fig. 14 Glucose, fructose, and galactose competition curves showing the normalized fluorescence intensity of DBA 3 versus saccharide concentration in a physiological buffer at neutral pH on an iDIOL 3 surface

6.2 *Implications of Binding Affinity Studies on Component Pair Selection*

Through these experiments, a selective glucose competition assay was established based on the binding affinities of DBAs for glucose and for an iDIOL surface. Additionally, our studies confirmed that candidate DBA:iDIOL pairs can be successfully screened for glucose sensitivity and selectivity. We have demonstrated that it is possible to use Keq values to compare the binding affinities of a DBA for glucose and of the same DBA for an iDIOL. This enabled us to qualitatively predict the glucose-competitive response of each DBA:iDIOL pair and to select candidate pairs that will generate reproducible glucose response curves with optimal sensitivity and selectivity. Intuitively, it can be assumed that component pairs that fall on either extreme of the Keq interaction graph will generate undesirable glucose response curves. On one end of the DBA:iDIOL affinity spectrum, the DBA binds too strongly to the iDIOL and glucose cannot effectively compete. On the other end of the affinity spectrum, the DBA binds too weakly to the iDIOL, which will not provide a useful dynamic range. With the capability of predicting glucose response curves based on the location of a Keq data interaction point, it was possible for us to quickly and efficiently eliminate component pair combinations that would be expected to perform in subsequent studies with low sensitivity and selectivity. Much to our advantage, this screening approach drastically limits the number of experiments that are required to select the best DBA:iDIOL combination, reducing time and cost investments. The results discussed above establish the validity of the Keq data interaction model for selection of candidate DBA:iDIOL pairs. The diversity of responses generated by each DBA:glucose:iDIOL system within our library ensures that we will be able to select DBA:iDIOL pairs with the appropriate physical and chemical properties necessary for analyzing glucose concentrations within the sensitivity and selectivity parameters required by the final device.

7 Concluding Remarks

We have designed and demonstrated a sensing system based on a DBA signaling component and immobilized saccharide mimic (iDIOL). Our materials ultimately do not require a fluorescent dye molecule to signal glucose concentration through DBA:glucose:iDIOL competition, as the device will function through a mass-sensitive signal transduction interface. In addition, the system components were synthesized with favorable aqueous solubility and stability characteristics. Each component was designed to include optimal structural motifs for the most favorable glucose sensitivity and selectivity. Faced with the challenge of sensing a range of physiologically relevant glucose concentrations in a complex matrix of potentially competing analytes, we developed a competitive binding model to expedite screening of our system components. Coordinated identification of DBA:iDIOL pairs that competitively interact with glucose was based on our evaluation of the Keq of a DBA for a diol versus the Keq of the DBA for glucose. A binding affinity model based on a three-component (DBA:diol:glucose) competitive assay was used to generate a scatter plot of interaction data. Based on the location of a representative data point on the interaction graph, the response of glucose and/or each diol for each DBA was evaluated for inclusion into the iDIOL screening process, which more closely mimic the format of the final device. The glucose sensing system described here provides a foundation for the integrated development of the final implantable device.

Acknowledgments We thank PositiveID Corporation for their partial financial support for this project.

References

1. Zimmet P, Alberti KGMM, Shaw J (2001) Global and societal implications of the diabetes epidemic. Nature 414:782–787
2. Centers for Disease Control and Prevention (2011) Successes and opportunities for population-based prevention and control at a Glance 2011. CDC. http://www.cdc.gov/chronicdisease/resources/publications/AAG/ddt.htm. Accessed 5 Nov 2011
3. World Health Organization (2011) Diabetes Fact Sheet Number 312. WHO. http://www.who.int/mediacentre/factsheet/fs312/en/. Accessed 5 Nov 2011
4. World Diabetes Foundation (2011) Diabetes facts. World Diabetes Foundation. http://www.worlddiabetesfoundation.org/compositive-35.htm. Accessed 5 Nov 2011
5. Karter AJ, Ferrara A, Darbinian JA et al (2000) Self-monitoring of blood glucose: language and financial barriers in a managed care population with diabetes. Diabetes Care 23:477–483
6. Gold AE, MacLeod KM, Frier BM (1994) Frequency of severe hypoglycemia in patients with type I diabetes with impaired awareness of hypoglycemia. Diabetes Care 17:697–703
7. Pickup J, McCartney L, Rolinski O et al (1999) In vivo glucose sensing for diabetes management: progress towards non-invasive monitoring. Br Med J 319:1289–1292
8. Novak BJ, Blake DR, Meinardi S et al (2007) Exhaled methyl nitrate as a noninvasive marker of hyperglycemia in type 1 diabetes. PNAS 104:15613–15618
9. Bindra DS, Zhang Y, Wilson GS et al (1991) Design and in vitro studies of a needle-type glucose sensor for subcutaneous monitoring. Anal Chem 63:1692–1696
10. Cronenberg C, Van Groen B, De Beer D et al (1991) Oxygen-independent glucose microsensor based on glucose oxidase. Anal Chim Acta 242:275–278
11. Ishikawa M, Schmidtke DW, Raskin P et al (1998) Initial evaluation of a 290-μm diameter subcutaneous glucose sensor: glucose monitoring with a biocompatible, flexible-wire, enzyme-based amperometric microsensor in diabetic and nondiabetic humans. J Diabetes Complications 12:295–301
12. Sternberg R, Barrau MB, Gangiotti L et al (1989) Study and development of multilayer needle-type enzyme-based glucose microsensors. Biosensors 4:27–40
13. Wilson R, Turner APF (1992) Glucose oxidase: an ideal enzyme. Biosens Bioelectron 7:165–185
14. McShane MJ (2002) Potential for glucose monitoring with nanoengineered fluorescent biosensors. Diabetes Technol Ther 4:533–538
15. Meadows D, Schultz JS (1988) Fiber-optic biosensors based on fluorescence energy transfer. Talanta 35:145–150
16. Moschou EA, Sharma BV, Deo SK et al (2004) Fluorescence glucose detection: advances toward the ideal in vivo biosensor. J Fluoresc 14:535–547

17. Pickup JC, Hussain F, Evans ND et al (2005) Fluorescence-based glucose sensors. Biosens Bioelectron 20:2555–2565
18. Russell RJ, Pishko MV, Gefrides CC et al (1999) A fluorescence-based glucose biosensor using concanavalin A and dextran encapsulated in a poly(ethylene glycol) hydrogel. Anal Chem 71:3126–3132
19. Arnold MA, Small GW (1990) Determination of physiological levels of glucose in an aqueous matrix with digitally filtered Fourier transform near-infrared spectra. Anal Chem 62:1457–1464
20. Gabriely I, Wozniak R, Mevorach M et al (1999) Transcutaneous glucose measurement using near-infrared spectroscopy during hypoglycemia. Diabetes Care 22:2026–2032
21. Haaland DM, Robinson MR, Koepp GW et al (1992) Reagentless near-infrared determination of glucose in whole blood using multivariate calibration. Appl Spectrosc 46:1575–1578
22. Heise HM, Marbach R, Koschinsky T et al (1994) Noninvasive blood glucose sensors based on near-infrared spectroscopy. Artif Organs 18:439–447
23. Stuart DA, Yonzon CR, Zhang X et al (2005) Glucose sensing using near-infrared surface-enhanced raman spectroscopy: gold surfaces, 10-day stability, and improved accuracy. Anal Chem 77:4013–4019
24. Cook A, Laughlin D, Moore M et al (2009) Differences in glucose values obtained from point-of-care glucose meters and laboratory analysis in critically ill patients. Am J Crit Care 18:65–71
25. Lacara T, Domagtoy C, Lickliter D et al (2007) Comparison of point-of-care and laboratory glucose analysis in critically ill patients. Am J Crit Care 16:336–346
26. Siegrist J, Kazarian T, Ensor C et al (2010) Continuous glucose sensor using novel genetically-engineered binding polypeptides towards in vivo applications. Sens Actuators B 149:51–58
27. Heo YJ, Shibata H, Okitsu T et al (2011) Long-term in vivo glucose monitoring using fluorescent hydrogel fibers. Proc Natl Acad Sci USA 108:13399–13403
28. Cote GL, McShane MJ, Pishko MV (2009) Fluorescence-based glucose biosensors. In: Tuchin VV (ed) Handbook of optical sensing of glucose in biological fluids and tissues. Taylor and Francis, New York
29. Oliver NS, Toumazou C, Cass AE et al (2009) Glucose sensors: a review of current and emerging technology. Diabet Med 26:197–210
30. Abadal G, Davis ZJ, Helbo B et al (2001) Electromechanical model of a resonating nano-cantilever-based sensor for high-resolution and high-sensitivity mass detection. Nanotechnology 12:100–104
31. Hansen KM, Thundat T (2005) Microcantilever biosensors. Methods 37:57–64
32. Sepaniak M, Datskos P, Lavrik N et al (2002) Microcantilever transducers: a new approach in sensor technology. Anal Chem 74:568A–575A
33. Ziegler C (2004) Cantilever-based biosensors. Anal Bioanal Chem 379:946–959
34. Chung W-Y, Yang C-H, Wang Y-F et al (2004) A signal processing ASIC for ISFET-based chemical sensors. Microelectron J 35:667–675
35. Liu X, Zheng YJ, Phyu MW et al (2010) A miniature on-chip multi-functional ECG signal processor with 30 μW ultra-low power consumption. Conf Proc IEEE Eng Med Biol Soc 2010:2577–2580
36. Ahn CH, Choi J-W, Beaucage G et al (2004) Disposable smart lab on a chip for point-of-care clinical diagnostics. Proc IEEE 92:154–173
37. Soin RS, Maloberti F, Franca J (1991) Analogue-digital ASICS: circuit techniques, design tools and applications. Peter Peregrinus Ltd, London
38. Abad E, Zampolli S, Marco S et al (2007) Flexible tag microlab development: gas sensors integration in RFID flexible tags for food logistic. Sens Actuators B 127:2–7
39. Opasjumruskit K, Thanthipwan T, Sathusen O et al (2006) Self-powered wireless temperature sensors exploit RFID technology. IEEE Pervasive Comput 5:54–61
40. Sample AP, Yeager DJ, Powledge PS et al (2008) Design of an RFID-based battery-free programmable sensing platform. IEEE Trans Instrum Meas 57:2608–2615
41. Leoni L, Desai T (2004) Micromachined biocapsules for cell-based sensing and delivery. Adv Drug Deliv Rev 56:211–229

42. Voskerician G, Shive MS, Shawgo RS et al (2003) Biocompatibility and biofouling of MEMS drug delivery devices. Biomaterials 24:1959–1967
43. Lieberzeit PA, Dickert FL (2008) Rapid bioanalysis with chemical sensors: novel strategies for devices and artificial recognition membranes. Anal Bioanal Chem 391:1629–1639
44. Sheehan AD, Quinn J, Daly S et al (2003) The development of novel miniaturized immunosensing devices: a review of a small technology with a large future. Anal Lett 36:511–537
45. Yi C, Li CW, Ji S et al (2006) Microfluidics technology for manipulation and analysis of biological cells. Anal Chim Acta 560:1–23
46. Baldrich E, O'Sullivan CK (2005) Ability of thrombin to act as molecular chaperone, inducing formation of quadruplex structure of thrombin-binding aptamer. Anal Biochem 341:194–197
47. Baldrich E, Acero JL, Reekmans G et al (2005) Displacement enzyme linked aptamer assay. Anal Chem 77:4774–4784
48. Baldrich E, Restrepo A, O'Sullivan CK (2004) Aptasensor development: elucidation of critical parameters for optimal aptamer performance. Anal Chem 76:7053–7063
49. Haupt K (2001) Molecularly imprinted polymers in analytical chemistry. Analyst 126:747–756
50. Jenkins AL, Yin R, Jensen JL (2001) Molecularly imprinted polymer sensors for pesticide and insecticide detection in water. Analyst 126:798–802
51. Moreno-Bondi M, Navarro-Villoslada F, Benito-Pena E et al (2008) Molecularly imprinted polymers as selective recognition elements in optical sensing. Curr Anal Chem 4:316–340
52. Kuhns WJ, Fernandez-Busquets X, Burger MM et al (1998) Hyaluronic acid-receptor binding demonstrated by synthetic adhesive proteoglycan peptide constructs and by cell receptors on the marine sponge microciona prolifera. Biol Bull 195:216–218
53. Schrader T, Hamilton AD (2005) Functional synthetic receptors. Wiley-VCH, Weinheim
54. Edwards NY, Sager TW, McDevitt JT et al (2007) Boronic acid based peptidic receptors for pattern-based saccharide sensing in neutral aqueous media, an application in real-life samples. J Am Chem Soc 129:13575–13583
55. Gamsey S, Suri JT, Wessling RA et al (2006) Continuous glucose detection using boronic acid-substituted viologens in fluorescent hydrogels: linker effects and extension to fiber optics. Langmuir 22:9067–9074
56. Horgan AM, Marshall AJ, Kew SJ et al (2006) Crosslinking of phenylboronic acid receptors as a means of glucose selective holographic detection. Biosens Bioelectron 21:1838–1845
57. James TD, Shinkai S (2002) Artificial receptors as chemosensors for carbohydrates. Top Curr Chem 218:159–200
58. Shoji E, Freund MS (2002) Potentiometric saccharide detection based on the pK_a changes of poly(aniline boronic acid). J Am Chem Soc 124:12486–12493
59. James TD, Shinmori H, Takeuchi M et al (1996) A saccharide 'sponge'. Synthesis and properties of a dendritic boronic acid. Chem Commun 6:705–706
60. Hall DG (2005) Boronic acids: preparation and applications in organic synthesis and medicine. Wiley-VCH, Weinheim
61. Yang W, Yan J, Springsteen G et al (2003) A novel type of fluorescent boronic acid that shows large fluorescence intensity changes upon binding with a carbohydrate in aqueous solution at physiological pH. Bioorg Med Chem 13:1019–1022
62. Adhikiri DP, Heagy MD (1999) Fluorescent chemosensor for carbohydrates which shows large change in chelation-enhanced quenching. Tetrahedron Lett 40:7893–7896
63. Arimori S, Ward CJ, James TD (2001) The first fluorescent sensor for boronic and boric acids with sensitivity at sub-micromolar concentrations—a cautionary tale. Chem Commun 2001:2018–2019
64. Cabell LA, Monahan M-K, Ansyln EV (1999) A competition assay for determining glucose-6-phosphate concentration with a tris-boronic acid receptor. Tetrahedron Lett 40:7753–7756
65. Eggert H, Frederiksen J, Morin C et al (1999) A new glucose-selective fluorescent bisboronic acid. First report of strong α-furanose complexation in aqueous solution at physiological pH. J Org Chem 64:3846–3852

66. James TD, Sandanayake KRAS, Iguchi R et al (1995) Novel saccharide-photoinduced electron transfer sensors based on the interaction of boronic acid and amine. J Am Chem Soc 117:8982–8987
67. James TD, Sandanayake KRAS, Shinkai S (1996) Saccharide sensing with molecular receptors based on boronic acid. Angew Chem Int Ed Engl 35:1910–1922
68. Wang W, Gao S, Wang B (1999) Building fluorescent sensors by template polymerization: the preparation of a fluorescent sensor for D-fructose. Org Lett 1:1209–1212
69. Wang W, Gao X, Wang B (2002) Boronic acid-based sensors. Curr Org Chem 6:1285–1317
70. Yang W, He H, Drueckhammer DG (2001) Computer-guided design in molecular recognition: design and synthesis of a glucopyranose receptor. Angew Chem Int Ed Engl 40:1714–1718
71. Yoon J, Czarnik AW (1992) Fluorescent chemosensors of carbohydrates. A means of chemically communicating the binding of polyols in water based on chelation-enhanced quenching. J Am Chem Soc 114:5874–5875
72. Fang H, Kaur G, Wang B (2004) Progress in boronic acid-based fluorescent glucose sensors. J Fluoresc 14:481–489
73. Springsteen G, Wang B (2002) A detailed examination of boronic acid-diol complexation. Tetrahedron 58:5291–5300
74. Yan J, Springsteen G, Deeter S et al (2004) The relationship among pK_a, pH, and binding constants in the interactions between boronic acids and diols-it is not as simple as it appears. Tetrahedron 60:11205–11209
75. Badugu R, Lakowicz JR, Geddes CD (2005) Boronic acid fluorescent sensors for monosaccharide signaling based on the 6-methyoxyquinolinium heterocyclic nucleus: progress toward noninvasive and continuous glucose monitoring. Bioorg Med Chem 13:113–119
76. Cao H, McGill T, Heagy MD (2004) Substituent effects on monoboronic acid sensors for saccharides based on N-phenyl-1,8-naphthalenedicarboximides. J Org Chem 69:2959–2966
77. Gamsey S, Miller A, Olmstead MM et al (2007) Boronic acid-based bipyridinium salts as tunable receptors for monosaccharides and α-hydroxycarboxylates. J Am Chem Soc 129:1278–1286
78. Alexeev VL, Das S, Finegold DN et al (2004) Photonic crystal glucose-sensing material for noninvasive monitoring of glucose in tear fluid. Clin Chem 50:2353–2360
79. DiCesare N, Lakowicz JR (2002) Charge transfer fluorescent probes using boronic acids for monosaccharide signaling. J Biomed Opt 7:538–545
80. Thoniyot P, Cappuccio FE, Gamsey S et al (2006) Continuous glucose sensing with fluorescent thin-film hydrogels. 2. Fiber optic sensor fabrication and in vitro testing. Diabetes Technol Ther 8:279–287
81. Hansen JS, Christensen JB, Solling TI et al (2011) Ortho-substituted aryl monoboronic acids have improved selectivity for D-glucose relative to D-fructose and L-lactate. Tetrahedron 67:1334–1340
82. Mulla HR, Agard NJ, Basu A (2004) 3-Methoxycarbonyl-5-nitrophenyl boronic: high affinity diol recognition at neutral pH. Bioorg Med Chem Lett 14:25–27
83. Tony JD, Phillips MD, Shinkai S (2006) Boronic acids in saccharide recognition. The Royal Society of Chemistry, Cambridge
84. Yang X, Lee M-C, Sartain F et al (2006) Designed boronate ligands for glucose-selective holographic sensors. Chem-Eur J 12:8491–8497
85. Jiang S, Escobedo JO, Kim KK et al (2006) Stereochemical and regiochemical trends in the selective detection of saccharides. J Am Chem Soc 128:12221–12228
86. Soundararajan S, Badawi M, Kohlrust CM et al (1989) Boronic acids for affinity chromatography: spectral methods for determinations of ionization and diol-binding constants. Anal Biochem 178:125–134
87. Kitano S, Hisamitsu I, Koyama Y et al (1991) Effect of the incorporation of amino groups in a glucose-responsive polymer complex having phenylboronic acid moieties. Polym Adv Technol 2:261–264
88. Linnane P, James TD, Shinkai S (1995) The synthesis and properties of a calixarene-based 'sugar bowl'. J Chem Soc Chem Commun 1995:1997–1998

89. Shiino D, Murata Y, Kubo A et al (1995) Amine containing phenylboronic acid gel for glucose-responsive insulin release under physiological pH. J Control Release 37:269–276
90. Bosch LI, Fyles TM, James TD (2004) Binary and ternary phenylboronic acid complexes with saccharides and lewis bases. Tetrahedron 60:11175–11190
91. Islam MT, Shi X, Balogh L (2005) HPLC separation of different generations of poly (amidoamine) dendrimers modified with various terminal groups. Anal Chem 77:2063–2070
92. Hong S, Bielinska AU, Mecke A et al (2004) Interaction of poly(amidoamine) dendrimers with supported lipid bilayers and cells: hole formation and the relation to transport. Bioconjug Chem 15:774–782
93. Tomalia DA, Naylor AM, Goddard WA (1990) Starburst dendrimers: molecular-level control of size, shape, surface chemistry, topology, and flexibility from atoms to macroscopic matter. Angew Chem Int Ed Engl 29:138–175
94. Majoros IJ, Thomas TP, Mehta CB et al (2005) Poly(amidoamine) dendrimer-based multifunctional engineered nanodevices for cancer therapy. J Med Chem 48:5892–5899
95. Kukowska-Latallo J, Candido KA, Cao Z et al (2005) Nanoparticle targeting of anticancer drug improves therapeutic response in animal model of human epithelial cancer. Cancer Res 65:5317–5324
96. Majoros IJ, Keszler B, Woehler S et al (2003) Acetylation of poly(amidoamine) dendrimers. Macromolecules 36:5526–5529
97. Arimori S, Ward CJ, James TD (2002) A D-glucose selective fluorescent assay. Tetrahedron Lett 43:303–305
98. Arimori S, Bosch AI, Ward CJ (2001) Fluorescent internal charge transfer (ICT) saccharide sensor. Tetrahedron Lett 42:4553–4555
99. Boduroglu S, El Khoury JM, Reddy V et al (2005) A colorimetric titration method for quantification of millimolar glucose in a pH 7.4 aqueous phosphate buffer. Bioorg Med Chem Lett 15:3974–3977
100. Musto CJ, Suslick KS (2010) Differential sensing of sugars by colorimetric arrays. Curr Opin Chem Biol 14:758–766

Evolving Trends in Transition Metal-Modified Receptor Design and Function

Paul A. Bertin

Abstract Nature elegantly invokes transition metal ions in a number of fundamental biomolecular processes from guiding protein folding and scaffolding tertiary structures to essential roles in information transmission through ligand binding, electron transfer, and catalysis. Often inspired by such natural systems, innovative transition metal-modified receptors are beginning to emerge with similar functions that hold substantial promise as components of next generation biosensors. This chapter aims to highlight some recent advances in the development of transition metal-modified synthetic receptors. Specifically, systems that incorporate organometallic, monometallic, and supramolecular coordination complexes are reviewed. The text will cover transition metal scaffolded peptide receptors, allosteric supramolecular enzyme mimics, and integrated monolayer-based electroactive receptor systems.

Keywords Allosteric supramolecular systems, Ferrocene bioconjugates, Metalloreceptor

Contents

P.A. Bertin (✉)
Ohmx Corporation, 1801 Maple Avenue, Suite 6143, Evanston, IL 60201, USA
e mail: paul.bertin2@gmail.com

S.A. Piletsky and M.J. Whitcombe (eds.), *Designing Receptors for the Next Generation of Biosensors*, Springer Series on Chemical Sensors and Biosensors (2013) 12: 239–260
DOI 10.1007/5346_2012_20, © Springer-Verlag Berlin Heidelberg 2012,
Published online: 27 June 2012

Abbreviations

Ala	Alanine
Arg	Arginine
Asp	Aspartic acid
Bpy	Bipyridine
BSA	Bovine serum albumin
Cys	Cysteine
DNA	Deoxyribonucleic acid
ELISA	Enzyme linked immunosorbent assay
ESI-MS	Electrospray ionization mass spectrometry
FRET	Fluorescence resonance energy transfer
Glu	Glutamic acid
His	Histidine
Leu	Leucine
NMR	Nuclear magnetic resonance
PCR	Polymerase chain reaction
Pro	Proline
RNA	Ribonucleic acid
SAM	Self-assembled monolayer
Ser	Serine
Trp	Tryptophan
Tyr	Tyrosine

1 Introduction

The term "biosensor" is used to describe a self-contained integrated device capable of transforming specific biochemical information from a sample into a useful quantitative or semiquantitative signal [1]. Biosensors comprise two principal components: (1) a recognition element or receptor capable of entering into a specific interaction with an analyte of interest, and (2) a closely linked physico-chemical transducer that translates this interaction into a physically measurable output (Fig. 1).

Since inception nearly 50 years ago with Professor Leland Clark's seminal address on how "to make electrochemical sensors more intelligent" by inclusion of "enzyme transducers" at electrodes for glucose detection [2], biosensing has evolved into a highly interdisciplinary analytical field that continues to proliferate from basic advances in biology, chemistry, materials science, and information technology. In many respects current biosensor research is becoming increasingly "molecularized" as how we think about overcoming problems in diagnostics is ever more reliant on an understanding of atomic-level structure and function along with

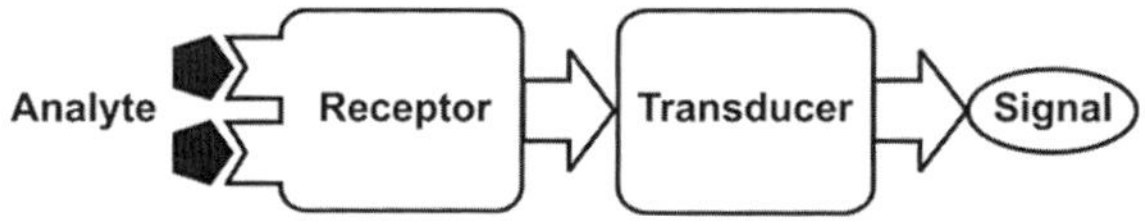

Fig. 1 Schematic diagram of a biosensor

an ability to design and synthesize complex molecular systems [3]. From this chemistry perspective, significant impetus exists to construct novel functional receptors with high performance characteristics to enhance and/or supplement conventional biomolecules (i.e., antibodies, enzymes, peptides, DNA). In particular, it is well known that transition metal ions serve many purposes in natural biological systems, from guiding protein folding and tertiary structure stabilization to crucial roles in information storage and retrieval through ligand binding, electron transfer, and catalysis. It follows then that innovative transition metal-modified receptors are beginning to emerge with similar functions that hold considerable promise as next generation biosensing elements.

This chapter aims to highlight some recent advances in the burgeoning field of transition metal-modified receptors. Specifically, systems that incorporate organometallic, monometallic, and supramolecular coordination complexes will be presented. The intention is to overview emerging concepts in synthetic receptor design and function that incorporate metal complexes through relevant examples from recent years rather than provide exhaustive and in-depth coverage of the field. The discussion will include transition metal-modified receptor scaffolds, allosteric supramolecular enzyme mimics, and integrated monolayer-based electroactive reporter systems.

2 Transition Metal Complexes as Scaffolds in Receptor Design

In biological systems, information is routinely stored in the size, shape, and electronic properties of molecules and transmitted by the way those molecules bind and react with each other. Beyond DNA, many of these information transfer tasks are accomplished using proteins. As distinct domains in multi-subunit protein building blocks fold and cooperate via noncovalent and directional bonding interactions, sophisticated nanoscale architectures are assembled that carry out complex life processes and chemical transformations.

Roughly one-third of all proteins require specific metal ion cofactors to assist in macromolecular folding and/or function [4]. Of the 20 naturally occurring amino acids programmed by the genetic code, only a relatively small number are frequently employed as metal ligands. These groups include the thiolate of Cys, the phenolate of Tyr, the imidazole nitrogens of His, and the carboxylates of Asp and Glu. Metal-binding sites in proteins are often classified into two categories: (1) preorganized environments, where metal binding only occurs if coordinating ligands are in the appropriate orientation, and (2) disordered environments, where protein folding is metal-directed. Zinc finger proteins are excellent examples of the latter as they contain relatively small structural domains stabilized by one or more Zn(II) cations [5]. The Zn(II) sites induce the correct folding of surrounding peptide fragments through tetrahedral coordination resulting in finger-like protrusions from the local protein structure that make tandem contacts with target molecules. Such zinc finger motifs commonly function as recognition modules that bind DNA, RNA, proteins, or small molecules. Hence, these proteins are involved in many fundamental cellular processes such as transcription and translation, replication and repair, cell proliferation and apoptosis, metabolism, and cell signaling [6].

2.1 Metal-Directed Peptide Assemblies for Molecular Recognition

Inspired by native metalloproteins, there is significant interest in utilizing metal-binding sites to prepare designed peptides and proteins as first steps toward novel biosensor receptors and de novo enzymes [7]. Early work by Ghadiri and coworkers [8, 9] demonstrated that α-helical secondary structures could be stabilized by selective metal complexation through judicious placement of either His or Cys residues one turn apart in relatively short amphiphilic peptide chains. In these examples, transition metal ions serve as intrapeptide crosslinks between i and $i + 4$ amino acid positions that stabilize helix formation by diminishing the entropy of the unfolded state relative to the folded state.

Furthermore, many groups have appended nonnatural synthetic ligands to peptide chains to allow for metal-directed assembly of stable tertiary structures (for a recent review, see [10]). For example, 2,2′-bpy ligands have gained prominence for many reasons: (1) they readily form tris-chelated octahedral complexes with a variety of transition metal precursors, (2) they serve as well-behaved spectroscopic probes, and (3) the resulting metal complexes often have high thermodynamic and kinetic stabilities. By covalently linking bpy ligands to the *N*-terminus of amphiphilic peptide sequences, triple helical coiled coil tertiary structures have been prepared upon coordination to Fe(II), Co(II), Ni(II), and Ru(II) precursors [11–14].

A schematic representation of such metal-directed trimeric coiled coils is shown in Fig. 2. The bpy-functionalized peptides are designed as short amphiphilic sequences that are relatively unstructured in solution. In the presence of a six-coordinate transition metal ion (M), a $[M(\text{bpy-peptide})_3]^{2+}$ complex is formed. As the metal complex forces the peptides into close proximity, α-helices are adopted, as characterized by circular dichroism. Stable assembly in the presence of metal ions is driven by burial of hydrophobic amino acid residues within the tertiary structure.

In the context of synthetic receptors, the $[M(\text{bpy-peptide})_3]^{2+}$ architecture is a well-defined, robust, and easily synthesized construct for molecular recognition studies. The 2,2′-bpy ligands can be linked to any helical peptide sequence via solid-phase peptide synthesis. Gochin et al. have employed $[M(\text{bpy-peptide})_3]^{2+}$ complexes using Fe(II) and Ni(II) ions to screen protein–peptide and protein–ligand interactions of the trimeric coiled coil subunit gp41 from the envelope glycoprotein

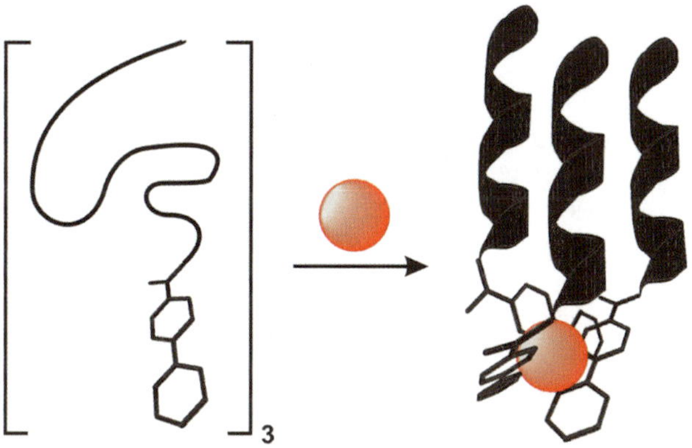

Fig. 2 Proposed model for metal-directed assembly of a trimeric coiled coil $[M(bpy\text{-}peptide)_3]^{2+}$. A six-coordinate metal ion (*sphere*) tethers three 2,2′-bpy ligands covalently linked to amphiphilic peptides (*left*), which drives hydrophobic collapse to yield the coiled coil structure (*right*). Adapted with permission from [10]. Copyright 2004 American Chemical Society

HIV-1 [15]. The $[M(bpy\text{-}peptide)_3]^{2+}$ coiled coil structures were shown to display receptor characteristics of the intact gp41 subunit by binding a small flanking peptide from the gp41 C-heptad repeat. By labeling this C-peptide with a fluorescent probe, compounds that bind the metal-directed coiled coil and displace the C-peptide were measured via FRET with a competitive inhibition assay. Importantly, this study suggests that $[M(bpy\text{-}peptide)_3]^{2+}$ motifs may serve as functional surrogates for full-length fusion proteins of class 1 enveloped viruses that are unsuitable for biophysical binding studies as the coiled coil regions of interest in the native proteins are masked by outer domains.

The ability to direct tertiary structure stabilization of designed peptide sequences through metal–ligand coordination provides biosensor researchers a powerful strategy for constructing receptors with defined rigid three-dimensional topologies. Further, it offers a facile route to synthetic receptors capable of discrete polyvalent binding interactions that may display increased affinities over their monovalent precursors. However, one potential limitation is the formation of multiple stereoisomers based on the coordination geometry of the metal complex which may complicate purification efforts. Regardless, it is anticipated that the molecular surfaces of such ordered peptides will more closely mimic naturally occurring protein recognition elements (i.e., ordered tertiary structures such as antigen-binding clefts in antibodies) and lead to synthetic receptors with increased specificity for a range of molecular targets.

Looking beyond oligopeptides, significant strides have also been made toward metal-directed protein self-assembly with the overarching goal to be able to master protein–protein interactions and access novel functional biomaterials [7, 16]. Ultimately, such designed protein interfaces coupled with designed metal-binding sites have the flexibility to provide unique specificity and reactivity for catalytic functions unknown in nature and may aid in the development of next generation biosensors.

2.2 Metalloreceptor Scaffolds in Differential Sensing

An emerging field within coordination chemistry is the use of synthetic metalloreceptors in array formats for the detection of small and large biomolecules. The biological processes of olfaction and taste, which use "differential" receptors biased toward specific classes of analytes, provide inspiration for this work. In contrast to traditional "lock-and-key" approaches aimed at identifying highly selective receptor–analyte interactions, each arrayed receptor for differential sensing is designed to be cross-reactive with diverse affinities for different targets of interest [17]. As a result, specific analytes and analyte mixtures produce fingerprint responses that can be extracted as unique diagnostic patterns using optical or chemometric tools. Synthetic receptor arrays are an exciting frontier in sensor development with applications in the detection of bioanalytes, pathogens in solution, environmental analysis, and medical diagnostics. While transition metal cores are not critical to the construction of such differential sensing receptor arrays, these subunits offer the potential advantage of target coordination over alternate covalent systems.

For example, Anslyn and coworkers [18] have designed a series of metalated synthetic receptor arrays as differential sensing platforms for a range of bioanalytes. A homogeneous array sensing scheme for the differentiation of small peptides and their phosphorylated analogues is shown in Fig. 3 [19]. Synthetic tripeptides derived from the amino acid sequence flanking Ser-129 of filamentous α-synuclein, a site of phosphorylation implicated in Parkinson's Disease, were used as model analytes to assess the discriminatory properties of the differential receptors for detecting protein phosphorylation. The platform involves a series of receptors created by appending two random peptide sequences to a central C_{3v}-symmetric metal complex core unit that selectively binds tetrahedral anions such as monophosphate esters. Using a combination of different metal ions and indicator dyes, receptors were screened optically for analyte binding using colorimetric/fluorescent indicator displacement protocols [20]. Patterns within the optical data sets were identified using pattern recognition algorithms.

A tris-[(2-pyridyl)methyl]amine core ligand subunit was selected due to its well-known capacity for binding several transition metals. Coordination of Cu(II) to this ligand framework substituted with guanidinium groups had previously been shown to produce a selective binding pocket for phosphate over other anions with different sizes, shapes, and charge [21, 22]. The high affinities reported for phosphate were proposed to proceed through a combination of ion-pairing interactions between guanidiniums and the Cu(II) center with the oxygens of the tetrahedral anion. This general ligand framework was functionalized with random tripeptide sequences via solid-phase synthesis to impart differential cross-reactivity toward the target peptides. Five different peptide side chains, three transition metal ions [Cu(II), Ni(II), Co(II)], and three indicator dyes were selected through extensive screening to produce an array of 45 receptor/metal/indicator combinations. Using linear discriminate analysis of the differential optical responses, 100% classification of the six peptide targets was achieved. Importantly, this general detection scheme

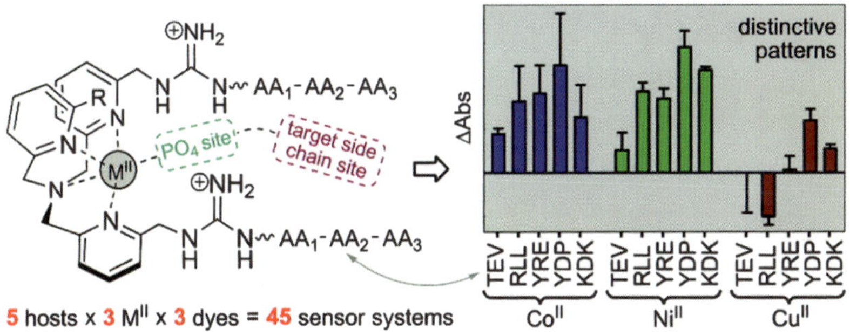

Fig. 3 A homogeneous array sensing scheme for the differentiation of tripeptides and their phosphorylated analogues. The array consists of five unique cross-reactive peptides coupled to tris-[(2-pyridyl)methyl]amine core ligands, three transition metal ions [Cu(II), Ni(II), Co(II)], and three indicator dyes totaling 45 indicator displacement assays. Linear discriminant analysis of optical data from this sensing ensemble resulted in 100% classification of the six target peptides. Reprinted with permission from [19]. Copyright 2009 American Chemical Society

may offer a more streamlined approach to screening specific kinase activity as it is not reliant on different selective receptors for each phosphorylated enzyme substrate.

Overall, this study highlights the utility of transition metal-modified receptor libraries biased toward a specific class of analytes as differential arrays for fingerprinting purposes. The receptor design strategy employs a single metal complex as both scaffold and analyte recognition element. This general approach should be broadly applicable to a range of bioanalytes and serve as a useful alternative to traditional, high affinity chemoselective sensing systems.

3 Functional Allosteric Supramolecular Receptors

The development of functional supramolecular coordination complexes has also been greatly influenced by biomimetic design principles over the past few decades. For instance, allosteric modulation of protein function is fundamental to many biological processes and occurs through binding of an effector, or regulatory molecule, to a protein site structurally distinct from the active site. Effector binding induces a conformational change in the protein that indirectly influences active site function. Effectors can either enhance or decrease the binding or catalytic efficiency of proteins, thus behaving as molecular "ON/OFF" switches. Inspired by such systems, chemists have recently evolved a number of strategies for endowing supramolecular systems with allosteric regulation [23, 24]. The overarching goal in this area is to develop abiotic synthetic systems that manifest the molecular recognition and catalytic properties of highly evolved natural proteins and enzymes.

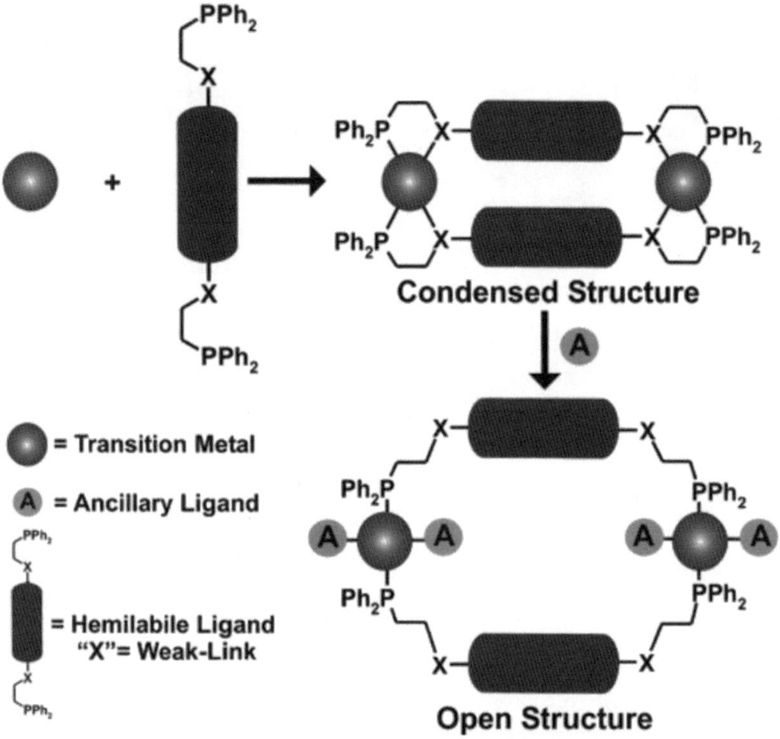

Fig. 4 General strategy of the WLA to macrocyclic coordination complexes. Reprinted with permission from [26]. Copyright 2005 American Chemical Society

Since their first report of an allosteric catalyst made possible through supramolecular chemistry [25], Mirkin and coworkers have remained pioneers in the development of synthetic enzyme mimics using the weak-link approach (WLA) to macrocyclic coordination complexes [26, 27]. The WLA allows one to construct multimetallic macrocycles with flexible hemilabile ligands that form both strong and weak coordination bonds with a transition metal center. Specifically, the WLA approach relies on condensed macrocyclic structures that contain at least two strategically placed strong (metal–phosphine) and weak (metal–X) bonds (Fig. 4). A critical feature of this approach is that the structural metals that direct the assembly process are available for further ligand substitution reactions without compromising the supramolecular structure. Thus, macrocycles can be toggled between well-defined architectures with different shapes and rigidities through ligand displacement reactions by selectively and reversibly breaking weak coordination bonds with small molecule effectors. This inherent property of all weak-link macrocycles makes the approach well suited for preparing functional systems that exhibit allosteric control through appropriately designed hemilabile ligands (i.e., transition metal catalysts, receptors).

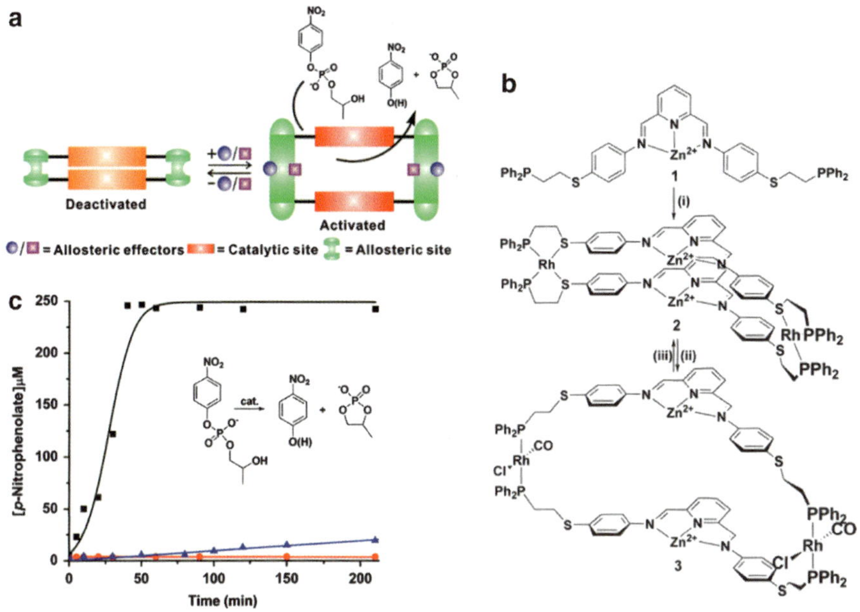

Fig. 5 (**a**) Supramolecular allosteric regulation via the WLA. (**b**) Synthesis of the supramolecular allosteric catalyst **3**. The AcO^- counterions ligands and counterions for the complexes are omitted for clarity. Reagents: (*i*) Rh(norbornadiene)$_2$BF$_4$; (*ii*) *n*-Bu$_4$NCl/CO; (*iii*) N_2 bubbling or 2 equiv of $AgBF_4$. (**c**) Product (*p*-nitrophenolate) concentration versus time for **2** (*circle*) and **3** (*square*) macrocyclic compounds. A control experiment with compound **1** (*triangle*) was carried out under the same reaction conditions. Reactions were monitored by UV–vis spectroscopy. Adapted with permission from [29]. Copyright 2007 American Chemical Society

Significantly, the WLA has been utilized to create structures that offer amplification in chemical detection where the effector molecule is the analyte and the catalytic reaction turned on by the effector generates a fluorescent surrogate molecule that provides an indirect signal output associated with the recognition event [28]. This completely synthetic system resembles the signal amplification properties of ELISAs commonly used in biosensing for a wide range of analytes. However, one drawback of this initial system was the small but measurable background catalysis exhibited by the closed "off" state of the macrocycle. In the context of sensor performance, higher background signal raises the lower limit of analyte detection and is therefore undesirable.

Recently, Mirkin et al. overcame this background issue with a tetrametallic macrocycle assembled via the WLA that operates as a completely reversible allosteric catalyst for the bimolecular hydrolysis of 2-(hydroxypropyl)-*p*-nitrophenyl phosphate (HPNP), a model substrate for RNA (Fig. 5a) [29]. Figure 5b shows the strategy for preparing the "open" active allosteric catalyst **3** from the "closed" condensed macrocycle **2** assembled from an equimolar ratio of hemilabile Zn(II)-functionalized ligand **1** and a Rh(I) precursor. The formation of compound **2** was characterized by ^{1}H and ^{31}P NMR, ESI-MS, and single-crystal X ray diffraction analysis. Notably, the proximity of the Zn(II) atoms in **2** precludes the possibility of any intramolecularly catalyzed HPNP hydrolysis reaction. In addition, the X-ray structure of **2** revealed that an acetate counterion bridges the two Zn atoms in a μ_2-fashion, further inhibiting the ability of HPNP to bind to the active site, which prevents intermolecular catalysis. Upon the addition of 2 equivalents of Cl^- and CO (1 atm) to a solution of **2**, open complex **3** rapidly forms as confirmed by ^1H and ^{31}P NMR, and ESI-MS. In contrast to **2**, open complex **3** has a suitable geometry for HPNP binding between the Zn(II) sites to enable catalytic bimolecular hydrolysis. Reversibility was demonstrated by bubbling N_2 into the solution to reform **2**.

The catalytic properties of ligand **1**, condensed macrocycle **2**, and open compound **3** were evaluated in the context of HPNP hydrolysis under pseudo-aqueous (mixed solvent) conditions and monitored by UV–vis spectroscopy as a function of time (Fig. 5c). Although ligand **1** exhibited slow but measurable catalytic activity, condensed macrocycle **2** was completely inactive under identical reaction conditions. In contrast, open complex **3** was extremely active and capable of quantitatively hydrolyzing all of the HPNP substrate in less than 40 min. The allosteric effect is generated by adjusting the distance between the catalytic Zn(II) metal centers using reversible coordination chemistry occurring at the Rh(I) metal centers. The system can be efficiently interconverted between completely "on" and "off" catalytic states through the use of small molecule regulators that alter the accessibility of substrate to the active binuclear Zn(II) sites.

In general, such synthetic allosteric systems represent the first step toward functional abiotic analyte detection strategies that take advantage of catalytic amplification. The attractive feature of a completely reversible allosteric catalyst with good turnover and catalytic rates is that it can become a central amplification motif used for many future systems where the regulatory sites are designed as receptors for different analytes with comparable amplification capabilities.

A particularly elegant and powerful extension of this technology is highlighted by the recent report of a supramolecular allosteric catalyst that exhibits a PCR-like reaction cascade [30]. PCR is an enzyme-mediated process where, in response to temperature cycling, a single target DNA molecule can be rapidly amplified into many billions of target molecules. Since its conception in 1985 [31], PCR has become a universal laboratory tool for most scientists involved in biochemical and molecular biological research. Indeed, effective PCR-based amplification is now the backbone of modern molecular diagnostics enabling the identification of genetic markers for disease with unmatched sensitivity and high reliability. A limitation of PCR, however, is that it only works with nucleic acid targets. Therefore, a significant challenge exists to create alternate synthetic systems that amplify the recognition events of non-nucleic acid targets. Such constructs would advance the development of ultrasensitive detection systems for a much wider class of analytes, including ones that are relevant to biosensing and beyond.

The catalyst system used to demonstrate PCR-like target amplification is shown in Fig. 6. In previous work [28, 32] catalyst **4** was opened in the presence of Cl^- and CO to generate **5**. The open cavity in **5** catalyzed an acyl transfer reaction between pyridyl carbinol and acetic anhydride to produce acetic acid which was signaled

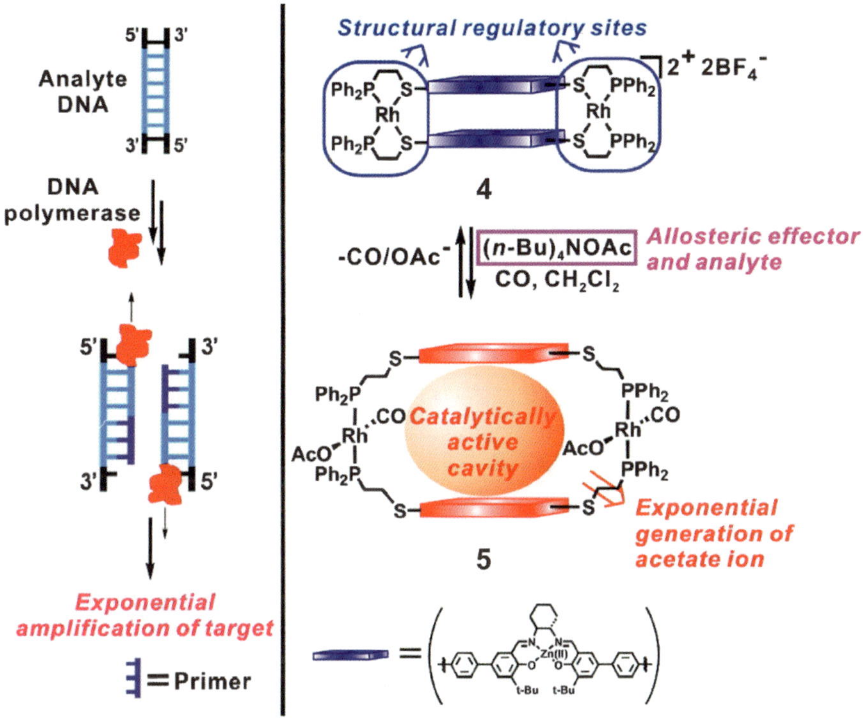

Fig. 6 Small molecule regulated target amplification using an allosteric enzyme mimic (*right*) with a reaction cascade analogous to PCR (*left*). Reprinted with permission from [30]. Copyright 2008 American Chemical Society

with a fluorescent pH indicator. By replacing the target effector Cl^- anion with acetate to generate a *trans* acetate and CO ligand at each Rh(I) site, the reaction was run under basic conditions to produce additional acetate from the same acyl transfer reaction. In the early stages of the reaction, only a minor amount of the catalyst is activated (in the form of **5**), but as the reaction proceeds, more acetate is generated, which leads to the formation of more **5** and progressively faster catalysis. The pH-sensitive fluorescent dye allows one to follow the amount of acetate generated during the catalytic cycles. This type of cascade reaction results in an exponential increase in the amount of initial target. As with PCR, slow induction followed by rapid exponential amplification, linear growth, and then eventual saturation was observed regardless of initial acetate concentrations. The time at which the exponential step turns on correlated well with the initial acetate target concentration. In principle, this approach can be operational in completely aqueous environments by incorporating water-soluble ligands [33] and extended to other analytes as long as one can envision catalytic reactions that generate an analyte of interest.

The above examples demonstrate that, through exquisite control over the synthesis of macrocyclic coordination complexes containing well-defined cavities, species can be designed that function in a specific and predictable fashion for the detection of analytes that behave in a manner similar to natural enzyme systems. Looking ahead, these abiotic enzyme mimics should inspire the next generation of functional synthetic metalloreceptors and may ultimately be rationally integrated into highly sensitive biosensing technologies. The challenge will be to design specific allosteric regulatory sites that are responsive to bioanalytes or surrogates of interest.

4 Monolayer Receptors with Integrated Metal-Based Reporters

The fundamental limitation that has prohibited more widespread success of commercial biosensors is the failure of most clinically relevant biomolecules to produce an easily measured signal upon receptor binding that is specific over the many potential interferants in a biological sample. Therefore, analytical approaches based on analyte–receptor recognition (i.e., immunoassays) almost always require burdensome, multistep procedures that are often limited to laboratory settings. Thus, a significant challenge remains to design functional receptor systems that can be integrated with physicochemical transducers and report on specific biomolecular interactions in complex matrices without intervention. Transition metal-modified receptors hold exciting promise in this regard due to their inherent capacity for participating in electron transfer reactions when integrated with signal transducing electrodes.

Indeed, nanoscale charge transfer processes represent a rapidly advancing frontier of fundamental science [34–36], with applications ranging from molecular electronics [37–40] and information storage [41] in addition to chemical and biological sensor fabrication [42]. Basic advances in our understanding of electron transfer continue to fuel growth in electrochemical biosensor research due to the tremendous commercial opportunity for fast, simple, scalable, and low-cost detection technologies that can be integrated with modern microelectronics. In particular, self-assembled monolayers (SAMs) of electroactive molecules adsorbed on noble metal electrodes have been intensely investigated as model systems for interfacial electron transfer events [43, 44]. The majority of electroactive SAMs studied to date comprise molecules with common design features, namely thiol-terminated organic bridges anchored to gold electrodes through gold-thiolate bonds with ω-functionalized redox-active head groups. Since the pioneering work of Chidsey and coworkers [45] in 1990, cyclic voltammetry (CV) studies of ferrocene-terminated SAMs have been extensively reported in the literature and the influence of bridge architecture, coadsorbed diluent molecules, and supporting electrolytes on the redox behavior of these SAMs has been well established [43].

In voltammetry studies of redox-active SAMs, current is measured as a function of potential between the SAM-modified working electrode and a reference electrode. It is well known that the kinetics and thermodynamics of such interfacial redox reactions are strongly affected by the nature of the medium in which they occur.

Currently, there is significant interest in coupling the electron transfer properties of surface-confined ferrocenes (and other redox probes) with the molecular recognition capabilities of designed receptors to prepare highly sensitive and selective integrated electrochemical sensors for a broad range of analytes.

Ferrocene scaffolds offer a number of advantages in the construction of integrated SAM-based electrochemical sensor platforms. These include (1) tunable [46] and fully reversible oxidation/reduction processes within a convenient electrochemical window for maintaining SAM integrity, (2) stability in aqueous media, and (3) commercial availability in many forms offering synthetic versatility and facile routes toward functionalization with anchoring organosulfur groups and/or recognition motifs such as DNA, peptides, and receptor ligands. In this section, recent examples of biosensing monolayers that incorporate ferrocene–receptor conjugates will be highlighted. In particular, emphasis will be placed on receptor architectures that enable reagentless electrochemical detection of different classes of analytes. For a more comprehensive treatment of ferrocene bioconjugates, the reader is directed to the recent reviews by Metzler-Nolte et al. [47] and Martic et al. [48].

4.1 Ferrocene–DNA Conjugates

A potentially general solution to the problem of signal detection in biosensors is based on the ability to program binding-induced "folding" responses of electrode-bound DNA probes. In particular, E-DNA sensors that utilize oligonucleotide receptors bound at one terminus to an electrode with the other covalently labeled with a ferrocene (or alternate) redox probe have become mainstay architectures for reagentless DNA detection [49]. In these devices, target DNA induced change in receptor strand conformation (i.e., folding, unfolding) and/or dynamics influences the electron transfer characteristics between the redox probe and the electrode leading to a detectable electrochemical signal.

The first reported E-DNA sensor system was based on a self-complementary ferrocene-functionalized DNA probe that formed a stable stem-loop structure [50]. A schematic illustration of the sensor is shown in Fig. 7. The sensor architecture closely resembles a surface-immobilized DNA hairpin molecular beacon [51]. In the absence of target single-stranded DNA, the stem-loop probe holds the ferrocene reporter in sufficiently close proximity to the electrode to enable efficient charge transfer during voltammetric scanning resulting in a large, quantifiable faradaic current response. Upon target DNA hybridization, the stem-loop structure is compromised and the ferrocene probe is displaced from frequent neighboring interactions with the electrode, significantly decreasing the current observed during scanning, thus providing a "signal-off" measurement for target DNA in the sample as low as 10 pM. Because the ferrocene moiety simply imparts redox activity to the DNA recognition sequence, it was envisaged that the platform could be translated to alternate electrochemical probes.

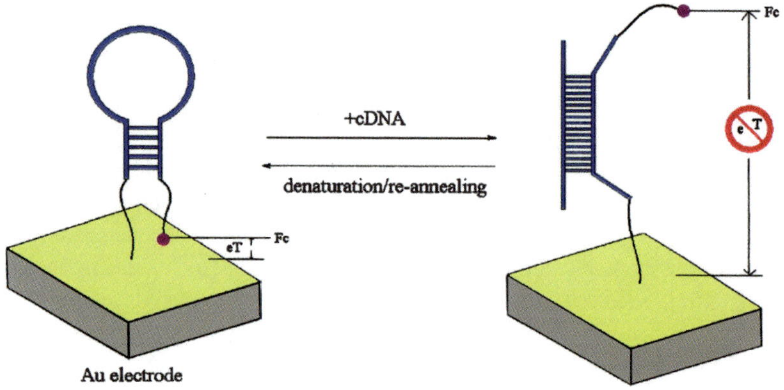

Fig. 7 A stem-loop oligonucleotide possessing terminal thiol and a ferrocene group is immobilized at a gold electrode through self-assembly. In the absence of target, the stem-loop structure holds the ferrocene tag into close proximity with the electrode surface, thus ensuring rapid electron transfer and efficient redox of the ferrocene label. On hybridization with the target sequence, a large change in redox currents is observed, presumably because the ferrocene label is separated from the electrode surface. Reprinted with permission from [50]. Copyright 2003 National Academy of Sciences, USA

Since this initial report, the Plaxco group has made a number of significant advances to the E-DNA platform [52]. Most notably, by inclusion of DNA aptamers as receptors, these folding-based sensors have been generalized to a wide range of specific protein and small molecule targets beyond nucleic acids. Further, because E-DNA signaling is associated with a target-induced change in the conformation and dynamics of the DNA probe and not simply target absorption to the electrode surface, the approach has been successfully utilized for analyte detection in complex matrices such as blood serum, saliva, urine, cell lysates, and environmental soil samples. Due to the high sensitivity of these systems in "dirty" samples coupled with their operational convenience, these E-DNA biosensors are well suited for potential commercial applications in pathogen detection, proteomics, metabolomics, and drug discovery.

4.2 *Peptide-Functionalized Metalloreceptors*

Alternatively, a number of ferrocene–peptide conjugates are beginning to emerge as functional receptors for reagentless electrochemical biosensing. Analogous to the E-DNA platforms, the peptide components in these systems invariably serve as analyte recognition elements, while the conjugated ferrocene groups impart redox activity to the receptor. Depending on the biosensor target, the electroactive receptors can be integrated with electrodes such that electron transfer properties are altered based on specific analyte–receptor interactions. Given the enormous challenge of developing effective label-free methods for protein detection, this section highlights some emerging sensor architectures that utilize ferrocene–peptide conjugates for this distinct purpose.

One class of proteins that has recently become an attractive target for biosensing is proteases. Proteolysis, the reaction catalyzed by these enzymes, is a simple hydrolytic process that cleaves the amide bond connecting adjacent amino acid residues in proteins. While certain proteases mediate nonspecific hydrolysis, many perform vital functions through selective and efficient cleavage of specific peptide substrates. In healthy individuals, the activity of proteases is tightly regulated and they are ubiquitous participants in maintaining fidelity in many critical biological processes such as apoptosis, matrix remodeling, and blood clotting. However, misregulation of protease activity has been implicated in an array of major life-threatening diseases such as cancer, AIDS, and infectious and neurodegenerative disorders. Thus, robust assay methods capable of quantifying target protease activity are in high demand.

Ferrocene–peptide conjugates are ideally suited as receptors for assaying protease activity when linked to interrogating electrodes. An example electrochemical proteolytic beacon for detecting matrix metalloproteinase-7 (MMP-7) activity is shown in Fig. 8 [53]. MMPs are a family of extracellular zinc proteases that have long been associated with tumor invasion and metastasis [54] and thus may serve as important cancer biomarkers. In this study, a known helix peptide substrate for MMP-7 (Arg-Pro-Leu-Ala-Leu-Trp-Arg-Ser) was modified with an *N*-terminal ferrocene and a Cys residue at the *C*-terminus to enable gold anchoring. The peptide was synthesized via traditional solid-phase synthesis and ferrocene acetic acid was conjugated to the *N*-terminus prior to cleavage from the resin. SAMs of the ferrocene–peptide conjugate were grown by immersing clean gold electrodes in ethanol solutions of the probe. In the initial "signal-on" state of the sensor, voltammetric scanning yields a well-defined peak with quantifiable charge due to efficient electron transfer between ferrocene and the underlying electrode. Upon exposure to MMP-7, the peptide probe is cleaved between the Ala-Leu residues and ferrocene is separated from the electrode. The removal of ferrocene by target protease produces a measurable "signal-off" decrease in current. At a constant incubation time, the amplitude of this signal decrease scaled with increasing concentrations of MMP-7 resulting in a detection limit of 3.4 pM. Specificity was confirmed in control experiments with different MMPs and BSA.

Importantly, similarly designed electrochemical protease biosensors have been employed for the detection of alternate disease markers such as caspase-3 [55], prostate-specific antigen (PSA) [56], plasmin [57], and trypsin and α-thrombin [58]. Hence, it is becoming increasingly more apparent that ferrocene–peptide conjugates may be used as effective alternatives to fluorescent-labeled peptide substrates for developing protease assays. This is significant as fluorescence-based assays are often complicated by intricate excitation and detection schemes to reduce background interference. It should be noted, however, that the electrochemical assays also require some degree of optimization as the signal response is influenced by the monolayer composition and probe densities. Looking ahead, one can easily envision multiplexing

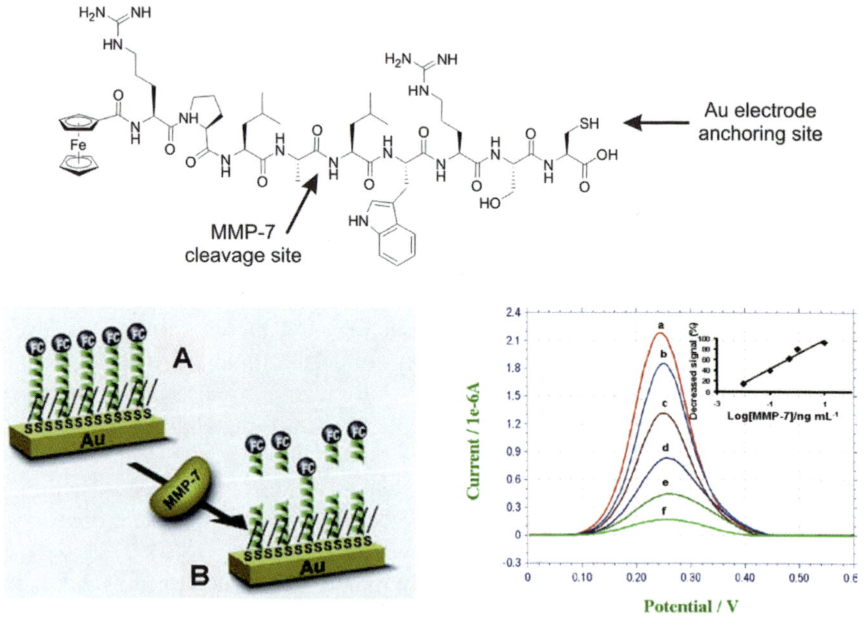

Fig. 8 (*Top*) Structure of the electrochemical proteolytic beacon for MMP-7 detection. (*Left*) Schematic of sensor construction and MMP detection: (*A*) self-assembling electrochemical ferrocene–peptide conjugate on a gold electrode; (*B*) cleavage of peptide substrate in the presence of MMP-7. (*Right*) Typical square wave voltammograms of ferrocene–peptide SAMs incubated with different concentrations of MMP-7 (from *a* to *f*: 0, 0.1, 0.5, 1, 5, and 10 ng mL^{-1}). All potentials are referenced to Ag/AgCl in 0.6M $NaClO_4$ supporting electrolyte. Adapted with permission from [53]. Copyright 2006 American Chemical Society

the electrochemical protease biosensor platform by incorporating multiple peptide substrates labeled with distinct nonoverlapping redox reporters in the same SAM or by exposing a sample to an array of electrodes each with uniquely modified peptide reporters.

In the development of protein biosensors, another general approach toward electroactive receptor design utilizes unsymmetrically substituted ferrocenes conjugated with both peptides and organosulfur anchoring moieties. Contrary to the protease systems discussed above where probe cleavage leads to a "signal-off" response, unsymmetrical ferrocene conjugates enable the more preferred "signal-on" sensing modality by exploiting the inherent sensitivity of the probe redox potential to its local environment (i.e., aqueous or protein). As a frame of reference, the ferrocene/ferrocenium (Fe(II)/Fe(III)) redox couple has been shown to shift positive 177 mV from its aqueous potential upon encapsulation in a hydrophobic protein matrix [59]. This higher redox potential may be rationalized by a destabilization of the cationic ferrocenium species in the protein microenvironment relative to a polar aqueous environment. Indeed, redox tuning of native transition metal complexes in metalloproteins is well known and a hallmark of bioinorganic chemistry.

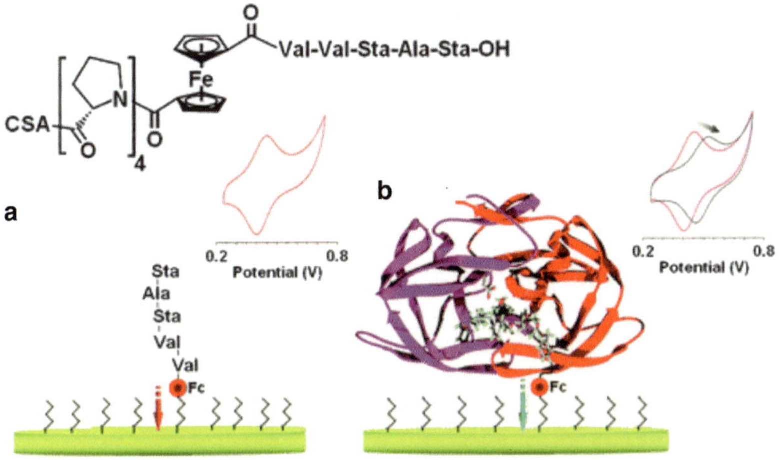

Fig. 9 (**a**) Structure of the ferrocene–pepstatin conjugate and schematic representation of the diluted self-assembled peptide film on a gold surface. The initial cyclic voltammogram of ferrocene probe SAM is in *red*. (**b**) The binding of HIV-1 PR to the ferrocene–pepstatin probe causes a significant shift in the formal potential and a decrease in the current intensity as shown in *blue*. Reprinted with permission from [61]. Copyright 2007 Royal Chemical Society

Kraatz and coworkers first reported integrated sensors with unsymmetric ferrocene architectures for the electrochemical detection of papain [60] and HIV type 1 protease (HIV-1 PR) [61] using conjugates of peptide enzyme inhibitors. Figure 9 shows the sensor for HIV-1 PR. In this example, the ferrocene probe is conjugated with a cystamine-terminated oligoproline anchoring group and the known aspartic protease inhibitor pepstatin. Mixed SAMs of the probe and hexanethiol diluent were prepared and the initial faradaic response characterized by CV. Upon exposure to target HIV-1 PR, the formal potential of the surface-bound ferrocene probe shifted to a higher potential suggesting protein–probe interaction. Increasing concentrations of target (from 40 to 100 nM) resulted in a linear increase in redox potential accompanied by progressively higher current loss. The analytical response was attributed to HIV-1 PR encapsulation of the surface-bound ferrocene probe resulting in ferrocenium destabilization and partial shielding of counteranion access during oxidation. Recent extension of this label-free biosensing platform to the detection of a panel of HIV proteins [62] using different peptide receptors suggests that the approach may be general.

Ultimately, the integrated electrochemical biosensors discussed above demonstrate that redox probe–receptor conjugates provide a means of signaling specific biomolecular recognition events. Our laboratory is pursuing a number of electroactive SAM building blocks [63] amenable to biofunctionalization to further explore similar modes of analyte-induced electronic signal transduction. The challenge of developing label-free, low cost, and multiplexed biosensor platforms provides significant inspiration and we anticipate significant advances in this area in the coming years.

5 Summary

On the road to next generation biosensors, the emerging field of transition metal-modified receptors has significantly evolved over the last decade. Inspired by the design and function of natural molecular recognition systems, several groups have constructed novel functional metalloreceptors with high performance characteristics that enhance and/or supplement conventional biomolecules such as proteins, enzymes, and DNA. In this chapter, three general roles for transition metals were discussed in the context of synthetic receptors: (1) metal as receptor scaffold, (2) metal as allosteric receptor for catalysis regulation, and (3) metal as integrated electroactive reporter. In each case, the versatility of transition metal coordination chemistry is exploited through rational design. It is anticipated that future work in these areas will remain a significant research focus.

References

1. Thevenot DR, Toth K, Durst RA, Wilson GS (2001) Electrochemical biosensors: recommended definitions and classification. Biosens Bioelectron 16(1–2):121–131
2. Clark LC Jr, Lyons C (1962) Electrode systems for continuous monitoring in cardiovascular surgery. Ann N Y Acad Sci 102:29–45
3. Wender P, Cummins CC, Poliakoff M, Kiessling L, Meijer EW, Alivisatos P, Wooley K, King D, Aizenberg J, Fleming G (2011) What lies ahead. Nature 469(7328):23–25
4. Rosenzweig AC (2002) Metallochaperones: bind and deliver. Chem Biol 9(6):673–677. doi:10.1016/S1074-5521(02)00156-4
5. Miller J, McLachlan AD, Klug A (1985) Repetitive zinc-binding domains in the protein transcription factor IIIA from *Xenopus oocytes*. EMBO J 4(6):1609–1614
6. Maret W, Li Y (2009) Coordination dynamics of zinc in proteins. Chem Rev 109 (10):4682–4707. doi:10.1021/cr800556u
7. Lu Y, Yeung N, Sieracki N, Marshall NM (2009) Design of functional metalloproteins. Nature 460(7257):855–862. doi:10.1038/nature08304
8. Ghadiri MR, Choi C (1990) Secondary structure nucleation in peptides – transition-metal ion stabilized alpha-helices. J Am Chem Soc 112(4):1630–1632
9. Ghadiri MR, Fernholz AK (1990) Peptide architecture – design of stable alpha-helical metallopeptides via a novel exchange-inert ruthenium(III) complex. J Am Chem Soc 112 (26):9633–9635
10. Doerr AJ, McLendon GL (2004) Design, folding, and activities of metal-assembled coiled coil proteins. Inorg Chem 43(25):7916–7925. doi:10.1021/ic0490573
11. Doerr AJ, Case MA, Pelczer I, McLendon GL (2004) Design of a functional protein for molecular recognition: specificity of ligand binding in a metal-assembled protein cavity probed by ^{19}F NMR. J Am Chem Soc 126(13):4192–4198. doi:10.1021/ja035798b
12. Ghadiri MR, Soares C, Choi C (1992) A convergent approach to protein design – metal ion-assisted spontaneous self-assembly of a polypeptide into a triple-helix bundle protein. J Am Chem Soc 114(3):825–831
13. Gochin M, Khorosheva V, Case MA (2002) Structural characterization of a paramagnetic metal-ion-assembled three-stranded alpha-helical coiled coil. J Am Chem Soc 124(37): 11018–11028. doi:10.1021/ja020431c

14. Lieberman M, Sasaki T (1991) Iron(II) organizes a synthetic peptide into 3-helix bundles. J Am Chem Soc 113(4):1470–1471
15. Gochin M, Guy RK, Case MA (2003) A metallopeptide assembly of the HIV-1 gp41 coiled coil is an ideal receptor in fluorescence detection of ligand binding. Angew Chem Int Ed 42(43):5325–5328. doi:10.1002/anie.200352006
16. Salgado EN, Radford RJ, Tezcan FA (2010) Metal-directed protein self-assembly. Acc Chem Res 43(5):661–672. doi:10.1021/ar900273t
17. Anslyn EV, Wright AT (2006) Differential receptor arrays and assays for solution-based molecular recognition. Chem Soc Rev 35(1):14–28. doi:10.1039/b505518k
18. Anslyn EV, Umali AP (2010) A general approach to differential sensing using synthetic molecular receptors. Curr Opin Chem Biol 14(6):685–692. doi:10.1016/j.cbpa.2010.07.022
19. Anslyn EV, Zhang TZ, Edwards NY, Bonizzoni M (2009) The use of differential receptors to pattern peptide phosphorylation. J Am Chem Soc 131(33):11976–11984. doi:10.1021/Ja9041675
20. Anslyn EV, Nguyen BT (2006) Indicator-displacement assays. Coordin Chem Rev 250 (23–24):3118–3127. doi:10.1016/j.ccr.2006.04.009
21. Anslyn EV, Tobey SL (2003) Energetics of phosphate binding to ammonium and guanidinium containing metallo-receptors in water. J Am Chem Soc 125(48):14807–14815. doi:10.1021/Ja030507k
22. Anslyn EV, Tobey SL, Jones BD (2003) C_{3v} symmetric receptors show high selectivity and high affinity for phosphate. J Am Chem Soc 125(14):4026–4027. doi:10.1021/Ja021390n
23. Kovbasyuk L, Kramer R (2004) Allosteric supramolecular receptors and catalysts. Chem Rev 104(6):3161–3187. doi:10.1021/cr030673a
24. Zhu L, Anslyn EV (2006) Signal amplification by allosteric catalysis. Angew Chem Int Ed 45(8):1190–1196. doi:10.1002/anie.200501476
25. Mirkin CA, Gianneschi NC, Bertin PA, Nguyen ST, Zakharov LN, Rheingold AL (2003) A supramolecular approach to an allosteric catalyst. J Am Chem Soc 125(35):10508–10509. doi:10.1021/Ja035621h
26. Mirkin CA, Gianneschi NC, Masar MS (2005) Development of a coordination chemistry-based approach for functional supramolecular structures. Acc Chem Res 38(11):825–837. doi:10.1021/Ar980101q
27. Mirkin CA, Holliday BJ (2001) Strategies for the construction of supramolecular compounds through coordination chemistry. Angew Chem Int Ed 40(11):2022–2043
28. Mirkin CA, Gianneschi NC, Nguyen ST (2005) Signal amplification and detection via a supramolecular allosteric catalyst. J Am Chem Soc 127(6):1644–1645. doi:10.1021/Ja0437306
29. Mirkin CA, Yoon HJ, Heo J (2007) Allosteric regulation of phosphate diester transesterification based upon a dinuclear zinc catalyst assembled via the weak-link approach. J Am Chem Soc 129(46):14182–14183. doi:10.1021/Ja077467v
30. Yoon HJ, Mirkin CA (2008) Pcr-like cascade reactions in the context of an allosteric enzyme mimic. J Am Chem Soc 130(35):11590–11591. doi:10.1021/Ja804076q
31. Saiki RK, Scharf S, Faloona F, Mullis KB, Horn GT, Erlich HA, Arnheim N (1985) Enzymatic amplification of beta-globin genomic sequences and restriction site analysis for diagnosis of sickle-cell anemia. Science 230(4732):1350–1354
32. Masar MS, Gianneschi NC, Oliveri CG, Stern CL, Nguyen ST, Mirkin CA (2007) Allosterically regulated supramolecular catalysis of acyl transfer reactions for signal amplification and detection of small molecules. J Am Chem Soc 129(33):10149–10158. doi:10.1021/Ja0711516
33. Wiester MJ, Mirkin CA (2009) Water-soluble macrocycles synthesized via the weak-link approach. Inorg Chem 48(17):8054–8056. doi:10.1021/Ic900983v
34. Adams DM, Brus L, Chidsey CED, Creager S, Creutz C, Kagan CR, Kamat PV, Lieberman M, Lindsay S, Marcus RA, Metzger RM, Michel-Beyerle ME, Miller JR, Newton MD, Rolison DR, Sankey O, Schanze KS, Yardley J, Zhu XY (2003) Charge transfer on the nanoscale: current status. J Phys Chem B 107(28):6668–6697. doi:10.1021/Jp0268462

35. Gray HB, Winkler JR (2005) Long-range electron transfer. Proc Natl Acad Sci U S A 102 (10):3534–3539. doi:10.1073/pnas.0408029102
36. McCreery RL (2004) Molecular electronic junctions. Chem Mater 16(23):4477–4496. doi:10.1021/Cm049517q
37. Flood AH, Stoddart JF, Steuerman DW, Heath JR (2004) Chemistry. Whence molecular electronics? Science 306(5704):2055–2056. doi:10.1126/science.1106195
38. Joachim C, Gimzewski JK, Aviram A (2000) Electronics using hybrid-molecular and mono-molecular devices. Nature 408(6812):541–548. doi:10.1038/35046000
39. Nitzan A, Ratner MA (2003) Electron transport in molecular wire junctions. Science 300 (5624):1384–1389. doi:10.1126/science.1081572
40. Yasutomi S, Morita T, Imanishi Y, Kimura S (2004) A molecular photodiode system that can switch photocurrent direction. Science 304(5679):1944–1947. doi:10.1126/science.1098489
41. Wei L, Padmaja K, Youngblood WJ, Lysenko AB, Lindsey JS, Bocian DF (2004) Diverse redox-active molecules bearing identical thiol-terminated tripodal tethers for studies of molecular information storage. J Org Chem 69(5):1461–1469. doi:10.1021/jo0349476
42. Privett BJ, Shin JH, Schoenfisch MH (2010) Electrochemical sensors. Anal Chem 82(12):4723–4741. doi:10.1021/Ac101075n
43. Eckermann AL, Feld DJ, Shaw JA, Meade TJ (2010) Electrochemistry of redox-active self-assembled monolayers. Coordin Chem Rev 254(15–16):1769–1802. doi:10.1016/j.ccr.2009.12.023
44. Love JC, Estroff LA, Kriebel JK, Nuzzo RG, Whitesides GM (2005) Self-assembled monolayers of thiolates on metals as a form of nanotechnology. Chem Rev 105(4): 1103–1169. doi:10.1021/Cr0300789
45. Chidsey CED, Bertozzi CR, Putvinski TM, Mujsce AM (1990) Coadsorption of ferrocene-terminated and unsubstituted alkanethiols on gold – electroactive self-assembled monolayers. J Am Chem Soc 112(11):4301–4306
46. Bertin PA, Meade TJ (2009) Novel redox active bifunctional crosslinkers from unsymmetrical 1,1′-disubstituted ferrocenes. Tetrahedron Lett 50(38):5409–5412. doi:10.1016/j.tetlet.2009.07.056
47. Metzler-Nolte N, Salmain M (2008) The bioorganometallic chemistry of ferrocene. In: Ferrrrocenes. Wiley, Chichester, pp 499–639. doi:10.1002/9780470985663.ch13
48. Martic S, Labib M, Shipman PO, Kraatz HB (2011) Ferrocene–peptido conjugates: from synthesis to sensory applications. Dalton Trans 40(28):7264–7290. doi:10.1039/C0dt01707h
49. Ricci F, Plaxco KW (2008) E-DNA sensors for convenient, label-free electrochemical detection of hybridization. Microchim Acta 163(3–4):149–155. doi:10.1007/s00604-008-0015-4
50. Fan CH, Plaxco KW, Heeger AJ (2003) Electrochemical interrogation of conformational changes as a reagentless method for the sequence-specific detection of DNA. Proc Natl Acad Sci U S A 100(16):9134–9137. doi:10.1073/pnas.1633515100
51. Du H, Disney MD, Miller BL, Krauss TD (2003) Hybridization-based unquenching of DNA hairpins on Au surfaces: prototypical "molecular beacon" biosensors. J Am Chem Soc 125 (14):4012–4013. doi:10.1021/Ja0290781
52. Lubin AA, Plaxco KW (2010) Folding-based electrochemical biosensors: the case for responsive nucleic acid architectures. Acc Chem Res 43(4):496–505. doi:10.1021/ar900165x
53. Liu G, Wang J, Wunschel DS, Lin Y (2006) Electrochemical proteolytic beacon for detection of matrix metalloproteinase activities. J Am Chem Soc 128(38):12382–12383. doi:10.1021/ja0626638
54. Egeblad M, Werb Z (2002) New functions for the matrix metalloproteinases in cancer progression. Nat Rev Cancer 2(3):161–174. doi:10.1038/nrc745
55. Xiao H, Liu L, Meng F, Huang J, Li G (2008) Electrochemical approach to detect apoptosis. Anal Chem 80(13):5272–5275. doi:10.1021/ac8005268
56. Zhao N, He YQ, Mao X, Sun YH, Zhang XB, Li CZ, Lin YH, Liu GD (2010) Electrochemical assay of active prostate-specific antigen (psa) using ferrocene-functionalized peptide probes. Electrochem Commun 12(3):471–474. doi:10.1016/j.elecom.2010.01.022

57. Ohtsuka K, Maekawa I, Waki M, Takenaka S (2009) Electrochemical assay of plasmin activity and its kinetic analysis. Anal Biochem 385(2):293–299. doi:10.1016/J.Ab.2008.11.006
58. Adjemian J, Anne A, Cauet G, Demaille C (2010) Cleavage-sensing redox peptide monolayers for the rapid measurement of the proteolytic activity of trypsin and alpha-thrombin enzymes. Langmuir 26(12):10347–10356. doi:10.1021/La100397g
59. Hwang HJ, Carey JR, Brower ET, Gengenbach AJ, Abramite JA, Lu Y (2005) Blue ferrocenium azurin: an organometalloprotein with tunable redox properties. J Am Chem Soc 127(44):15356–15357. doi:10.1021/ja054983h
60. Mahmoud KA, Kraatz HB (2007) A bioorganometallic approach for the electrochemical detection of proteins: a study on the interaction of ferrocene-peptide conjugates with papain in solution and on Au surfaces. Chem Eur J 13(20):5885–5895. doi:10.1002/chem.200601878
61. Kerman K, Mahmoud KA, Kraatz HB (2007) An electrochemical approach for the detection of HIV-1 protease. Chem Commun (37):3829–3831. doi: 10.1039/B707140j
62. Kerman K, Kraatz HB (2009) Electrochemical probing of HIV enzymes using ferrocene-conjugated peptides on surfaces. Analyst 134(12):2400–2404. doi:10.1039/b912083a
63. Bertin PA, Ahrens MJ, Bhavsar K, Georganopoulou D, Wunder M, Blackburn GF, Meade TJ (2010) Ferrocene and maleimide-functionalized disulfide scaffolds for self-assembled monolayers on gold. Org Lett 12(15):3372–3375. doi:10.1021/Ol101180r

Index

S.A. Piletsky and M.J. Whitcombe (eds.), *Designing Receptors for the Next Generation of Biosensors*, Springer Series on Chemical Sensors and Biosensors (2013) 12: 261–264
DOI 10.1007/5346_2012, © Springer-Verlag Berlin Heidelberg 2013

Printed by Printforce, the Netherlands